Intelligent Control and Learning Systems

Volume 26

The Springer book series Intelligent Control and Learning Systems addresses the emerging advances in intelligent control and learning systems from both mathematical theory and engineering application perspectives. It is a series of monographs and contributed volumes focusing on the in-depth exploration of learning theory in control such as iterative learning, machine learning, deep learning, and others sharing the learning concept, and their corresponding intelligent system frameworks in engineering applications. This series is featured by the comprehensive understanding and practical application of learning mechanisms. This book series involves applications in industrial engineering, control engineering, and material engineering, etc.

The Intelligent Control and Learning System book series promotes the exchange of emerging theory and technology of intelligent control and learning systems between academia and industry. It aims to provide a timely reflection of the advances in intelligent control and learning systems. This book series is distinguished by the combination of the system theory and emerging topics such as machine learning, artificial intelligence, and big data. As a collection, this book series provides valuable resources to a wide audience in academia, the engineering research community, industry and anyone else looking to expand their knowledge in intelligent control and learning systems.

Chen Yang · Yuanqing Xia

Set Theory-Based Spacecraft Dynamics and Control

Springer

Chen Yang
School of Astronautics
Beihang University
Beijing, China

Yuanqing Xia
School of Automation
Beijing Institute of Technology
Beijing, China

ISSN 2662-5458 ISSN 2662-5466 (electronic)
Intelligent Control and Learning Systems
ISBN 978-981-95-9178-7 ISBN 978-981-95-9179-4 (eBook)
https://doi.org/10.1007/978-981-95-9179-4

This Springer imprint is published by the registered company Springer Nature Singapore Pte Ltd.
The registered company address is: 152 Beach Road, #21-01/04 Gateway East, Singapore 189721, Singapore

Acknowledgements

This book is supported by the National Natural Science Foundation of China (Nos. 12472058 and 11972355), Beijing Natural Science Foundation (No. L258044), and the Fundamental Research Funds for the Central Universities.

Introduction

The advancement of spacecraft technology has significantly propelled human explorations of the universe, driven scientific and technological innovation, and profoundly impacted our daily lives. As space missions grow in complexity and scale, the demands on spacecraft performance continue to become increasingly stringent. For instance, spacecraft must maintain precise attitudes to orient themselves toward Earth or the Sun for navigation and solar power generation, as well as perform orbital adjustments. Additionally, spacecraft must control the vibrations caused by various factors to prevent structural damage and local instability. Given the critical importance of attitude/vibration dynamics and control for spacecraft safety and functionality, effective spacecraft dynamics and control systems are indispensable. Numerous nominal dynamics and control methods have been developed and applied in spacecraft. However, these methods often fall short when dealing with the multi-source uncertainties inherent to space environments, spacecraft systems, etc. Such uncertainties arise from spacecraft structural errors, extreme space conditions, and discrepancies between ground test platforms and actual space environments. These factors render traditional dynamics and control methods inadequate for achieving precise spacecraft dynamics and control. Compared to probabilistic and statistical methods that require a large number of statistical samples to handle uncertainty, set theory-based theory has a significant advantage in dealing with small-sample uncertainty, as it only needs to obtain the uncertainty bounds to achieve accurate quantification of uncertainty parameters. Set theory-based uncertainty theory is an effective way to deal with the spacecraft dynamics and control system. The main techniques throughout the book can be summarized mainly into three parts.

Set Theory-Based Reduced-Order Model

This part provides a set theory-based uncertain modeling method for spacecraft dynamics. In the field of spacecraft dynamics system design, the development of effective reduced-order modeling methods is crucial due to the computational limitations imposed by high-order spacecraft dynamics. By introducing the unknown-but-bounded parameters, interval and convex-set uncertainty methods enable the construction of simplified models that represent the core system dynamics. These methods not only ensure that the reduced-order model accurately represents the original nominal system but also quantify the uncertainties introduced in the full order dynamics. Such quantification is vital for dynamic model validation and subsequent controller optimization steps, providing a solid foundation for dynamical characteristics and control system design. For set theory-based uncertainty theory, interval and convex-set uncertainty methods provide an accurate and efficient and reliable approach for reduced-order model in spacecraft dynamics with uncertainties. They enable the retention of essential dynamic features while significantly reducing computational consumption, thus underpinning the efficient design and optimization of spacecraft control systems.

Set Theory-Based Attitude Determination and Control

This part establishes the set theory-based methods for uncertain spacecraft dynamics and control. Spacecraft faces challenges in attitude control due to uncertainties like inertia deviation and complex space environments. These can disrupt the orientation of the spacecraft, impacting mission success. Set theory-based uncertainty methods provide an accurate description of the uncertain spacecraft attitude control by designing a control system containing interval or convex-set uncertainty analysis. In this part, set theory-based uncertainty attitude determination, linear quadratic regulator (LQR), and sliding mode control methods are established. The proposed design method based on set theory aims to design the uncertainty component of the controller to achieve uncertainty control of spacecraft systems. These methods enhance spacecraft attitude control stability and precision while reducing computational consumption with reliability evaluation. The proposed methods ensure accurate, stable, and rapid attitude control for spacecraft in environments with multi-source uncertainties.

Set Theory-Based Control for Rigid-Flexible Coupling Spacecrafts

This part provides theories of set theory-based uncertainty analysis in complex dynamics and control systems. Similar to rigid spacecraft, which face attitude control challenges from uncertainties, rigid-flexible coupling spacecraft encounters additional difficulties in managing both rigid-body motions and flexible deformations. Traditional control methods often fall short in maintaining stability and precision when these two dynamic types interact. In this part, for rigid-flexible coupling spacecraft operating under uncertain conditions, uncertainty analysis methods based on set theory for rigid-flexible coupling dynamics and control system are introduced. Applying these set theory-based methods to attitude and vibration control of proposed spacecraft will provide more accurate responses for the entire control system and simultaneously improve the control performance of the spacecraft in the space environment.

By applying these advanced methods, spacecraft can achieve an accurate, stable, and rapid control effect, also with optimum performance, low costs, and high safety in uncertain conditions. Therefore, this book offers comprehensive theoretical foundations for the dynamics and control of the spacecraft in complex space environments with uncertainties, along with practical guidance for optimizing spacecraft dynamics and control systems.

Contents

About the Authors

Chen Yang is Associate Professor at Beihang University. He received his B.S., M.S., and Ph.D. degrees from Beihang University, Beijing, China, and his main research interests are spacecraft dynamics and control, inverse problems in mechanics, uncertainty, and reliability-based control system. Dr. Yang has won Marie Skłodowska-Curie Actions (MSCA) Fellowships. Dr. Yang has been awarded 'whole career' and 'single year' in the World's Top 2% Scientists list by Stanford University and the Elsevier database for 3 (2023–2025) and 6 (2020–2025) consecutive years, respectively. Dr. Yang has published more than 100 papers, including more than 70 SCI papers as First or Corresponding Author, and three monographs as First Author. Dr. Yang has presided over more than 30 projects, including three projects of National Natural Science Foundation of China, etc. Dr. Yang is Associate Editor/Subject Editor of seven SCI journals, including IEEE Transactions on Systems, Man, and Cybernetics, IEEE Transactions on Aerospace and Electronic Systems, IEEE Sensors Journal, Applied Mathematical Modeling, Journal of Vibration and Control, Journal of Vibration Engineering and Technologies, and International Journal of Aeronautical and Space Sciences. He has been serving as Reviewer for more than 70 SCI journals.

Yuanqing Xia is IEEE Fellow and received the grant from the National Science Foundation for Distinguished Young Scholars of China and the Cheung Kong Scholars Programme of China. He has been awarded 'Highly Cited Chinese Researchers' published by the Elsevier database. He received the Ph.D. degree in control theory and control engineering from Beihang University, Beijing, China, in 2001. He was Research Fellow in several academic institutions from 2002 to 2008, including the National University of Singapore and the University of Glamorgan, UK. Since 2004, he has been with the Beijing Institute of Technology, China, where he is currently Full Professor. His research interests include flight control, cloud control systems for flight, networked control systems, robust control and signal processing, and active disturbance rejection control. He is Director of the specialized committee on cloud control and decision of the Chinese Institute of Command and Control (CICC), Member of the 8th Disciplinary Review Group of the Academic Degrees

Committee of the State Council, Member of the Big Data Expert Committee of the Chinese Computer Society, and Vice Chairperson of the Internet of Things Working Committee of the Chinese Institute of Instrumentation. He is Deputy Editor of the Journal of Beijing Institute of Technology, Associate Editor of Acta Automatica Sinica, International Journal of Automation and Computing, Gyroscopy and Navigation, and IET Control Theory and Applications.

Chapter 1
Introduction

Abstract This chapter provides a comprehensive introduction to the dynamics and control of spacecraft in complex space environments with multi-source uncertainties. It reviews critical challenges in the space environment, dynamic modeling, and control systems, followed by a literature review on spacecraft dynamics, control methodologies, and uncertainty-based approaches. The chapter highlights the advantages of set theory-based methods in handling small-sample uncertainty quantification, and outlines the book's contributions in reduced-order modeling, attitude determination and control with reliability constraints, and rigid-flexible coupling spacecraft control.

1.1 Background of Spacecraft Dynamics and Control

Spacecraft have become essential tools for human space exploration, which are currently the only reliable means of communication and operation available to us in space. Spacecraft continuously perform on-orbit service tasks such as data transmission, space telescope operations, space station construction, planetary sampling, asteroid defense, and advanced manufacturing, while continuously expanding the scope of our exploration of space. Therefore, spacecraft have established their status as a key platform for cutting-edge scientific research, which requires continuous technological advancements to ensure that each generation of spacecraft is more advanced and capable than the previous one. As the complexity of space tasks continues to increase, from Earth observation to deep space exploration, the performance requirements for spacecraft have reached new heights, imposing critical demands for high precision, high autonomy, long lifespan, and high reliability in their operations (National Academies of Sciences, Medicine, Division on Engineering, Physical Sciences, Aeronautics, Space Engineering Board, & Committee on NASA Technology Roadmaps 2016). Thus, as shown in Fig. 1.1, a current trend in spacecraft development is the gradual evolution toward large-scale, complex, time-varying systems, which impose increasingly stringent demands on the accuracy, rapid response, and reliability of dynamics and control systems. However, meeting

C. Yang and Y. Xia, *Set Theory-Based Spacecraft Dynamics and Control*, Intelligent Control and Learning Systems 26, https://doi.org/10.1007/978-981-95-9179-4_1

Fig. 1.1 Large-scale spacecraft

these demands requires precise and rational dynamic modeling and control system design optimization of spacecraft systems, tailored to the complex space environment. The rest of this section will discuss the demands and challenges in spacecraft dynamics and control from four perspectives: the space environment, dynamic modeling, control systems, and their design optimization.

First, the extreme environment in space, characterized by extreme temperatures, particle radiation, microgravity, space debris, solar radiation pressure, and gravitational perturbations from celestial bodies, forms a multi-source, time-varying, strongly coupled, and extremely complex environment that may interfere with spacecraft electronic equipment and damage its structural integrity. As a result, the uncertainties generated by these factors can cause significant interference with spacecraft dynamics and control, which is illustrated in Fig. 1.2. The direct impact is that such multi-source uncertainties can degrade the performance of spacecraft dynamics and control and affect spacecraft attitude and lifespan. These multi-source uncertainties caused by complex unknown factors in the space environment may cause spacecraft to deviate from their predefined orbits and attitudes, increase the difficulty of spacecraft dynamics and control, and pose numerous challenges to the successful implementation of space tasks.

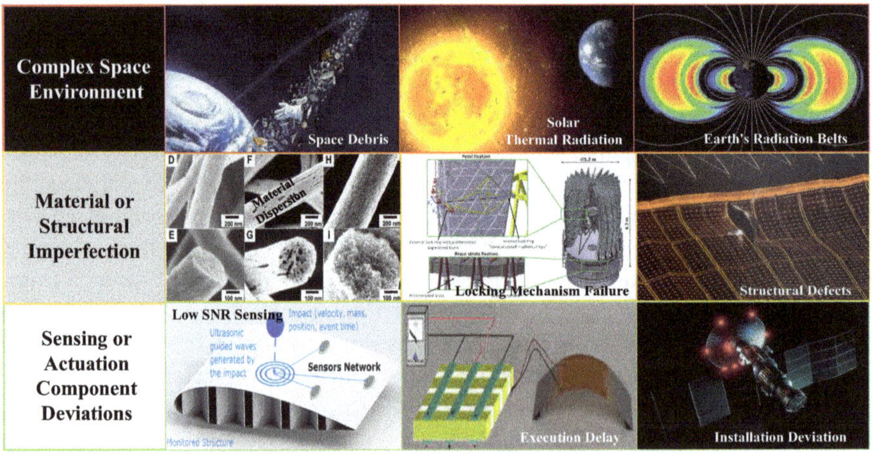

Fig. 1.2 Multi-source uncertainties affecting spacecraft

Additionally, the accuracy, rapid response, and reliability of spacecraft dynamics and control systems often depend on the establishment of accurate spacecraft models. However, spacecraft system models are generally highly complex and are easily influenced by factors such as multi-physics coupling, structural flexibility and scaling, and system parameter uncertainty. Attitude dynamics are deeply coupled with other physical processes such as structural thermal deformation, fuel sloshing, and electromagnetic interference, forming a highly nonlinear coupled system that makes precise modeling of the system challenging. The widespread application of large lightweight solar arrays, deployable antennas, and space manipulators has resulted in spacecraft exhibiting significant structural flexibility, inducing nonlinear low-frequency structural vibrations that severely impair the effectiveness of high-precision spacecraft control.

Based on the dynamic modeling, accurate, stable, and rapid spacecraft control is the technical prerequisite and core guarantee for achieving the aforementioned task objectives. The precision of spacecraft control is critical for completing various specific tasks, particularly those involving precise pointing, energy acquisition, payload operations, orbital maneuvers, and safe rendezvous and docking. In addition, effective spacecraft control ensures the safety performance of the spacecraft, enabling it to maintain its original attitude characteristics under a complex environment. The ability to reliably and precisely execute these tasks depends on the effectiveness of the control systems of the spacecraft (Hasan et al. 2022). These systems are critical to spacecraft operations and are key to ensuring that spacecraft operate as intended in the uncertain space environment, thereby achieving safe and effective space tasks. Developing advanced control methods with the capability to execute predefined tasks is an important frontier in aerospace engineering and will profoundly influence future space exploration and utilization. However, due to the influence of multiple factors, the characteristic parameters of spacecraft systems exhibit significant uncertainty, which often simultaneously causes degradation of the actuators, making it difficult for the spacecraft controller to attain specified performance. Besides, existing control methods are unable to precisely and rapidly quantify the impact of multi-source uncertainties on system responses, consequently failing to achieve accurate, stable, and rapid control for spacecraft with uncertain dynamics in complex space environments.

Nevertheless, the technical chain linking reliability analysis, uncertainty-based dynamics and control, and the reliability-based design optimization (RBDO) for the entire control system has not yet been fully established in spacecraft dynamics and control research. While there has been extensive research on spacecraft dynamics and control in the past, there have been relatively few studies on dynamics and control methods for uncertain systems, making it challenging to achieve dynamics and control of spacecraft systems that involve complex uncertainties. Additionally, due to the existence of multiple sources of uncertainties within the system, the responses of the spacecraft often do not fully comply with predefined requirements, leading to dynamics and control system failures. Currently, there is limited research on the reliability analysis of spacecraft dynamics and control systems, making it difficult to establish a research framework suitable for the uncertain dynamics and control systems. Furthermore, spacecraft dynamics and control systems often require multi-objective design, aiming to achieve high reliability and predefined dynamics and control performance while ensuring stability. However, there is a lack of suitable RBDO methods in this research field.

Essentially, spacecraft dynamics and control is not only a technical consideration but also a crucial factor in the accomplishment of space tasks in the scientific field. With the continued growth in demand for space exploration, the importance of spacecraft dynamics and control systems will further increase. In modern engineering, when designing spacecraft dynamics and control systems, it is particularly important to consider the complex space environment and specific space-related scenarios, the complexity and limitations of the spacecraft system, and the multi-source uncertainty.

1.2 Literature Review of Spacecraft Dynamics and Control

The primary challenge in achieving highly efficient spacecraft dynamics and control lies in the contradiction between high-precision modeling of spacecraft dynamics and the computational efficiency of the control system. The large flexible structures and the strong coupling of multi-physics fields lead to a significant increase in model complexity, resulting in excessive time consumption for model development and response calculations. Therefore, finding a reasonable model reduction method to effectively reduce computational consumption while maintaining a certain level of model accuracy is an urgent priority. In the field of control, existing methods struggle to effectively analyze and control spacecraft control systems with multi-source uncertainties. Therefore, it is hoped that optimal control methods can be used to establish a reasonable uncertainty-based control system for parameters sensitive to perturbations, thereby achieving a high-precision characterization of the uncertain control responses. Additionally, uncertainty analysis methods for spacecraft operating in complex space environments are urgently needed. Probabilistic methods are constrained by the small sample size dilemma in space tasks, severely limiting their application in such scenarios and thereby constraining system performance. Finally, reliability assessment has long relied on probabilistic models, which cannot effectively quantify the performance of spacecraft control systems in small sample scenarios, resulting in a lack of effective matching with uncertainty-based spacecraft control systems that can achieve high precision of the uncertain response quantification. Therefore, this section will review and summarize the existing technologies for spacecraft uncertain dynamics and control from the perspectives of spacecraft dynamics, control, uncertainty, and reliability methods.

1.2.1 Spacecraft Dynamics

Precise control of spacecraft dynamics is the key to the success of space tasks. Whether spacecraft involves precise orientation toward the Sun or Earth, scientific observations, effective load orientation, correction of deviated spacecraft orbits, and precise rendezvous and docking processes, the quality of the spacecraft control system directly impacts the effectiveness and efficiency of various space tasks. However, the actual control performance of spacecraft in real-world engineering often depends on the establishment and computational complexity of the simulation models. Therefore, a highly accurate dynamic model constructed with a rigid-body spacecraft model will facilitate further spacecraft control system design. Besides, considering the rigid-flexible coupling (RFC) dynamics of spacecraft essentially aims to implement corresponding motion or dynamic control for attitude changes caused by the flexible components of the spacecraft. However, in general, the models of spacecraft can be converted into mathematically describable nonlinear differential equations by considering their kinematic and dynamic characteristics. Such models often have high-dimensional characteristics, significantly increasing system complexity. Therefore, model order reduction techniques are crucial for the efficient control of spacecraft. Thus, the review of spacecraft dynamics will be sequentially developed from the following aspects: first, the kinematic modeling methods for spacecraft attitude models; second, studies on modeling the dynamic system of the spacecraft to be controlled; and finally, reviews on reduced-order models (ROMs) for complex spacecraft dynamic systems.

Kinematics-based spacecraft attitude modeling methods mainly focus on the mathematical description of dynamics parameters and coordinate system transformations, and generally do not involve the effects of forces and torques. These methods are characterized by their ability to clearly reflect changes in attitude angles during spacecraft motion. Based on differences in mainstream methods, these can be separated into three main varieties: methods based on Euler angle (Bai et al. 2022), methods based on quaternion (Yang 2019), and methods based on Lie group (Lee et al. 2012). Pezent et al. (2022) formulated the kinematic and dynamic behavior described by the Euler equations of a rigid object that is asymmetric and under the action of torque aligned or counter-aligned with the angular momentum. Subsequently, by combining the solutions of parallel and anti-parallel torque, the static and directional maneuvers of any asymmetric spacecraft can be analyzed. Guan et al. (2022) established a nonlinear ordinary differential equation for variable speed control by combining the Euler method with virtual prototype simulation, and analyzed the control framework of variable speed control. Yefymenko et al. (2022) proposed models to describe the dynamic angular motion of spacecraft, which are described using second-order quaternion differential equations, and explored methods for synthesizing spacecraft control algorithms based on these models. Li et al. (2023) studied the rigid spacecraft orbit and attitude tracking subject to actuation constraints. Within the Lie group SE(3) framework, Li et al. (2023) described the six DOF maneuver of the spacecraft and subsequently developed a nonlinear model predictive control strategy using the derived discrete-time formulation. The aforementioned research provides a wealth of methods for modeling spacecraft kinematics. And after establishing the kinematics model of the spacecraft, dynamics modeling and control system design can be conducted. To conclude, kinematics-based spacecraft dynamic modeling methods have been extensively studied, based on which the dynamic models of spacecraft can be established.

Dynamic models of spacecraft form the foundation for achieving spacecraft control. These models incorporate parameters such as torque and inertia to describe the mechanical mechanisms of motion. Although these methods involve techniques such as Newton mechanics and analytical mechanics, and their mathematical models exhibit subtle differences, they all clearly illustrate the mechanical characteristics of the spacecraft, particularly their attitude and vibration responses. The mainstream spacecraft dynamics modeling methods based on the model method can be categorized into the following three approaches: the Newton–Euler method (Hooker and Margulies 1965), the Lagrange method (Meirovitch 2010), and the Kane method (Kane et al. 1983). Liu et al. (2024) designed a novel anti-roll device combining three harpoons and a nanosatellite, and further utilized this control strategy in the dynamics and control issues. Considering external disturbances and changes in inertial parameters, they established a high-precision dynamics model of the orbital-attitude coupling of the combined system in the de-spin phase. Based on Jourdain's principle of velocity variation and the forward recursive construction formula of flexible multibody dynamics, Wang et al. (2022) established the dynamics of an RFC satellite, proposed an improved attitude dynamic model, and designed an attitude controller based on this model. In the design of the spacecraft attitude tracker, Newton–Euler rotation equations were converted into dynamic equations based on selected inertial coordinate systems, and then the Lyapunov method was used to design a controller to ensure asymptotic convergence of the attitude toward the target direction. Iannelli et al. (2022) established a nonlinear on-orbit dynamics model of a flexible Soil Moisture Active Passive spacecraft using the Newton–Euler method, and designed a model predictive control strategy based on the model to achieve momentum compensation and triaxial attitude control of the spin element, and control of the motor for the spin mechanism as well. Xie et al. (2023a) constructed the dynamic model of a rigid spacecraft and proposed a predefined-time attitude constraint control scheme under a backstepping framework for rigid spacecraft with external disturbances. Li et al. (2022a) studied the modeling and high-precision attitude control of large spacecraft. To describe the complex constraint relationships between rigid bodies in a multi-body system, they established a dynamic model of a flexible spacecraft by applying Lagrange's equations. If the captured object is comparable in size to or larger than the spacecraft, and there is significant modeling uncertainty, it will generate a large residual disturbance torque. Bandyopadhyay et al. (2016) ensured global exponential convergence of attitude tracking error and finite gain stability while considering modeling uncertainty and disturbances, and reduced the generated disturbance torque. Yang et al. (2024) used Stoneking's implementation of Kane's method to derive a symbolic multibody nonlinear model of the spacecraft body. Since the analytical linearization method is difficult to implement manually, this symbolic nonlinear model is linearized using symbolic functions. Since the Kane method allows ideal constraints to be ignored in dynamic modeling, Han (2016) studied the dynamics and control of robotic arm installed on a spacecraft as it freely flies around a space target in a circular Earth orbit, and verified the model through numerical simulation. Therefore, model-driven dynamic modeling methods—including Newton–Euler, Lagrange, and Kane formulations—provide the mathematical basis for describing spacecraft dynamical behavior. These models enable the design of high-precision and robust controllers by systematically addressing nonlinearity, coupling, and uncertainty inherent in spacecraft systems.

Model-driven approaches for spacecraft dynamics require extensive knowledge of complex prior models and are not suitable for some kinds of spacecraft that lack a modeling foundation. With the rise of artificial intelligence technologies, some data-based modeling methods have been applied to the establishment of spacecraft models in recent years. The advantages of data-based modeling methods are that they are model-free methods that do not require a large amount of computing resources to identify accurate model parameters, and they are also more accurate when constructing spacecraft attitude models. Since accurate dynamic models are highly complex, nonlinear, and costly for online identification, they make subsequent controller design impractical. To address this issue, Liu et al. (2022a) used the Q-learning algorithm to directly obtain spacecraft attitude data from system input/output measurements without relying on an explicit dynamic model. Based on the Clohessy-Wiltshire equation, Yang et al. (2023) constructed a Markov decision process model of spacecraft dynamics. Yu et al. (2025) addressed the challenges in data-driven dynamic modeling of spacecraft equipment under low signal-to-noise conditions by developing a novel method based on model frequency analysis and Monte Carlo simulation. Qiao et al. (2024) address attitude estimation for uncooperative known space objects by proposing DSOAE-Net, a fast and accurate direct-attitude-estimation network based on a novel pose representation method, which achieves an optimal balance between speed and generalization on weakly textured datasets. Mansell et al. (2021) presented a hybrid method that integrates one-class support-vector machines with long short-term memory neural networks, enhanced by transfer learning, to achieve automated, robust fault diagnosis for spacecraft attitude dynamic systems. Silvestrini and Lavagna (2021) addressed spacecraft relative dynamics modeling under unmodeled perturbations by employing a nonlinear autoregressive exogenous recurrent neural network for online, adaptive dynamics reconstruction, enabling accurate trajectory prediction for formation-flying spacecraft. In conclusion, data-driven modeling methods offer a promising alternative to traditional model based approaches by learning spacecraft dynamics directly from operational data, thereby reducing dependency on complex prior knowledge and enabling adaptive control without extensive parameter identification. These model-free techniques, such as reinforcement learning (RL), enhance adaptability and practicality for spacecraft with uncertain or poorly characterized dynamics.

Although the precise dynamic models of spacecraft are well-established and lay the foundations for spacecraft control, models for precise control of spacecraft are often computationally complex, posing challenges for efficient control system calculations. When such models are linear or weakly nonlinear, after linearization or other processes, they can be simplified to reduce their complexity, thereby enabling efficient simulation calculations. The following three methods for model order reduction are briefly introduced: modal reduction analysis (Hurty 1965), balanced truncation (Moore 2003), and the modal cut-back method (Craig and Bampton 1968). Nicassio et al. (2022) proposed a technique for deriving low-order models of large deformable spacecraft. The modeling method is based on a hybrid Newton-Lagrange method and uses a hypothetical modal method to estimate the modal characteristics of the main deformable structures based on their actual characteristics to achieve model order reduction. Renganathan (2020) transformed the control equations to a higher-dimensional space, resulting in a linearly underdetermined equation, simplifying the identification of the system matrix compared to the original nonlinear formulation. To achieve model reduction, the underdetermined system is then closed through a set of model-specific nonlinear constraints. By combining principal component decomposition with the discrete empirical interpolation method and the trajectory segmented linear method, Qian et al. (2015) proposed a subspace projection-based ROM modeling method and an integrated framework for automatically generating ROMs for spacecraft analysis. Similarly, combining the characteristics of classical model reduction methods and fully exploiting the advantages of data-based methods in handling high-dimensional uncertainties and achieving full-process adaptive order reduction, data-based spacecraft model reduction methods have also advanced rapidly in the past few years. To reduce the computational cost of flexible multi-body system simulation, Peng et al. (2022) developed a novel model reduction strategy based on the concept of data-driven modeling. Snapshot matrices were obtained from the empirical data set of the full-order model, and the cotangent enhancement approach was applied to convert the high-dimensional Simpson matrix into a low-dimensional form through reduced-order basis vectors. Jang et al. (2024) proposed a data-driven ROM for lunar orbiters based on principal component analysis and artificial neural networks, thereby improving convergence speed and training performance. To address the computational complexity inherent in high-fidelity spacecraft models, model order reduction techniques such as modal reduction, balanced truncation, and data-driven approaches provide effective means to derive simplified yet representative low-order representations. These methods enable efficient simulation and control design while preserving the essential dynamic characteristics of the original dynamic system.

In general, high-precision and relatively low-complexity modeling of spacecraft dynamics is a critical precondition for the success of space tasks, as its performance directly depends on the accuracy of the dynamics and computational efficiency. Spacecraft dynamics modeling describes the evolution of multidimensional states through the analysis of system energy and motion relationships, providing a physical foundation for controller design. However, it requires the multi-physics coupling mechanisms of spacecraft attitude, yet current research still faces challenges posed by strong nonlinearity and uncertainty. Additionally, driven by the advancement of artificial intelligence technology, data-driven modeling and model order reduction methods are gradually addressing the drawbacks of model-based methods that require precise model parameters. Finally, model order reduction techniques are indispensable in dynamics systems with high-dimensional models, as they can significantly enhance the computational efficiency of the entire control system.

In the future, it is important to achieve multi-scale intelligent fusion of models by combining physical information neural networks with mechanism-based models, and enhance the generalization capability of rigid-flexible-fluid-thermal coupling models. This will enable the establishment of more accurate spacecraft dynamics models, thereby achieving better control performance. Additionally, it is significant to further address the trade-off between model accuracy, computational complexity, and engineering feasibility by developing spacecraft modeling and model order reduction methods that balance physical consistency, real-time performance, and robustness, thereby providing more precise and reliable model-based control for spacecraft.

1.2.2 Spacecraft Control

While precise and efficient models are indispensable for analyzing spacecraft dynamics, they alone are insufficient for mission execution. The transition from theoretical modeling to actual task completion fundamentally relies on the design and implementation of effective control systems. However, in space tasks, spacecraft often rely on dynamics-based control for attitude maneuvers and vibration suppression to perform a variety of predefined tasks, including orientation, orbital correction, and docking. To achieve dynamic evolution in accordance with engineering requirements, effective control is necessary, including control of the primary attitude changes caused by rigid body models and vibrations caused by flexible components such as flexible rods (Zhang and Duan 2023). Therefore, it is crucial to achieve precise spacecraft control of spacecraft to meet the requirements of space tasks and ensure the success of the entire mission. Spacecraft control has unique requirements compared to other control applications, including optimal performance, multiple constraints, high uncertainty, and high reliability. The development of spacecraft control methods has been a long process, evolving alongside advancements in control theory. Based on the chronological order of application, logical sequence of development, and algorithm complexity, this section categorizes spacecraft control methods into the following four categories: classical control methods, optimal control methods, advanced control methods, and intelligent control methods.

Classic control methods often use closed-loop feedback control to optimize errors and achieve spacecraft control. When the attitude of the spacecraft deviates, sensors can obtain current attitude information and calculate control inputs through negative feedback. This control method does not rely on an accurate spacecraft model and is easy to combine with drivers to achieve the desired control effect in engineering. It was widely used in the early stages of space tasks to realize the preset performance of the spacecraft. The large-angle control of spacecraft is affected by nonlinear coupling and uncertainty. By modifying the traditional PID controller with additional control actions, Bang et al. (2003) improved the control design of a fixed-gain proportional-integral-derivative (PID) controller based on quaternion attitude variables for large-angle spacecraft reorientation, thereby achieving improved performance. To address the spacecraft pointing stability problem, Li et al. (2012) proposed an optimal PID-type controller capable of handling continuous inequality constraints on angular velocity and control inputs, along with terminal constraints on attitude and angular velocity. Sari et al. (2019) developed a fuzzy adaptive PID controller for the control of geostationary satellites' attitude, which incorporates a single-input fuzzy inference engine and a priority fuzzy inference engine. To achieve precise tracking control of the spacecraft's Euler angle, a corresponding cost function was considered to establish a sampling data adaptive control algorithm. Based on this, Hao et al. (2025) developed an adaptive PID controller that employs full-form dynamic linearization technology. Compared with rigid spacecraft, flexible spacecraft are more susceptible to interference and more difficult to control. Therefore, for flexible spacecraft, it is particularly necessary to provide attitude control with prescribed performance boundaries. Under the assumption that attitude and angular velocity measurement errors, as well as modal state estimation errors, are bounded, Xiao et al. (2020) designed an effective and computationally simple PD + controller, then analyzed and proved the robustness of the PD + controller and its capability to deliver non-conservative and guaranteed performance on attitude control. Considering the flexibility of the appendages attached to the spacecraft, under actuator saturation conditions, the implicit oscillations in the PID-controlled spacecraft attitude system were located and their parameters determined. Kuznetsov et al. (2022) focused on adopting reaction wheels for spacecraft attitude control. The combination of stiffness and flexibility has a huge effect on the control accuracy of spacecraft with large flexible appendages during attitude maneuvers. In response to this, Zhong et al. (2022) proposed an improved PD controller that effectively improved system performance. Classic control methods, therefore, have evolved from simple fixed-gain PID implementations to sophisticated variants, effectively addressing challenges such as nonlinearity, flexibility, and performance constraints in spacecraft dynamics and control.

Although classical control is renowned for its computational simplicity and ease of implementation, it can effectively achieve the objective of spacecraft control. However, due to the limited internal volume of spacecraft and the resulting constraints on fuel capacity, spacecraft often cannot afford to perform spacecraft control at the expense of excessive fuel consumption. Therefore, if optimal system control performance is to be further considered, optimal control methods represent a superior choice. Optimal control methods involve constructing an optimization problem and solving it to obtain the optimal control rate for the response, thereby achieving optimal control performance for the spacecraft. The key lies in solving the optimal problem that has been constructed. Jin et al. (2024) designed the cost function of the optimal control problem by integrating a prescribed performance control framework and a barrier Lyapunov function, enabling simultaneous solution of spacecraft attitude stabilization and tracking under performance constraints, angular velocity limits, and actuator saturation. Wang et al. (2022) settled the attitude tracking challenge of large-angle deviation of spacecraft attitude, generated optimized reference commands that satisfy the principles of low energy consumption and short turning time, and developed a robust inverse optimal control design based on the generation of optimized attitude commands. Wang et al. (2024) established a nonlinear optimal control problem for real-time generation of optimal control commands for solar sail spacecraft subject to attitude constraints in coplanar circular-to-circular interplanetary transfer, and derived the necessary conditions for optimality through Pontryagin's minimum principle. By developing a design method for attitude control laws and an objective function for optimization problems, Hu et al. (2023) proposed a distributed optimization approach for flexible spacecraft vibration suppression and pointing control. When all redundant variables converge to zero through an intelligent consensus algorithm, all actuators obtain the optimal solution. To conclude, optimal control methods transcend classical approaches in some scenarios involving explicit performance metrics and multiple constraints—such as large-angle attitude maneuvers, vibration suppression, and fuel-limited orbital transfers—by systematically formulating and solving constrained optimization problems, enabling precise and relatively low-cost spacecraft control even under complex dynamical and environmental constraints.

While optimal control provides a rigorous framework for performance optimization under idealized assumptions, its practical applications in spacecraft systems often face challenges in addressing real-time and complex dynamics. To better achieve the high adaptability and comprehensive performance optimization capabilities of complex dynamic systems in spacecraft, and with the maturation of advanced control theory, some advanced control methods have been proposed, which have also been further applied to spacecraft. Advanced control methods are often capable of addressing multi-constraint problems that classical control methods and optimal control methods struggle to handle. They also employ techniques such as back-stepping and sliding mode variable structure to resolve strongly nonlinear issues that previous methods found difficult to handle accurately. Building on this foundation, these methods can further ensure the convergence of control errors under complex disturbances during space tasks. To address the system becoming under-driven due to actuator failure, Santos et al. (2024) utilized extended model predictive control to point the satellite in the nadir direction in the event of actuator failure. The controller optimizes control variables within a finite prediction time domain through the system model to obtain an approximate optimal solution. Since the control reliability of model predictive control for satellite attitude is closely related to the accuracy of the predictive model describing satellite dynamics, Zhou et al. (2025) developed a new nonlinear scheme based on a multivariate radial basis function autoregressive model to overcome the problems of satellite parameter uncertainty and model mismatch caused by external disturbances during in-orbit control. Pecorilla et al. (2024) proposed a reliable and effective solution for attitude determination and control systems in Earth observation satellites. By combining H–∞ optimal output feedback control and an extended Kalman filter, this framework significantly improves the stability of pointing of satellite systems. By using the derivative of the actual torque as the virtual control law to be designed, Guo et al. (2023) studied a robust continuous controller for solving the challenge of achieving finite-time stability in attitude control of rigid spacecraft systems affected by multiple disturbances. For a flexible communication spacecraft, based on a deep-coupled sliding mode disturbance observer, Bao et al. (2025) proposed a composite attitude control scheme to achieve accurate and rapid suppression of flexible vibration disturbances. Based on the fixed-time terminal sliding mode generating the appropriate angular velocity in the MPC part of the outer layer of the composite controller, Khodaverdian et al. (2023) proposed a dual-loop control approach for spacecraft simulators by integrating the strengths of SMC and model predictive control, enabling simultaneous attitude and reaction wheel momentum control. In this framework, a fixed-time terminal sliding mode surface is employed in the outer MPC loop to generate the desired angular velocity trajectory. To effectively solve the attitude tracking control problem and ensure near-optimal cost, Huang et al. (2024) designed a robust attitude tracking controller for dumbbell-shaped flexible spacecraft affected by actuator saturation and attitude maneuvering speed constraints based on adaptive dynamic programming. Hu et al. (2021) used the same set of actuators to simultaneously achieve attitude control and vibration suppression. The desired input torque for attitude control was derived from an advanced control law, subsequently assigned to actuators via a low-level control allocation strategy that minimizes vibration of flexible modes. In summary, advanced control methods demonstrate a superior capability to handle complex dynamics, and strict constraints, thereby enhancing performance in spacecraft dynamics and control under increasingly demanding operational conditions.

As spacecraft missions grow more complex, the limitations of model-dependent advanced control methods in highly uncertain environments have become increasingly apparent, prompting a shift toward intelligent, data-driven control methods. Owing to significant developments in artificial intelligence technology, intelligent control methods relying on multi-agent systems, neural networks, and other learning-based control methods have been further improved. Compared to traditional control methods, the advantage of intelligent control methods lies in their ability to achieve intelligent decision-making in control systems through data-driven approaches, which enables autonomous adaptation of complex systems and collaborative optimization of multiple task objectives. The application of intelligent control methods in spacecraft control follows the same principle, reducing reliance on manual adjustments of controller parameters in engineering and enabling autonomous optimization of multiple performance metrics. To address the problem of attitude control for model-free fault-tolerant spacecraft, a data-driven adaptive control scheme based on fuzzy logic, lazy adaptation constant inflection point inference rules was proposed by Sun et al. (2023), which is suitable for spacecraft with actuator failures and performance constraints. By combining a type II fuzzy neural network controller, a PD controller, and a robust controller in parallel, a new controller was designed by Liu et al. (2023) to improve the stability of the orbital plane oscillation angle in the attitude stabilization of a magnetically rigid satellite. By using the simulation results of the rocket's nonlinear six-degree-of-freedom model as the objective function value, Fan et al. (2024) employed a genetic algorithm to adaptively adjust parameters within a given range to achieve global optimal attitude control performance. They also analyzed the characteristics of the algorithm's convergence to the global optimal value under specified conditions. Considering external disturbances, inertial uncertainties, and actuator saturation, a self-adaptive neural finite-time command filtering method is employed to design virtual control laws and actual controllers. Zhao et al. (2023) introduced an adaptive finite-time output feedback controller for attitude tracking on the basis of a neural network. The estimator employs an iterative learning algorithm based on a network with radial basis activation functions. Jia et al. (2024) developed a nonlinear learning neural network estimator that achieves fault tolerance control for satellite attitude tracking based on actuator error reconstruction. For spacecraft with actuator failures, performance constraints, and external disturbances, Xiao et al. (2023) converted the attitude stabilization error subject to predefined performance constraints into unconstrained variables and introduced an RL-based fixed-time optimal control strategy to address the attitude stabilization problem. By solving the model via dynamic programming, they derived an optimal policy in the form of a decision table. This policy was subsequently employed as input information to train a neural network, enabling compact and efficient function approximation. Under the premise that spacecraft attitude control is used as a benchmark, Elkins et al. (2022) demonstrated the experimental results of using RL to train neural network spacecraft attitude controllers, thereby promoting its application in online learning scenarios. Mansell et al. (2021) proposed a method for diagnosing time-dependent faults in the attitude determination and control subsystem of a simulated cube satellite using a single-class support vector machine to train a long short-term memory neural network. The results are robust to missing data, false positives, modeling errors, conflicting information, and multiple faults. To address the issue of thruster fault diagnosis in satellite attitude control systems, Xiao et al. (2021) first converted this complex problem into a binary image classification problem, and subsequently proposed a deep learning–based, data-driven diagnostic framework. A key advantage of this approach is its model-free nature: it operates solely on raw measurement data, requiring no prior knowledge of the thruster dynamics or the attitude control system model. Thus, intelligent control methods, by leveraging data-driven learning, adaptive neural structures, and AI-based optimization, enable autonomous and fault-tolerant spacecraft control even under complex and evolving mission conditions.

In summary, for both rigid body and RFC models, spacecraft control methods for spacecraft have made significant progress, evolving from error feedback control methods dependent on system models to considering system optimal performance, and finally to intelligently handling dynamic problems such as attitude deviations in complex situations without physical models. Although this section has reviewed a large number of studies related to spacecraft control, most of these studies have only focused on nominal controller design for spacecraft attitude systems or other related nominal dynamical systems, without explicitly considering the impact of disturbances in the complicated space conditions and uncertain resources existing in the system itself on the entire spacecraft control system. Meanwhile, the safety issues caused by control system responses deviating from nominal responses of the spacecraft dynamic model due to uncertainties are not included in the above research.

1.2.3 Uncertainty-Based Dynamics and Control for Spacecraft

Nominal control methods often yield satisfactory dynamics and control performance for ideal nominal spacecraft systems, as the desired dynamics and control performance of such systems can be effectively achieved through precise pre-simulation calculations or real-time planning and control. However, system uncertainties caused by factors such as perturbations in spacecraft structural parameters, control actuator deviations and degradations, complex space environments, and simplification errors in spacecraft models often significantly impact the dynamics and control accuracy and performance of spacecraft dynamics and control systems. In severe cases, these uncertainties can lead to system failures, resulting in task failures, posing challenges to space task safety, and causing substantial economic losses. Given that uncertainty is inherent in control systems of spacecraft, spacecraft dynamics and control approaches considering uncertainties play a crucial role in improving spacecraft dynamics and control theory, robustness, and adaptability. Numerous uncertainty-based methods derived from uncertainty analysis techniques have been developed for application in spacecraft dynamics and control tasks. Among these, uncertainty modeling serves as the foundation for system and equipment uncertainty design, and a series of corresponding uncertainty-based control methods have been developed based on uncertainty models. Uncertainty-based spacecraft control algorithms can be broadly classified according to the underlying uncertainty modeling framework, including probability theory-based methods, fuzzy theory-based methods, set theory-based methods, etc.

The probability theory-based dynamics and control methods for spacecraft is a kind of widely used methods, which utilizes the probabilistic and statistical characteristics of the data of the relevant task obtained from the ground experimental environment of the spacecraft as well as previous space tasks to realize the uncertainty quantification and uncertainty propagation process of the spacecraft system and control system of it, to derive the entire control system uncertainty response. The advantage of the control method based on probability theory is that it can realize an accurate uncertainty quantification through huge samples, so that the spacecraft control system's uncertain responses can be more accurately obtained. By proposing an adaptive SMC method that ensures stability in practical scenarios of the system even under uncertainty, Hu et al. (2022) explored the attitude tracking problem of a spacecraft-borne imaging platform under system uncertainty. To address control trajectory errors and torque disturbances, Golpashin et al. (2020) proposed an uncertain spacecraft attitude system model that reproduces the stochastic behavior inherent to continuous low-thrust propulsion and achieves better uncertainty control compared to deterministic optimal control. Zhong et al. (2019) applied the extended fault detection scheme based on Hi/H∞ optimization to the satellite attitude control system, and proposed a recursive algorithm to realize online fault detection for probabilistic analysis of satellite attitude control system fault diagnosis. Cao et al. (2020) proposed a flexible spacecraft attitude system considering external disturbances, actuator uncertainties, and inertial parameters uncertainties, and developed a fixed-time robust control strategy for attitude stabilizing, enabling attitude stability that is robust to inertial and actuator disturbances and uncertainties. Lu et al. (2023) investigated the fixed-time controller on a special orthogonal group for spacecraft orientation tracking, and ensured that the spacecraft dynamics under external disturbances and parameter uncertainties can achieve both the attitude constraints and the fixed-time control performance. To better handle disturbances, including unknown nonlinear dynamics, parameter uncertainties, and flexible vibrations, Liu et al. (2022b) developed a hybrid control architecture for spacecraft attitude and position tracking by integrating the SMC and disturbance observer-based control, which further improves attitude-control effectiveness under operational conditions. The aforementioned researches demonstrate the broad application of probability theory in model-based control system. Probability theory-based dynamics and control methods, by systematically quantifying and propagating uncertainties through probabilistic modeling, provide a robust framework for enhancing the precision and reliability of spacecraft attitude systems under multi-source uncertainties.

In addition, some learning-based uncertain dynamics and control methods can also be classified as probability theory-based dynamics and control methods, which compensate for the uncertainty of the spacecraft system by learning some uncertain characteristics of the spacecraft control system to achieve more precise control effects. Meng et al. (2022a) simplified the attitude control system to a nominal form containing synthetic fault and uncertainty terms by adopting incremental nonlinear control technology, thereby studying the optimal fault-tolerant attitude control problem of rigid spacecraft under model uncertainty, control-effector failures, and exogenous disturbance torques. Luo et al. (2025) proposed an online RL algorithm to solve the optimal rotational control issue pertaining to rigid-body satellites' attitude subject to dynamic uncertainty. The method simultaneously derives an estimate of the optimal control policy while ensuring closed-loop system stability and algorithmic convergence. Meng et al. (2022b) addressed the adaptive tracking control problem for a two-link manipulator with rigid and flexible segments, subject to parametric uncertainty, subject to limits on oscillation magnitude. By integrating radial basis function neural networks with SMC, they designed an adaptive controller that achieves precise position control through motion planning and online adaptation. Wang et al. (2023) investigated an intelligent SMC strategy for satellite dynamics under orbital transfer, where system uncertainties arise from an uncertain inertia matrix and disturbance torques induced by the propulsion system. By integrating a neural network to approximate unknown dynamics and an adaptive disturbance observer to estimate perturbations, the method addresses dynamic uncertainty through neural learning and real-time estimation. The control method proposed by

Guan et al. (2025) using the neural network approximation method, the Nussbaum gain technique, and the positive integral function is able to achieve asymptotic attitude tracking under inertial parametric imprecision, exogenous disturbances, and thruster malfunctions, and is rigorously verified by the Lyapunov stability theory. Learning-based probabilistic control methods adaptively compensate for system unknowns through real-time data, enabling more precise and robust spacecraft control under dynamic uncertainties, actuator faults, and environmental disturbances.

Although the uncertainty-based dynamics and control method using probability theory can achieve a more precise characterization of the indeterminate transient performance of the spacecraft control system, the more accurate calculation results often need to rely on a large number of data samples, which results in excessive computational consumption. Even though the computation time can be reduced by adopting more advanced processors, in space task scenarios, it is often hard to obtain those accurate and significant amounts of sample data, which has an effect on the accuracy of the spacecraft uncertainty-based dynamics and control system applying probabilistic theory. Among non-probabilistic approaches, fuzzy theory-based control methods are widely adopted. These methods, rooted in the principles of fuzzy mathematics, construct a linguistic rule base derived from expert knowledge to model and control complex or uncertain systems, and handle nonlinearity and uncertainty by means of the affiliation function to realize robust control that does not rely on precise mathematical models. Sun et al. (2019) addressed the spacecraft attitude tracking problem under the superposition of control saturation, unknown external perturbations, parameter uncertainty, and measurement uncertainty by developing a saturated adaptive hierarchical fuzzy controller designed within a backstepping framework. Considering external disturbances, unknown system uncertainties, control input constraints, and actuator faults, by introducing a fuzzy-inference framework for estimating the uncertainty in the dynamic model, Zou et al. (2011) studied the spacecraft attitude control problem and were able to successfully attain precise pointing accuracy using the proposed controller. Huo et al. (2017) proposed a stable fuzzy-adaptive fault-tolerant scheme for six-DOF attitude tracking of rigid-body spacecraft in the existence of disturbances, unmeasured velocities, actuator failures, and input saturation. By employing a fuzzy logic system to reconstruct uncertain nonlinear behaviors and designing an adaptive fuzzy-based state reconstructor to infer the unmeasurable angular velocity, the method ensures reliable attitude tracking without requiring full state feedback. Focusing on the adaptive fuzzy predefined-time attitude tracking problem of rigid-body spacecraft with state constraints, inertial uncertainty, and external disturbances, Xie et al. (2023b) used fuzzy logic to reconstruct the unmodeled nonlinear behavior and introduced a quadratic fractional Lyapunov function to achieve effective tracking of the spacecraft's attitude in the controller design. Addressing the attitude stabilization of networked flexible spacecraft during large-angle slewing maneuvers, Dong et al. (2009) to compensate for network-induced delays and overcome the negative effects of uncertainty, thus proposing an adaptive fuzzy sliding mode controller for addressing dynamic models that have network-induced delays, nonlinearities, and uncertain parameters. Based on the T–S fuzzy modeling method, Li et al. (2022b) studied the problem of event-triggered integral sliding-mode attitude-tracking control for a nonlinear elastic spacecraft system with the unknown actuator dead-zone and a redundant reaction wheel over digital communication channels. In summary, fuzzy theory-based control methods provide an effective alternative for spacecraft dynamics and control under limited data and model imprecision by utilizing expert knowledge to handle nonlinearity and uncertainty.

Dynamics and control methods based on fuzzy theory are indeed effective in some specific scenarios, but their disadvantages are also more obvious. The disadvantages come from the fundamental drawbacks of fuzzy methods, i.e., the design of the rule base is highly dependent on the experience of experts and lacks a systematic approach, and it is difficult to strictly guarantee the optimality and closed-loop equilibrium retention. The dynamics and control method based on set theory can further make up for the disadvantages of the fuzzy theory-based control method while considering non-probabilistic advantages, so as to accomplish a better uncertainty-based dynamics and control effect of the spacecraft. Because the control method based on set theory only considers the uncertain bound of uncertain parameters when dealing with uncertainty, and thus quantifies the uncertainty based on the set rules, the uncertain responses of the spacecraft can be gained more accurately and quickly by relying on a small number of samples. Zhang et al. (2024) introduce a reliable state feedback control approach for linear systems with uncertainties and a terminal reliability constraint. The method models uncertainties from system parameters and disturbances using multiple probabilistic intervals whose total probability equals one. These intervals are then propagated through the system dynamics to approximately estimate corresponding probabilistic intervals of the control responses. Qiu and Elishakoff (1998) developed a non-probabilistic, interval modeling of uncertain-but-non-random parameters for structures for anti-optimization analysis, consisting of determining the least favorable responses. The uncertain-but-non-random parameter is considered to be a deterministic variable belonging to a set modeled as an interval. Qiu et al. (1996) proposed a solution for problems involving uncertain but non-random interval stiffness and/or mass matrices. If one views the deviation amplitude of the interval matrix as a perturbation around the nominal value of the interval matrix, one can solve the standard eigenvalue problem of the interval matrix by applying the interval extension to the matrix perturbation method. For structures with uncertain-but-bounded parameters, Qiu (2003) established a computational framework based on non-probabilistic convex models and finite element analysis to compute static displacements through numerical and interval methods. Qiu and Wang (2005) employed the parameter perturbation method to estimate the bounds of dynamic responses in structures with uncertain-but-bounded parameters. The uncertainty were set as an interval vector. The first order perturbation quantitics of responses of the perturbed system were obtained through the parameter perturbation method, and then, taking advantage of interval mathematics, a new algorithm to estimate the response interval was presented. Yang et al. (2025) investigate the vibration steady-state behaviors of a non-probabilistic uncertain dual-rotor system with/without rub-impact fault between turbine blades and casings. The effects of individual uncertain parameters and combinations of multiple uncertain parameters are described by the Chebyshev interval algorithm. Set theory-based dynamics and control methods thus provide a computationally efficient framework for spacecraft dynamics and control under uncertainty, overcoming the reliance on expert knowledge in fuzzy approaches while delivering precise and sample-efficient uncertainty quantification through unknown-but-bounded (UBB) parameter sets.

It can be easily seen that various types of uncertainty-based spacecraft dynamics and control methods have produced better results for uncertain dynamics and control tasks compared to nominal methods. At the same time, the number of studies focusing on these methods remains relatively limited compared to the nominal dynamics and control methods, and a large part of them are still in the preliminary stage of research, which has a huge potential for future research as well as development.

1.2.4 Reliability Assessment of Safe Flight for Spacecraft

The complex environments often have a significant impact on spacecraft systems, particularly in space environments, where the existence of a variety of multi-source uncertainties within the system often poses challenges to the reliability of spacecraft dynamics and control systems. System reliability refers to the probability of a system continuously performing its intended functions without failure within specified operating conditions and timeframes (Bartko and Carpenter 1976). To put it simply, reliability for spacecraft means that the spacecraft can first maintain its basic operations stably throughout its designated service life span, while ensuring that the entire control system can precisely execute predefined tasks such as spacecraft attitude control during this period. Additionally, the spacecraft must demonstrate the ability to adapt to environmental disturbances. Typically, in spacecraft dynamics and control, i.e., attitude adjustment scenarios in orbit, due to the influence of those uncertainties mentioned above, the spacecraft dynamics and control system response often fails to reach the previously settled state. Once the response exceeds the safety threshold, the dynamics and control of the spacecraft is deemed to have failed, and such situations are defined as unreliable; conversely, they are considered reliable. Therefore, reliability analysis methods can be proposed to quantify the safety performance of spacecraft dynamics and control systems.

Reliability analysis methods generally follow their definitions and use probability methods. System reliability analysis with probability methods is a process of evaluating the ability of an entire system to operate without failure under specified conditions through mathematical probability models and their probabilistic or statistical characteristics. Its core lies in quantifying the reliability parameters of the entire control responses of the spacecraft dynamics and control system to predict the overall system performance. The confidence-limit assessment for feedback control hardware in spacecraft is typically conducted using probabilistic models to assess the impact of component failure probabilities on the overall system reliability. Tools such as Markov chains and Monte Carlo simulations (MCSs) are commonly employed to calculate the reliability of the system over a specific time period. The spacecraft dynamics and control subsystem is the main driver of spacecraft unreliability, accounting for a large proportion of all spacecraft failures. Therefore, Wayer et al. (2013) provided nonparametric and parametric results for attitude-loop dependability margin, including a mixed Weibull distribution, supplemented by a multi-state failure analysis of the subsystem. Bohlouri et al. (2019) used a reliability-based robust optimization design approach and applied system optimization concepts to the detailed engineering design for the dynamics and control system of spacecraft, improving the robustness of open-loop attitude control under conditions of uncertainty while limiting the probability of failure of orientation regulation under environmental torques. Damircheli et al. (2020) quantified the dependability margin of the attitude control loop of flexible spacecraft in low Earth orbit. A reliability block diagram was developed on the basis of the design structure of the actuator of the attitude subsystem. Based on a simulation approach, a new concept, the fault assessment logic model, was proposed. Liang et al. (2007) proposed passive and active reliable control design schemes and explored the problem of variable structure reliable control in a series of second-order non-affine dynamics and their utilization in spacecraft orientation control and its application in attitude control to achieve flexible, robust, and most effective spacecraft attitude control methods. Hassan et al. (2008) proposed a probabilistic reliability design optimization method on the basis of a composition of genetic algorithms and Monte Carlo sampling for satellite dynamics and control system design. This method uses confidence constraints to ensure that system reliability requirements are met with high probability and further reduces the computational complexity of reliability analysis for satellite attitude control systems. Patton et al. (2008) proposed a diagnostic and fault-decoupling routine for the onboard gyroscope sensors and thrusters in the spacecraft attitude control subsystem on the basis of the Mars Express satellite. The design and optimization are achieved through robust observers or filters. Subsequently, the robustness and reliability characteristics of the robust residual generator are simulated and verified by selecting appropriate performance indicators and combining them with Monte Carlo analysis. Therefore, probabilistic reliability assessment methods provide a rigorous quantitative framework for predicting and enhancing the operational dependability of spacecraft dynamics and control systems, serving as a critical tool for design optimization under uncertainty.

In addition to the classical probabilistic reliability analysis methods for spacecraft dynamics and control systems mentioned above, there are also novel reliability analysis methods proposed based on the characteristics of the uncertainty parameters of the corresponding systems, such as reliability analysis methods with fuzzy theory and reliability analysis methods based on neural networks. These methods often yield highly accurate reliability results for specific types of problems. However, due to limitations in fuzzy rule design and sample data quantity/quality, they do have certain limitations. Nevertheless, these methods are still promising and worthy of further research. Considering the mixed random and fuzzy uncertainties in the system input, Wang et al. (2018) used fuzzy parameters to quantify the cognitive uncertainty associated with expert opinions, thereby proposing a new numerical method for reliability analysis and reliability-based optimization. You et al. (2022) designed a series of reliable fuzzy controllers to achieve exponential attitude stability under random fault conditions, thereby exploring the control problem of non-periodic sampling data for flexible spacecraft with random actuator faults, to show the feasibility and performance of the introduced fuzzy control method. To address the difficulty of solving reliability allocation problems caused by insufficient data and uncertainty in the relationships between components in mechanical systems, Wu et al. (2012) propose a hybrid framework merging linguistic scoring with fuzzy-rule-based Petri net inference to achieve reliability allocation for solar cell arrays. To properly analyze reliability issues in spaceflight, Chen et al. (2018) proposed a space environment-based human reliability analysis method for quantifying the probability of human error in space missions by constructing a Bayesian network (BN) to model these space-based performance-shaping factors and their dependencies. During orbit operation, spacecraft often experience abnormal events that affect reliability due to factors such as the space environment and performance degradation. Yang et al. (2021) developed an improved deviation-from-norm identification method using for reliability analysis of spacecraft telemetry data. Wen et al. (2025) developed a machine learning (ML) architecture to promote the reliability of ML systems and developed a BN model for the corresponding ML system to assess the reliability of outputs. The results indicated that the introduced method outperformed the baseline model considering prediction performance. Therefore, it can be concluded that, the above methods, by systematically design fuzzy expert rules or fully exploring the reliability logic behind massive amounts of data, have certain advantages over model-driven reliability assessment in some application scenarios.

While significant advances have been achieved in the reliability analysis of spacecraft control systems through statistical and probabilistic methods, these approaches are often hindered by high computational costs and limited effectiveness in handling epistemic or non-probabilistic uncertainties. Therefore, there is an urgent need to incorporate non-probabilistic uncertainty models into the design of spacecraft control systems. While uncertain state responses—such as attitude angles, angular velocities, and flexible structural deformations—can reflect dynamic fluctuations, they do not inherently quantify system safety or reliability. This limitation becomes particularly critical when states approach or exceed predefined thresholds. By integrating safety thresholds with indeterminate state trajectories, reliability approaches can quantitatively characterize system behavior and capture the evolution of reliability over time. A reliability model based on stochastic process theory—particularly first-passage failure criteria—has been developed to assess the probability that system states remain within safe bounds throughout the mission. This approach has been successfully applied to spacecraft such as satellites, enabling dynamic reliability evaluation under uncertainty. Additionally, the application of probabilistic reliability methods is often constrained by the requirement for large amounts of sample data, which are typically unavailable in spacecraft applications due to high costs and operational uniqueness. In contrast, set theory-based analysis techniques are well-suited for engineering problems with limited or imprecise data. Consequently, a new class of non-probabilistic reliability models has been developed based on these methods to evaluate the safety and robustness of spacecraft control systems under uncertainty. To deal with bounded uncertainty that affects feedback-loop quality indices during the design process of the LQR controller, Li et al. (2018) extended the concept of non-probabilistic reliability to determine the weight matrix selection of the LQR controller, thereby proposing an approach for analyzing the propagation of uncertainty in active control systems. Tang et al. (2022) proposed a non-probability positioning accuracy reliability measure of a robot end-effector, based on the non-probability theory and method, combined with the prescribed permission interval and error interval, and different states of reliability can be judged according to the position relationship. The non-probability reliability properties are outlined, and the positioning accuracy reliability assessment model is established. Bai et al. (2014) developed a reliability analysis approach for complex engineering problems involving black-box limit state functions by combining response surface technology with convex modeling methods. The introduced method was integrated with the iHL-RF algorithm to compute approximate reliability indices. An iterative response surface updating procedure was implemented to refine the surrogate model, thereby enhancing the accuracy and convergence of the reliability analysis. Given the inadequacy of uncertainty information in actual engineering projects, Wang et al. (2017) developed a non-probabilistic time-dependent reliability (NPTDR) method that merges vibration suppression principles with set-membership uncertainty tools. This approach enables the effective assessment of time-dependent safety margins for feedback-controlled structures subject to interval-valued parametric deviations. Previous works typically assume that uncertain parameters follow a specific probability distribution, often neglecting the time-dependent nature of uncertainties and system responses. Gao et al. (2024) introduced an NPTDR analysis approach for evaluating the motion accuracy of mechanisms, achieving the efficiency and accuracy of local optimization algorithms in non-probabilistic models. In summary, non-probabilistic reliability methods offer a computationally efficient and data-sparse research idea for assessing spacecraft system safety under bounded uncertainty, effectively complementing traditional probabilistic approaches.

The methods listed above are all reliability analysis methods. While these methods can indeed be used to quantitatively evaluate the reliability performance of the spacecraft control system, the analysis process alone is insufficient for the entire on-orbit service task of the spacecraft. Some studies have begun to utilize the reliability results of spacecraft dynamics and control systems as corresponding performance metrics or constraints to construct RBDO for spacecraft (Meng et al. 2021). Such optimization problem construction methods enable spacecraft dynamics and control systems to achieve the desired control performance objectives while satisfying preset safety performance requirements, thereby enabling the accomplishment of highly reliable space tasks. Allen et al. (2004) used a first-order reliability method to approximate system reliability and proposed an analysis and design method that combines RBDO with high-fidelity aeroelastic simulation for the analysis and design of aeroelastic structures under given uncertainty conditions. Nam et al. (2018) proposed a multi-stage RBDO method. This design optimization method considers the subsequent correction decisions triggered by the partial realization values of random parameters selected at certain stages during the engineering design process, thereby further

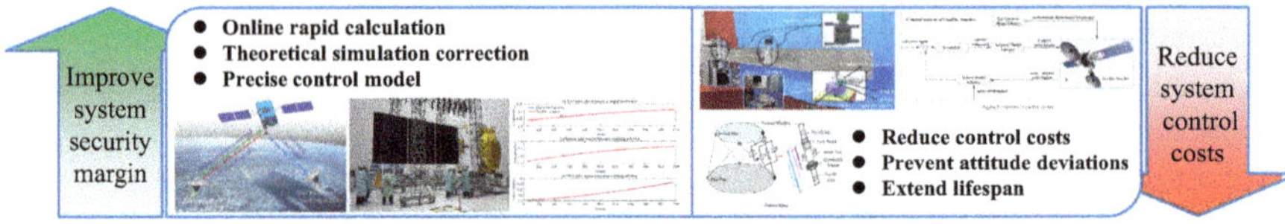

Fig. 1.3 Reliability-based optimization for spacecraft dynamics and control system

enhancing the practicality of RBDO. Paiva et al. (2014) proposed an architecture for simultaneous analysis, robustness, and reliability calculations in aircraft wing design optimization using a Kriging substitution model, avoiding foreseeable implementation issues that may arise in uncertainty-based designs while minimizing the impact of robustness/reliability assessments on performance. Nikbay et al. (2015) further optimized the deorbiting system using deterministic design and RBDO techniques, proposing a design method and developing a dependable end-of-mission drag device for 3USAT that secures swift, low-risk orbit lowering. To perform reliability design, Lü et al. (2022) created reliability constraints using confidence measures related to the inherent characteristics of the system and proposed a reliability-based robust design optimization method for handling uncertainty problems involving imprecise information. Consequently, the aforementioned reliability analysis can be integrated into RBDO, which transforms quantified safety metrics into actionable design constraints, enabling the spacecraft dynamics and control systems to balance high performance with high reliability.

In conclusion, reliability analysis is a vital technique for ensuring the safe operation of the spacecraft controller in orbit. It quantifies the safety performance of spacecraft dynamics and control systems by assessing critical safety thresholds and dynamic failure probabilities. Reliability analysis can also directly avoid serious task failures during spacecraft operation by estimating the results and further optimizing dynamics and control systems based on safety performance. Based on reliability analysis, optimization of spacecraft dynamics and control system can be conducted to both improve system security margin and reduce system control costs, which is illustrated in Fig. 1.3.

1.3 Contents of This Book

Building upon the preceding review of advances in spacecraft dynamics and control with uncertainty quantification and reliability assessment, current methodologies exhibit fundamental limitations in addressing multi-source uncertainty disturbances in space environments and the constraint of limited sample data, thereby impeding the realization of accurate, stable, and rapid control for spacecraft dynamics. To overcome these challenges, this book aims to develop set theory-based frameworks for spacecraft uncertainty dynamics and control.

The contents of this book are organized as follows: In Part I, which contains Chaps. 2 and 3, model reduction is implemented for spacecraft dynamic models with UBB uncertain parameters, thereby improving computational efficiency for complex models and enabling precise characterization of system uncertain responses. In Part II, which contains Chaps. 4–8, set theory-based attitude determination and control methods are introduced. A set theory-based uncertain spacecraft attitude determination method is proposed in Chap. 4, which achieves uncertainty analysis and optimization design for a kind of classical attitude determination problem. In Chaps. 5–8, considering the impact of UBB uncertain parameters on spacecraft attitude dynamics, a series of spacecraft control methods based on set theory are proposed to accurately characterize the uncertain attitude response of the spacecraft system. Set theory-based NPTDR analysis methods are introduced to gauge the safety margin indices of the entire control system and achieve subsequent RBDO of the satellite attitude stabilizing system. In Part III, which contains Chaps. 9 and 10, spacecraft with uncertain attitude-vibration response characteristics due to the RFC model of the spacecraft are investigated, and a series of uncertainty-based control methods are proposed. RBDO is implemented for the entire uncertainty-based control system to simultaneously enhance the safety performance and control performance of the spacecraft systems. Collectively, this book establishes a systematic framework for spacecraft dynamics and control, including model reduction, set-theoretic control design, and RBDO, to advance the precision, efficiency, and robustness of spacecraft dynamics and control under multi-source uncertainty with limited data.

References

Allen M, Maute K (2004) Reliability-based design optimization of aeroelastic structures. Struct Multidisc Optim. 27:228–242. https://doi.org/10.1007/s00158-004-0384-1

Bai YC, Han X, Jiang C, Bi RG (2014) A response-surface-based structural reliability analysis method by using non-probability convex model. Appl Math Model 38:3834–3847. https://doi.org/10.1016/j.apm.2013.11.053

Bai Q, Shehata M, Nada A (2022) Review study of using Euler angles and Euler parameters in multibody modeling of spatial holonomic and non-holonomic systems. Int J Dynam Control 10:1707–1725. https://doi.org/10.1007/s40435-022-00913-9

Bandyopadhyay S, Chung S-J, Hadaegh FY (2016) Nonlinear attitude control of spacecraft with a large captured object. J Guid Control Dyn 39:754–769. https://doi.org/10.2514/1.g001341

Bang H, Tahk M-J, Choi H-D (2003) Large angle attitude control of spacecraft with actuator saturation. Control Eng Pract 11:989–997. https://doi.org/10.1016/s0967-0661(02)00216-2

Bao Z, Zhu Y, Teng H, Cui Y, Guo L (2025) Deep-coupled sliding mode disturbance observer-based composite attitude control for flexible communication spacecrafts. IEEE Trans Ind Electron 72:833–844. https://doi.org/10.1109/TIE.2024.3398670

Bartko JJ, Carpenter WT (1976) On the methods and theory of reliability. J Nerv Ment Dis 163:307–317

Bohlouri V, Jalali-Naini SH (2019) Application of reliability-based robust optimization in spacecraft attitude control with PWPF modulator under uncertainties. J Braz Soc Mech Sci Eng 41:449. https://doi.org/10.1007/s40430-019-1955-9

Cao L, Xiao B, Golestani M (2020) Robust fixed-time attitude stabilization control of flexible spacecraft with actuator uncertainty. Nonlinear Dyn 100:2505–2519. https://doi.org/10.1007/s11071-020-05596-5

Chen J, Zhou D, Lyu C, Zhu X (2018) A method of human reliability analysis and quantification for space missions based on a Bayesian network and the cognitive reliability and error analysis method. Qual Reliab Eng Int 34:912–927. https://doi.org/10.1002/qre.2300

Craig RR Jr, Bampton MC (1968) Coupling of substructures for dynamic analyses. AIAA J 6:1313–1319

Damircheli M, Fakoor M, Yadegari H (2020) Failure assessment logic model (FALM): A new approach for reliability analysis of satellite attitude control subsystem. Reliab Eng Syst Saf 198:106889. https://doi.org/10.1016/j.ress.2020.106889

Mansell JR, Spencer DA (2021) Deep learning fault diagnosis for spacecraft attitude determination and control J Aerosp Inf Syst 18 102 115 https://doi.org/10.2514/1.I010881

Dong C, Xu L, Chen Y, Wang Q (2009) Networked flexible spacecraft attitude maneuver based on adaptive fuzzy sliding mode control. Acta Astronaut 65:1561–1570. https://doi.org/10.1016/j.actaastro.2009.04.004

Dos Santos PH, Chagas RAJ (2024) Satellite attitude control using extended model predictive control (EMPC) under actuator failures. Adv Space Res 74:1314–1326. https://doi.org/10.1016/j.asr.2024.05.010

Elkins JG, Sood R, Rumpf C (2022) Bridging reinforcement learning and online learning for spacecraft attitude control. J Aerosp Inf Syst 19:62–69. https://doi.org/10.2514/1.i010958

Fan X, Bai X, Jiang Z, Zhang S (2024) An optimization method of attitude control parameters based on genetic algorithm for the boost-glide rocket. IEEE Trans Aerosp Electron Syst 60:4356–4376. https://doi.org/10.1109/TAES.2024.3374712

Gao X, Chen L, Chen Z, Yang L (2024) Non-probabilistic time-dependent reliability analysis of mechanism motion accuracy based on extreme value response. J Mech Sci Technol 38:3379–3390. https://doi.org/10.1007/s12206-024-0612-0

Golpashin AE, Yeong HC, Ho K, Sri Namachchivaya N (2020) Spacecraft attitude control: a consideration of thrust uncertainty. J Guid Control Dyn 43:2349–2365. https://doi.org/10.2514/1.G005280

Guan R, Li C, Lv Q (2022) VSCMG system model based on Euler kinematics equation. In Hung JC, Chang J-W, Pei Y, Wu W-C (eds) Innovative computing. Springer Nature, Singapore, pp 409–416

Guan T, Li B, Song Y, Duan G-R (2025) Unified performance control of spacecraft attitude tracking with relaxed quaternion conditions. IEEE Trans Circuits Syst Regul Pap 72:3718–3728. https://doi.org/10.1109/TCSI.2024.3484515

Guo Z, Wang Z, Ding S, Li S, Wang X (2023) Global finite-time stabilization of spacecraft attitude with disturbances via continuous nonlinear controller. IEEE Trans Aerosp Electron Syst 59:2608–2620. https://doi.org/10.1109/TAES.2022.3217199

Han Y (2016) Kane method based dynamics modeling and control study for space manipulator capturing a space target. Int J Aerosp Eng:1–13. https://doi.org/10.1155/2016/7375685

Hao R, Zheng R, Yu H, Tang L, Guan X, Shi D (2025) Design of sampled-data PID controllers with adaptive updating rules for spacecraft attitude angle tracking. IEEE Trans Aerosp Electron Syst:1–16. https://doi.org/10.1109/TAES.2025.3584906

Hasan MN, Haris M, Qin S (2022) Fault-tolerant spacecraft attitude control: A critical assessment. Prog Aerosp Sci 130:100806. https://doi.org/10.1016/j.paerosci.2022.100806

Hassan R, Crossley W (2008) Spacecraft reliability-based design optimization under uncertainty including discrete variables. J Spacecr Rockets 45:394–405. https://doi.org/10.2514/1.28827

Hooker W, Margulies G (1965) The dynamical attitude equations for n-body satellite. J Astronaut Sci

Hu Y, Geng Y, Biggs JD (2021) Simultaneous spacecraft attitude control and vibration suppression via control allocation. J Guid Control Dyn 44:1853–1861. https://doi.org/10.2514/1.G005834

Hu Q, Chen W, Guo L, Biggs JD (2022) Adaptive fixed-time attitude tracking control of spacecraft with uncertainty-rejection capability. IEEE Trans Syst Man Cybern Syst 52:4634–4647. https://doi.org/10.1109/TSMC.2021.3100903

Hu Y, Bernelli-Zazzera F, Geng Y (2023) Distributed optimization method for spacecraft attitude control and vibration suppression. J Guid Control Dyn 46:752–760. https://doi.org/10.2514/1.G007097

Huang W, Ran G, Wang B, Li D, Dong W (2024) Robust attitude tracking control based on adaptive dynamic programming for flexible dumbbell-shaped spacecraft. IEEE Trans Aerosp Electron Syst 60:2394–2406. https://doi.org/10.1109/TAES.2024.3354686

Huo B, Xia Y, Yin L, Fu M (2017) Fuzzy adaptive fault-tolerant output feedback attitude-tracking control of rigid spacecraft. IEEE Trans Syst Man Cybern Syst 47:1898–1908. https://doi.org/10.1109/TSMC.2016.2564918

Hurty WC (1965) Dynamic analysis of structural systems using component modes. AIAA J 3:678–685

Iannelli P, Angeletti F, Gasbarri P (2022) A model predictive control for attitude stabilization and spin control of a spacecraft with a flexible rotating payload. Acta Astronaut 199:401–411. https://doi.org/10.1016/j.actaastro.2022.07.024

Jang B, Lee W, Lee J-J, Jin H (2024) Artificial neural network-based temperature prediction of a lunar orbiter in thermal vacuum test: Data-driven reduced-order models. Aerosp Sci Technol 145:108867. https://doi.org/10.1016/j.ast.2023.108867

Jia Q, Zhang C, Ahn CK, Wu J, Liu M (2024) Reconfigurable fault-tolerant attitude tracking for spacecraft with unknown nonlinear dynamics using neural network estimators with learning-type weight updating. Nonlinear Dyn 112:8213–8227. https://doi.org/10.1007/s11071-024-09502-1

Jin K, Huang G, Lei R, Wei C, Wang Y (2024) Multi-constrained predictive optimal control for spacecraft attitude stabilization and tracking with performance guarantees. Aerosp Sci Technol 155:109599. https://doi.org/10.1016/j.ast.2024.109599

Kane TR, Likins PW, Levinson DA (1983) Spacecraft dynamics

Khodaverdian M, Malekzadeh M (2023) Fixed-time constrained model predictive sliding mode control of spacecraft simulator. IEEE Trans Ind Electron 70:2739–2747. https://doi.org/10.1109/TIE.2022.3167150

Kuznetsov NV, Andrievsky B, Kudryashova EV, Kuznetsova OA (2022) Stability and hidden oscillations analysis of the spacecraft attitude control system using reaction wheels. Aerosp Sci Technol 131:107973. https://doi.org/10.1016/j.ast.2022.107973

Lee D, Springmann J, Spangelo S, Cutler J (2012) Satellite dynamics simulator development using lie group variational integrator. In: AIAA modeling and simulation technologies conference. American Institute of Aeronautics and Astronautics

Li C, Teo KL, Li B, Ma G (2012) A constrained optimal PID-like controller design for spacecraft attitude stabilization. Acta Astronaut 74:131–140. https://doi.org/10.1016/j.actaastro.2011.12.021

Li Z, Wang H, Xu S (2023) Nonlinear model predictive control of a 6DOF spacecraft on lie group SE3. In: 2023 2nd conference on fully actuated system theory and applications. CFASTA, pp 960–965

Li A, Ahn CK, Liu M (2022b) T–s fuzzy-based event-triggering attitude-tracking control for elastic spacecraft with quantization. IEEE Trans Aerosp Electron Syst 58:124–139. https://doi.org/10.1109/TAES.2021.3097952

Li Y, Li S, Xin M (2022a) Dynamic modeling and attitude control of large-scale flexible parallel multibody spacecraft. J Guid Control Dyn 45:2304–2317. https://doi.org/10.2514/1.G007137

Li Y, Xu M, Chen J, Wang X (2018) Nonprobabilistic reliable LQR design method for active vibration control of structures with uncertainties. AIAA J 56:2443–2454. https://doi.org/10.2514/1.J056510

Liang Y-W, Xu S-D, Tsai C-L (2007) Study of VSC reliable designs with application to spacecraft attitude stabilization. IEEE Trans Contr Syst Technol 15:332–338. https://doi.org/10.1109/tcst.2006.883186

Liu X, Zhao T, Cao J, Li P (2023) Design of an interval type-2 fuzzy neural network sliding mode robust controller for higher stability of magnetic spacecraft attitude control. ISA Trans 137:144–159. https://doi.org/10.1016/j.isatra.2023.01.011

Liu Y, Liu X, Cai G, Xu F (2024) Shengyong Tang: Orbit-attitude coupled dynamics modeling and adaptive sliding mode control for detumbling large space debris. Aerosp Sci Technol 151:109269. https://doi.org/10.1016/j.ast.2024.109269

Liu Y, Ma G, Lyu Y, Wang P (2022a) Neural network-based reinforcement learning control for combined spacecraft attitude tracking maneuvers. Neurocomputing 484:67–78. https://doi.org/10.1016/j.neucom.2021.07.099

Liu Z, Zhu Y, Qiao J (2022b) Composite anti-disturbance position and attitude control for spacecrafts with parametric uncertainty and flexible vibration. Chin J Aeronaut 35:242–252. https://doi.org/10.1016/j.cja.2022.01.005

Lü H, Mao H, Huang X, Yin H, Shangguan W-B (2022) An effective approach for reliability-based robust design optimization of uncertain powertrain mounting systems involving imprecise information. Eng Comput 38:1193–1214. https://doi.org/10.1007/s00366-020-01266-7

Lu X, Jia Y, Fu Y, Matsuno F (2023) Constrained attitude control of uncertain spacecraft with appointed-time control performance. IEEE Trans Syst Man Cybern Syst 53:178–190. https://doi.org/10.1109/TSMC.2022.3174845

Luo C, Ning X, Tang R (2025) Online optimal attitude stabilization via reinforcement learning for rigid spacecraft with dynamic uncertainty. IEEE Trans Aerosp Electron Syst:1–12. https://doi.org/10.1109/TAES.2025.3564579

Mansell JR, Spencer DA (2021) Deep learning fault diagnosis for spacecraft attitude determination and control J Aerosp Inf Syst 18 102 115 https://doi.org/10.2514/1.I010881

National Academies of Sciences, Medicine, Division on Engineering, Physical Sciences, Aeronautics, Space Engineering Board, & Committee on NASA Technology Roadmaps (2016) NASA space technology roadmaps and priorities revisited. National Academies Press

Meirovitch L (2010) Methods of analytical dynamics. Courier Corporation

Meng Z, Li G, Wang X, Sait SM, Yildiz AR (2021) A comparative study of metaheuristic algorithms for reliability-based design optimization problems. Arch Comput Methods Eng 28:1853–1869. https://doi.org/10.1007/s11831-020-09443-z

Meng Q, Lai X, Yan Z, Su C-Y, Wu M (2022b) Motion planning and adaptive neural tracking control of an uncertain two-link rigid–flexible manipulator with vibration amplitude constraint. IEEE Trans Neural Netw Learn Syst 33:3814–3828. https://doi.org/10.1109/TNNLS.2021.3054611

Meng Q, Yang H, Jiang B (2022a) Fault-tolerant optimal spacecraft attitude maneuver: An incremental model approach. J Guid Control Dyn 45:1676–1691. https://doi.org/10.2514/1.G006417

Moore B (2003) Principal component analysis in linear systems: Controllability, observability, and model reduction. IEEE Trans Autom Control 26:17–32

Najafizadeh Sari N, Jahanshahi H, Fakoor M (2019) Adaptive fuzzy PID control strategy for spacecraft attitude control. Int J Fuzzy Syst 21:769–781. https://doi.org/10.1007/s40815-018-0576-2

Nam T, Mavris DN (2018) Multistage reliability-based design optimization and application to aircraft conceptual design. J Aircr 55:2022–2036. https://doi.org/10.2514/1.C032099

Nicassio F, Fattizzo D, Giannuzzi M, Scarselli G, Avanzini G (2022) Attitude dynamics and control of a large flexible space structure by means of a minimum complexity model. Acta Astronaut 198:124–134. https://doi.org/10.1016/j.actaastro.2022.05.047

Nikbay M, Acar P, Aslan AR (2015) Development of a nanosatellite de-orbiting system by reliability based design optimization. Acta Astronaut 117:469–483. https://doi.org/10.1016/j.actaastro.2015.09.002

Paiva RM, Crawford C, Suleman A (2014) Robust and reliability-based design optimization framework for wing design. AIAA J 52:711–724. https://doi.org/10.2514/1.J052161

Patton RJ, Uppal FJ, Simani S, Polle B (2008) Reliable fault diagnosis scheme for a spacecraft attitude control system. Proc Inst Mech Eng Part O J Risk Reliab. https://doi.org/10.1243/1748006XJRR98

Pecorilla M, Stesina F (2024) Novel extended Kalman filter + infty optimal output feedback control configuration for small satellites with high pointing stability requirements. Int J Robust Nonlinear Control 34:2991–3010. https://doi.org/10.1002/rnc.7119

Peng H, Song N, Kan Z (2022) Data-driven model order reduction with proper symplectic decomposition for flexible multibody system. Nonlinear Dyn 107:173–203. https://doi.org/10.1007/s11071-021-06990-3

Pezent JB, Sood R (2022) New solution to Euler's equations for asymmetric bodies: Applications to spacecraft reorientation. J Guid Control Dyn 45:2289–2303. https://doi.org/10.2514/1.g006747

Qian J, Wang Y, Song H, Pant K, Peabody H, Ku J, Butler CD (2015) Projection-based reduced-order modeling for spacecraft thermal analysis. J Spacecr Rockets 52:978–989. https://doi.org/10.2514/1.a33117

Qiao S, Zhang H, Xie F, Jiang Z (2024) Deep-learning-based direct attitude estimation for uncooperative known space objects. IEEE Trans Aerosp Electron Syst 60:2526–2541. https://doi.org/10.1109/TAES.2023.3325801

Qiu Z (2003) Comparison of static response of structures using convex models and interval analysis method. Int J Numer Methods Eng 56:1735–1753. https://doi.org/10.1002/nme.636

Qiu Z, Chen S, Elishakoff I (1996) Bounds of eigenvalues for structures with an interval description of uncertain-but-non-random parameters. Chaos Solitons Fractals 7:425–434. https://doi.org/10.1016/0960-0779(95)00065-8

Qiu Z, Elishakoff I (1998) Antioptimization of structures with large uncertain-but-non-random parameters via interval analysis. Comput Methods Appl Mech Eng 152:361–372. https://doi.org/10.1016/S0045-7825(96)01211-X

Qiu Z, Wang X (2005) Parameter perturbation method for dynamic responses of structures with uncertain-but-bounded parameters based on interval analysis. Int J Solids Struct 42:4958–4970. https://doi.org/10.1016/j.ijsolstr.2005.02.023

Renganathan SA (2020) Koopman-based approach to nonintrusive reduced order modeling: Application to aerodynamic shape optimization and uncertainty propagation. AIAA J 58:2221–2235. https://doi.org/10.2514/1.j058744

Silvestrini S, Lavagna M (2021) Neural-based predictive control for safe autonomous spacecraft relative maneuvers. J Guid Control Dyn 44:2303–2310. https://doi.org/10.2514/1.G005481

Sun L, Zheng Z (2019) Saturated adaptive hierarchical fuzzy attitude-tracking control of rigid spacecraft with modeling and measurement uncertainties. IEEE Trans Ind Electron 66:3742–3751. https://doi.org/10.1109/TIE.2018.2856204

Sun X, Shen Q, Wu S (2023) Fuzzy supervised learning-based model-free adaptive fault-tolerant spacecraft attitude control with deferred asymmetric constraints. IEEE Trans Aerosp Electron Syst 59:8884–8900. https://doi.org/10.1109/TAES.2023.3312363

Tang Z, Peng J, Sun J, Meng X, Tang Z, Peng J, Sun J, Meng X (2022) Non-probabilistic reliability analysis of robot accuracy under uncertain joint clearance. Machines 10. https://doi.org/10.3390/machines10100917

Wang B, Liu Z, Zheng P (2022) Rigid-flexible coupling dynamic modeling and analysis of dumbbell-shaped spacecraft. Aerosp Sci Technol 126:107641. https://doi.org/10.1016/j.ast.2022.107641

Wang C, Matthies HG, Xu M, Li Y (2018) Hybrid reliability analysis and optimization for spacecraft structural system with random and fuzzy parameters. Aerosp Sci Technol 77:353–361. https://doi.org/10.1016/j.ast.2018.03.014

Wang K, Lu F, Chen Z, Li J (2024) Real-time optimal control for attitude-constrained solar sailcrafts via neural networks. Acta Astronaut 216:446–458. https://doi.org/10.1016/j.actaastro.2024.01.026

Wang L, Wang X, Li Y, Lin G, Qiu Z (2017) Structural time-dependent reliability assessment of the vibration active control system with unknown-but-bounded uncertainties: Time-variant reliability for controlled system with UBB uncertainties. Struct Control Health Monit 24:1965. https://doi.org/10.1002/stc.1965

Wang P, Zhang X (2022) Optimized Bézier-curve-based command generation and robust inverse optimal control for attitude tracking of spacecraft. Aerosp Sci Technol 121:107183. https://doi.org/10.1016/j.ast.2021.107183

Wang X, Xu B, Pan Y (2023) Neural network-based sliding mode control for satellite attitude tracking. Adv Space Res 71:3565–3573. https://doi.org/10.1016/j.asr.2022.04.047

Wayer JK, Castet J-F, Saleh JH (2013) Spacecraft attitude control subsystem: Reliability, multi-state analyses, and comparative failure behavior in LEO and GEO. Acta Astronaut 85:83–92. https://doi.org/10.1016/j.actaastro.2012.12.003

Wen Q, Machida F (2025) Reliability modeling for three-version machine learning systems through Bayesian networks. Reliab Eng Syst Saf 260:111016. https://doi.org/10.1016/j.ress.2025.111016

Wu J, Yan S, Xie L, Gao P (2012) Reliability apportionment approach for spacecraft solar array using fuzzy reasoning Petri net and fuzzy comprehensive evaluation. Acta Astronaut 76:136–144. https://doi.org/10.1016/j.actaastro.2012.02.023

Xiao Y, Ruiter J, AH, Y, D, S, Z (2020) Attitude tracking control for rigid-flexible coupled spacecraft with guaranteed performance bounds. J Guid Control Dyn 43:327–337. https://doi.org/10.2514/1.G004535

Xiao B, Yin S (2021) A deep learning based data-driven thruster fault diagnosis approach for satellite attitude control system. IEEE Trans Ind Electron 68:10162–10170. https://doi.org/10.1109/TIE.2020.3026272

Xiao B, Zhang H, Chen Z, Cao L (2023) Fixed-time fault-tolerant optimal attitude control of spacecraft with performance constraint via reinforcement learning. IEEE Trans Aerosp Electron Syst 59:7715–7724. https://doi.org/10.1109/TAES.2023.3292809

Xie S, Chen Q, He X (2023a) Predefined-time approximation-free attitude constraint control of rigid spacecraft. IEEE Trans Aerosp Electron Syst 59:347–358. https://doi.org/10.1109/TAES.2022.3183550

Xie S, Chen Q, Yang Q (2023b) Adaptive fuzzy predefined-time dynamic surface control for attitude tracking of spacecraft with state constraints. IEEE Trans Fuzzy Syst 31:2292–2304. https://doi.org/10.1109/TFUZZ.2022.3223253

Yang L, Ma Y, Zeng F, Peng X, Liu D (2021) Improved deep learning based telemetry data anomaly detection to enhance spacecraft operation reliability. Microelectron Reliab 126:114311. https://doi.org/10.1016/j.microrel.2021.114311

Yang Y (2019) Spacecraft modeling, attitude determination, and control quaternion-based approach. CRC Press, Boca Raton, FL

Yang Y, Bentz W, Lewis L (2024) A systematic methodology for modeling and attitude control of multibody space telescopes. IEEE Trans Aerosp Electron Syst 60:5359–5372. https://doi.org/10.1109/TAES.2024.3390648

Yang Y, Ma T, Zeng J, Chen G, Song S, Li C (2025) An integrated simulation and experimental study on the dynamics of a non-probabilistic uncertain dual rotor system. Nonlinear Dyn 113:25503–25529. https://doi.org/10.1007/s11071-025-11507-3

Yang Z, Xing L, Gu Z, Xiao Y, Zhou Y, Huang Z, Xue L (2023) Model-based reinforcement learning and neural-network-based policy compression for spacecraft rendezvous on resource-constrained embedded systems. IEEE Trans Ind Inf. 19:1107–1116. https://doi.org/10.1109/tii.2022.3192085

Yefymenko N, Kudermetov R (2022) Quaternion models of a rigid body rotation motion and their application for spacecraft attitude control. Acta Astronaut 194:76–82. https://doi.org/10.1016/j.actaastro.2022.01.029

You Z, Yan H, Sun J, Zhang H, Li Z (2022) Reliable control for flexible spacecraft systems with aperiodic sampling and stochastic actuator failures. IEEE Trans. Cybern. 52:3434–3445. https://doi.org/10.1109/TCYB.2020.3008045

Yu C, Luo H, Luo Z, He F (2025) Data-driven dynamic modeling method for spacecraft equipment based on nondestructive vibration tests. AIAA J 63:4465–4478. https://doi.org/10.2514/1.J065629

Zhang F, Duan G (2023) Coupled dynamics and integrated control for position and attitude motions of spacecraft: a survey. IEEECAA J Autom Sin 10:2187–2208. https://doi.org/10.1109/JAS.2023.123306

Zhang J, Wang E, Qiao J, Qiu Z (2024) Reliable state feedback control for uncertain linear systems with terminal reliability constraint. IEEE Trans Syst Man Cybern Syst 54:2833–2845. https://doi.org/10.1109/TSMC.2024.3351121

Zhao L, Yu J, Chen X (2023) Neural-network-based adaptive finite-time output feedback control for spacecraft attitude tracking. IEEE Trans Neural Netw Learn Syst 34:8116–8123. https://doi.org/10.1109/TNNLS.2022.3144493

Zhong M, Liu C, Zhou D, Li W, Xue T (2019) Probability analysis of fault diagnosis performance for satellite attitude control systems. IEEE Trans Ind Inf 15:5867–5876. https://doi.org/10.1109/TII.2019.2907382

Zhong R, Zhao Y, Wang H (2022) Rigid-flexible coupling identification and attitude control based on deep neural networks. Adv Space Res 69:1538–1549. https://doi.org/10.1016/j.asr.2021.10.057

Zhou Y, Hu Y, Ling K-V, Ding F (2025) Hybrid two-stage identification-based nonlinear MPC strategy for satellite attitude control. IEEE Trans Aerosp Electron Syst:1–12. https://doi.org/10.1109/TAES.2025.3543466

Zou A-M, Kumar KD (2011) Adaptive fuzzy fault-tolerant attitude control of spacecraft. Control Eng Pract 19:10–21. https://doi.org/10.1016/j.conengprac.2010.08.005

Part I
Set Theory-Based Reduced-Order Model

Chapter 2
Interval-Based Reduced-Order Model Method

Abstract This chapter introduces an interval-based reduced-order model (IROM) framework designed to handle high-dimensional linear systems subject to unknown-but-bounded (UBB) parameter uncertainties. In contrast to conventional probabilistic techniques that require extensive sampling of data, the proposed method leverages interval analysis to ensure robustness in data-sparse scenarios. The formulation begins by establishing interval state-space equations, which are then converted to a balanced realization form to preserve system invariants. To accurately quantify the uncertainty propagation, bounds of the observability and controllability Gramians, and the Hankel singular values (HSVs), are derived using interval Lyapunov equations and a customized interval perturbation-based singular value decomposition (SVD) technique. Furthermore, addressing the issue of dense interval distributions of HSV in the system, a novel truncation criterion based on interval possibility theory is developed to objectively determine the optimal reduced order. Numerical validations involving mechanical systems demonstrate that the IROM approach achieves approximation accuracy comparable to Monte Carlo simulations (MCSs) while offering superior computational efficiency.

2.1 Introduction

In the realm of modern engineering, the increasing complexity of dynamics systems has brought high-dimensional problems to the forefront, demanding efficient solutions. While advancements in high-performance computing have boosted the ability to manage these large-scale challenges, the need for further optimization persists. This is where reduced-order models (ROMs) come into play, offering a streamlined approach that balances accuracy and computational efficiency. ROMs are designed to approximate the response of a full-order system with a simpler, lower-order model, significantly reducing the computational resources required while maintaining essential performance characteristics (Strazzullo and Vicini 2023; Ivagnes et al. 2023; Zhang et al. 2024).

C. Yang and Y. Xia, *Set Theory-Based Spacecraft Dynamics and Control*, Intelligent Control and Learning Systems 26, https://doi.org/10.1007/978-981-95-9179-4_2

Since their inception by Mullis and Roberts (1976) and later developments by Moore (1981), ROMs have become a cornerstone in control theory, facilitating numerous applications across dynamic systems (Farlessyost and Singh 2022), vibration control (Zhang et al. 2020), inverse problems (Xu et al. 2024), and model predictive control (Chen et al. 2022a). The ability to estimate and manage the error between reduced and full-order models has allowed these techniques to maintain consistency and reliability, even in complex scenarios.

However, the practical application of control systems often involves uncertainties, which can complicate the modeling process (Pu et al. 2023; Chen et al. 2022b; Shen et al. 2021). Traditional probabilistic methods, though powerful, can be resource-intensive and less effective when dealing with limited data. This has resulted in the exploration of interval-based approaches as a promising alternative, offering robustness in uncertain environments without the need for extensive probabilistic analysis (Lee et al. 2024; Zhu et al. 2022; Mierlo et al. 2022; Wang et al. 2022).

Despite the potential of IROMs, current research in this area remains limited. This chapter is dedicated to addressing these issues. It is structured as the following three parts: First, we establish the foundational interval state-space equation and explore its solutions in the absence of high-order uncertainties. Next, we delve into the construction of a balanced interval-valued state realization, followed by the introduction of an interval-based SVD method for uncertainty bounds estimation. Finally, we present the interval order truncation and discuss optimization techniques for state selection.

Based on the interval perturbation method, this chapter develops a tool for uncertainty analysis in a control system. The proposed IROM framework provides an efficient, non-probabilistic framework for uncertainty quantification, extending the application of IROM in modern control systems.

2.2 Interval Uncertainty-Based State-Space Equation

2.2.1 First Type of Set Theory: Interval Analysis Method

Although probability and statistics methods have achieved significant progress and are widely applied in dynamics and control theory and related areas, they generally rely on a large amount of sample data. The acquisition of such samples is often costly, making probabilistic and statistical approaches less suitable when only limited data are available.

In contrast, interval analysis methods are advantageous for problems with small sample sizes. Most uncertain parameters can be treated as UBB uncertainties, which makes interval analysis particularly effective for quantitative characterization. After determining the upper bounds (UB) and lower bounds (LB) of uncertain variables, their corresponding intervals can be defined precisely. Accordingly, interval parameters can be expressed as

$$\underline{\mathbf{b}} \le \mathbf{b} \le \overline{\mathbf{b}}, \quad \underline{b_k} \le b_k \le \overline{b_k}, \quad k = 1, 2, \cdots, m \tag{2.1}$$

Here $\mathbf{b}$ represents the interval vector, m indicates the count of interval variables, b_k indicates the kth element of $\mathbf{b}$, and represent the corresponding LB/UB. Thus, Eq. (2.1) can be rewritten as

$$\mathbf{b} = \mathbf{b}^{\mathrm{c}} + \Delta\mathbf{b}^{\mathrm{I}} = \mathbf{b}^{\mathrm{c}} + \Delta\mathbf{b}\xi \in \mathbf{b}^{\mathrm{I}} = \left[\underline{\mathbf{b}}, \overline{\mathbf{b}}\right] = \left[\mathbf{b}^{\mathrm{c}} - \Delta\mathbf{b}, \mathbf{b}^{\mathrm{c}} + \Delta\mathbf{b}\right] \tag{2.2}$$

where $\xi = [-1, 1]$. $\mathbf{b}^{\mathrm{I}}$ represents the interval vector, with the central vector $\mathbf{b}^{\mathrm{c}}$ and the radius vector $\Delta\mathbf{b}$ defined as

$$\mathbf{b}^{\mathrm{c}} = \left\{b_k^{\mathrm{c}}\right\} = \frac{1}{2}\left(\overline{\mathbf{b}} + \underline{\mathbf{b}}\right), \quad \Delta\mathbf{b} = \left\{\Delta b_k\right\} = \frac{1}{2}\left(\overline{\mathbf{b}} - \underline{\mathbf{b}}\right) \tag{2.3}$$

Thus, the component-wise interval description is given by

$$\begin{aligned} & b_k = b_k^{\mathrm{c}} + \Delta b_k^{\mathrm{I}} = b_k^{\mathrm{c}} + \Delta b_k \xi \in b_k^{\mathrm{I}} = \left[\underline{b_k}, \overline{b_k}\right] = \left[b_k^{\mathrm{c}} - \Delta b_k, b_k^{\mathrm{c}} + \Delta b_k\right] \\ & b_k^{\mathrm{c}} = \left(\overline{b_k} + \underline{b_k}\right)/2, \quad \Delta b_k = \left(\overline{b_k} - \underline{b_k}\right)/2 \end{aligned} \tag{2.4}$$

2.2.2 *Framework of Interval Uncertainty-Based State-Space Equation*

Consider a linear time-invariant (LTI) state-space system:

$$\begin{cases} \dot{\mathbf{x}} = \mathbf{A}\mathbf{x} + \mathbf{B}\mathbf{u} & \mathbf{x} \in \mathbb{R}^n, \mathbf{u} \in \mathbb{R}^p \\ \mathbf{y} = \mathbf{C}\mathbf{x} + \mathbf{D}\mathbf{u} & \mathbf{y} \in \mathbb{R}^q \end{cases} \tag{2.5}$$

Here $\mathbf{x}$, $\mathbf{u}$, $\mathbf{y}$ is the state, input, and output vectors, respectively; $\mathbf{A}$, $\mathbf{B}$, $\mathbf{C}$, $\mathbf{D}$ represent the four matrices that respectively represent the system matrix, input matrix, output matrix, and transfer matrix.

Equation (2.5) describes only the deterministic model, without considering uncertainties. In practice, uncertainties, modeling errors, and incomplete information are always present. To account for these, the deterministic model in Eq. (2.5) can be reformulated using interval analysis with UBB uncertainties as

$$\begin{cases} \dot{\mathbf{x}}^{\mathrm{I}} = \mathbf{A}^{\mathrm{I}}\mathbf{x}^{\mathrm{I}} + \mathbf{B}^{\mathrm{I}}\mathbf{u}^{\mathrm{I}} & \mathbf{x}^{\mathrm{I}} \in \mathbb{R}^n, \mathbf{u}^{\mathrm{I}} \in \mathbb{R}^p \\ \mathbf{y}^{\mathrm{I}} = \mathbf{C}^{\mathrm{I}}\mathbf{x}^{\mathrm{I}} + \mathbf{D}^{\mathrm{I}}\mathbf{u}^{\mathrm{I}} & \mathbf{y}^{\mathrm{I}} \in \mathbb{R}^q \end{cases} \text{or} \left(\begin{array}{c|c} \mathbf{A}^{\mathrm{I}} & \mathbf{B}^{\mathrm{I}} \\ \hline \mathbf{C}^{\mathrm{I}} & \mathbf{D}^{\mathrm{I}} \end{array}\right) \tag{2.6}$$

where ^{I} indicates the interval uncertainties, expressed as

$$\Xi^{\mathrm{I}}(\mathbf{b})=\Xi^{\mathrm{c}}(\mathbf{b})+\Delta\Xi^{\mathrm{I}}(\mathbf{b})=\Xi^{\mathrm{c}}(\mathbf{b})+\Delta\Xi(\mathbf{b})\xi=\Xi^{\mathrm{c}}(\mathbf{b})+\left[\sum_{k=1}^{m}\left(\partial\Xi(\mathbf{b})/\partial b_k\right)\Delta b_k\right]\xi \tag{2.7}$$

where Ξ can stand for the control system in Eq. (2.5). Therefore, Eq. (2.6) can be expanded as

$$\begin{cases} \dot{\mathbf{x}}^{\mathrm{c}}+\Delta\dot{\mathbf{x}}^{\mathrm{I}} \approx \left(\mathbf{A}^{\mathrm{c}}+\Delta\mathbf{A}^{\mathrm{I}}\right)\left(\mathbf{x}^{\mathrm{c}}+\Delta\mathbf{x}^{\mathrm{I}}\right)+\left(\mathbf{B}^{\mathrm{c}}+\Delta\mathbf{B}^{\mathrm{I}}\right)\left(\mathbf{u}^{\mathrm{c}}+\Delta\mathbf{u}^{\mathrm{I}}\right) \\ \mathbf{y}^{\mathrm{c}}+\Delta\mathbf{y}^{\mathrm{I}} \approx \left(\mathbf{C}^{\mathrm{c}}+\Delta\mathbf{C}^{\mathrm{I}}\right)\left(\mathbf{x}^{\mathrm{c}}+\Delta\mathbf{x}^{\mathrm{I}}\right)+\left(\mathbf{D}^{\mathrm{c}}+\Delta\mathbf{D}^{\mathrm{I}}\right)\left(\mathbf{u}^{\mathrm{c}}+\Delta\mathbf{u}^{\mathrm{I}}\right) \end{cases} \tag{2.8}$$

Keep only the linear term and discard all higher-order contributions. Thus, the interval state-space model is obtained:

$$\begin{cases} \dot{\mathbf{x}}^{\mathrm{c}} = \mathbf{A}^{\mathrm{c}}\mathbf{x}^{\mathrm{c}} + \mathbf{B}^{\mathrm{c}}\mathbf{u}^{\mathrm{c}} \\ \mathbf{y}^{\mathrm{c}} = \mathbf{C}^{\mathrm{c}}\mathbf{x}^{\mathrm{c}} + \mathbf{D}^{\mathrm{c}}\mathbf{u}^{\mathrm{c}} \end{cases} \tag{2.9}$$

and

$$\begin{cases} \Delta\dot{\mathbf{x}}^{\mathrm{I}} = \mathbf{A}^{\mathrm{c}}\Delta\mathbf{x}^{\mathrm{I}}+\Delta\mathbf{A}^{\mathrm{I}}\mathbf{x}^{\mathrm{c}}+\mathbf{B}^{\mathrm{c}}\Delta\mathbf{u}^{\mathrm{I}}+\Delta\mathbf{B}^{\mathrm{I}}\mathbf{u}^{\mathrm{c}} \\ \Delta\mathbf{y}^{\mathrm{I}} = \mathbf{C}^{\mathrm{c}}\Delta\mathbf{x}^{\mathrm{I}}+\Delta\mathbf{C}^{\mathrm{I}}\mathbf{x}^{\mathrm{c}}+\mathbf{D}^{\mathrm{c}}\Delta\mathbf{u}^{\mathrm{I}}+\Delta\mathbf{D}^{\mathrm{I}}\mathbf{u}^{\mathrm{c}} \end{cases} \tag{2.10}$$

Remark 2.1 Aiming at solving the interval state-space model as expressed in Eqs. (2.9) and (2.10), the nominal and interval parts of the input/output vectors are, respectively, combined as

$$\begin{cases} \begin{bmatrix} \dot{\mathbf{x}}^{\mathrm{c}} \\ \Delta\dot{\mathbf{x}}^{\mathrm{I}} \end{bmatrix} = \begin{bmatrix} \mathbf{A}^{\mathrm{c}} & \mathbf{0} \\ \Delta\mathbf{A}^{\mathrm{I}} & \mathbf{A}^{\mathrm{c}} \end{bmatrix} \begin{bmatrix} \mathbf{x}^{\mathrm{c}} \\ \Delta\mathbf{x}^{\mathrm{I}} \end{bmatrix} + \begin{bmatrix} \mathbf{B}^{\mathrm{c}} & \mathbf{0} \\ \Delta\mathbf{B}^{\mathrm{I}} & \mathbf{B}^{\mathrm{c}} \end{bmatrix} \begin{bmatrix} \mathbf{u}^{\mathrm{c}} \\ \Delta\mathbf{u}^{\mathrm{I}} \end{bmatrix} \\ \begin{bmatrix} \mathbf{y}^{\mathrm{c}} \\ \Delta\mathbf{y}^{\mathrm{I}} \end{bmatrix} = \begin{bmatrix} \mathbf{C}^{\mathrm{c}} & \mathbf{0} \\ \Delta\mathbf{C}^{\mathrm{I}} & \mathbf{C}^{\mathrm{c}} \end{bmatrix} \begin{bmatrix} \mathbf{x}^{\mathrm{c}} \\ \Delta\mathbf{x}^{\mathrm{I}} \end{bmatrix} + \begin{bmatrix} \mathbf{D}^{\mathrm{c}} & \mathbf{0} \\ \Delta\mathbf{D}^{\mathrm{I}} & \mathbf{D}^{\mathrm{c}} \end{bmatrix} \begin{bmatrix} \mathbf{u}^{\mathrm{c}} \\ \Delta\mathbf{u}^{\mathrm{I}} \end{bmatrix} \end{cases} \tag{2.11}$$

which can be further reconstructed as follows:

$$\begin{cases} \dot{\mathcal{X}} = \mathcal{A}\mathcal{X} + \mathcal{B}\mathcal{U} & \dot{\mathcal{X}} \in \mathbb{R}^{2n}, \mathcal{U} \in \mathbb{R}^{2p} \\ \mathcal{Y} = \mathcal{C}\mathcal{X} + \mathcal{D}\mathcal{U} & \mathcal{Y} \in \mathbb{R}^{2q} \end{cases} \tag{2.12}$$

where $\mathcal{A} = \begin{bmatrix} \mathbf{A}^{\mathrm{c}} & \mathbf{0} \\ \Delta\mathbf{A}^{\mathrm{I}} & \mathbf{A}^{\mathrm{c}} \end{bmatrix}$ and $\dot{\mathcal{X}} = \left[(\dot{\mathbf{x}}^{\mathrm{c}})^{\mathrm{T}}\ (\Delta\dot{\mathbf{x}}^{\mathrm{c}})^{\mathrm{T}} \right]^{\mathrm{T}}$. The operator ^{T} indicates transpose.

Therefore, the interval order-expanded state-space model is solved in the classical case, yielding the uncertain vectors $\Delta\mathbf{x}^{\mathrm{I}}$.

$$\Delta\mathbf{x}^{\mathrm{I}}=\sum_{k=1}^{m}\left(\partial\mathbf{x}(\mathbf{b})/\partial b_k\right)\Delta b_k \tag{2.13}$$

Thus, the radius can be calculated as

$$\Delta\mathbf{x}=\sum_{k=1}^{m}\left|\partial\mathbf{x}(\mathbf{b})/\partial b_k\right|\Delta b_k \tag{2.14}$$

Furthermore, the LB/UB of the interval state vector are

$$\begin{cases}\underline{\mathbf{x}}=\mathbf{x}^{\mathrm{c}}(\mathbf{b})-\sum_{k=1}^{m}\left|\partial\mathbf{x}(\mathbf{b})/\partial b_k\right|\Delta b_k\\ \overline{\mathbf{x}}=\mathbf{x}^{\mathrm{c}}(\mathbf{b})+\sum_{k=1}^{m}\left|\partial\mathbf{x}(\mathbf{b})/\partial b_k\right|\Delta b_k\end{cases} \tag{2.15}$$

The same approach applies to obtain the uncertainty radii of $\Delta\dot{\mathbf{x}}$ and $\Delta\mathbf{y}$. Therefore, both uncertain states and outputs can be determined using the interval state-space model.

2.2.3 Solution of Interval Uncertainty-Based State-Space Equation

As the foundation for the uncertainty equation, the classical nominal state-space equation solution is as follows:

$$\mathbf{x}(t)=\exp(\mathbf{A}t)\mathbf{x}(0)+\int_0^t\exp[\mathbf{A}(t-\tau)]\mathbf{B}\mathbf{u}(\tau)\mathrm{d}\tau \tag{2.16}$$

The interval state-space is developed through Taylor expansion as

$$\mathbf{x}^{\mathrm{c}}(t)=\exp(\mathbf{A}^{\mathrm{c}}t)\mathbf{x}^{\mathrm{c}}(0)+\int_0^t\exp(\mathbf{A}^{\mathrm{c}}(t-\tau))\mathbf{B}^{\mathrm{c}}\mathbf{u}^{\mathrm{c}}(\tau)\mathrm{d}\tau \tag{2.17}$$

$$\begin{aligned}\Delta\mathbf{x}^{\mathrm{I}}(t)&=\exp(\mathbf{A}^{\mathrm{c}}t)\Delta\mathbf{x}^{\mathrm{I}}(0)+\Delta\exp\left(\mathbf{A}^{\mathrm{I}}t\right)\mathbf{x}^{\mathrm{c}}(0)\\&\quad+\int_0^t\exp(\mathbf{A}^{\mathrm{c}}(t-\tau))\left[\mathbf{B}^{\mathrm{c}}\Delta\mathbf{u}^{\mathrm{I}}(\tau)+\Delta\mathbf{B}^{\mathrm{I}}\mathbf{u}^{\mathrm{c}}(\tau)\right]\mathrm{d}\tau\\&\quad+\int_0^t\Delta\exp\left(\mathbf{A}^{\mathrm{I}}(t-\tau)\right)\mathbf{B}^{\mathrm{c}}\mathbf{u}^{\mathrm{c}}(\tau)\mathrm{d}\tau\end{aligned} \tag{2.18}$$

where $\mathbf{x}^c(0)$ and $\Delta\mathbf{x}^I(0)$ denote the initial deterministic and uncertain states. By integrating Eqs. (2.17) and (2.18), the Eq. (2.12) corresponding solution is

$$\begin{bmatrix} \mathbf{x}^c(t) \\ \Delta\mathbf{x}^I(t) \end{bmatrix} = \begin{bmatrix} \exp(\mathbf{A}^c t) & \mathbf{0} \\ \Delta\exp(\mathbf{A}^I t) & \exp(\mathbf{A}^c t) \end{bmatrix} \begin{bmatrix} \mathbf{x}^c(0) \\ \Delta\mathbf{x}^I(0) \end{bmatrix} + \int_0^t \begin{bmatrix} \exp(\mathbf{A}^c(t-\tau))\mathbf{B}^c & \mathbf{0} \\ \exp(\mathbf{A}^c(t-\tau))\Delta\mathbf{B}^I + \Delta\exp(\mathbf{A}^I(t-\tau))\mathbf{B}^c & \exp(\mathbf{A}^c(t-\tau))\mathbf{B}^c \end{bmatrix} \begin{bmatrix} \mathbf{u}^c(\tau) \\ \Delta\mathbf{u}^I(\tau) \end{bmatrix} d\tau \tag{2.19}$$

Using the earlier definitions, Eq. (2.19) can be written in compact form:

$$\mathcal{X}(t) = \hat{\mathcal{A}}(t)\mathcal{X}(0) + \int_0^t [\hat{\mathcal{B}}(t,\tau)\mathcal{U}(\tau)]d\tau, \mathcal{X} \in \mathbb{R}^{2n} \tag{2.20}$$

Here, the matrix $\hat{\mathcal{A}}(t)$ and $\hat{\mathcal{B}}(t,\tau)$ can be respectively expressed as

$$\hat{\mathcal{A}}(t) = \begin{bmatrix} \exp(\mathbf{A}^c t) & 0 \\ \Delta\exp(\mathbf{A}^I t) & \exp(\mathbf{A}^c t) \end{bmatrix} \tag{2.21}$$

and

$$\hat{\mathcal{B}}(t,\tau) = \begin{bmatrix} \exp(\mathbf{A}^c(t-\tau))\mathbf{B}^c & \mathbf{0} \\ \exp(\mathbf{A}^c(t-\tau))\Delta\mathbf{B}^I + \Delta\exp(\mathbf{A}^I(t-\tau))\mathbf{B}^c & \exp(\mathbf{A}^c(t-\tau))\mathbf{B}^c \end{bmatrix} \tag{2.22}$$

Therefore, the interval state response $\mathbf{x}^I(t) = \mathbf{x}^c(t) + \Delta\mathbf{x}^I(t)$ can be obtained.

2.3 Interval Balance System

Direct reduction of a control system is generally inadvisable because truncating the model may change the behavior of the initial high-dimensional model. Balanced realizations are widely used in ROM since they preserve key invariance properties of the full system during reduction. If the given full system is not balanced, it should first be transformed into a balanced form and then reduced.

Lemma 2.1 (Mullis and Roberts 1976; Moore 1981): *For an $n \times n$-dimensional controllable and observable balanced state-space system:*

$$\tilde{\mathbb{P}} = \tilde{\mathbb{G}} \tag{2.23}$$

Here $\tilde{\mathbb{P}}$ and $\tilde{\mathbb{G}}$ denote the controllability and observability Gramian energy matrices of the balanced system. The master coordinate balancing condition is satisfied when:

$$\sigma(\tilde{\mathbb{P}}) = \sigma(\tilde{\mathbb{G}}) = \tilde{\Sigma} \tag{2.24}$$

where $\tilde{\Sigma}$ is a diagonal matrix consisting of the HSVs.

Lemma 2.2 (Mullis and Roberts 1976; Moore 1981): *The transformation matrix* $\mathbf{T}$ *from the initial full system to the balanced system is given by*

$$\mathbf{T} = \mathbf{\Sigma}^{1/2}\mathbf{Z}^{\mathrm{T}}\mathbf{V}^{-1} \tag{2.25}$$

where $\mathbb{P} = \mathbf{V}\mathbf{V}^{\mathrm{T}}$. $\mathbf{Z}$ contains the eigenvectors of $\mathbb{G}$, and the controllability and observability Gramians of the regular system are denoted as $\mathbb{P}$ and $\mathbb{G}$, respectively, which are different from the $\tilde{\mathbb{P}}$ and $\tilde{\mathbb{G}}$. The notation $^{-1}$ denotes the inverse of a matrix. Therefore, the balanced system is equivalent to the original system as

$$\tilde{\Sigma}=\left(\begin{array}{c|c} \tilde{\mathbf{A}} & \tilde{\mathbf{B}} \\ \hline \tilde{\mathbf{C}} & \end{array}\right) \tag{2.26}$$

where $\tilde{\mathbf{A}}=\mathbf{T}\mathbf{A}\mathbf{T}^{-1}$, $\tilde{\mathbf{B}}=\mathbf{T}\mathbf{B}$, $\tilde{\mathbf{C}}=\mathbf{C}\mathbf{T}^{-1}$, and $\mathbf{T}^{-1} = \mathbf{V}\mathbf{Z}^{\mathrm{T}}\mathbf{\Sigma}^{-1/2}$.

Thus, $\tilde{\Sigma}$ can be expressed by combining the nominal and uncertain parts:

$$\tilde{\Sigma}^{\mathrm{c}}+\Delta\tilde{\Sigma}^{\mathrm{I}}=\tilde{\Sigma}^{\mathrm{I}}=\left(\begin{array}{c|c} \tilde{\mathbf{A}}^{\mathrm{I}} & \tilde{\mathbf{B}}^{\mathrm{I}} \\ \hline \tilde{\mathbf{C}}^{\mathrm{I}} & \end{array}\right) = \left(\begin{array}{c|c} \tilde{\mathbf{A}}^{\mathrm{c}} & \tilde{\mathbf{B}}^{\mathrm{c}} \\ \hline \tilde{\mathbf{C}}^{\mathrm{c}} & \end{array}\right) + \left(\begin{array}{c|c} \Delta\tilde{\mathbf{A}}^{\mathrm{I}} & \Delta\tilde{\mathbf{B}}^{\mathrm{I}} \\ \hline \Delta\tilde{\mathbf{C}}^{\mathrm{I}} & \end{array}\right) \tag{2.27}$$

Using the interval natural expansion, the uncertainty radius $\Delta\tilde{\Sigma}$ is solved:

$$\Delta\tilde{\Sigma} = \left(\begin{array}{c|c} \Delta\tilde{\mathbf{A}} & \Delta\tilde{\mathbf{B}} \\ \hline \Delta\tilde{\mathbf{C}} & \end{array}\right) \tag{2.28}$$

To reduce the balanced system $\tilde{\Sigma}^{\mathrm{I}} = \mathrm{diag}(\tilde{\sigma}_1, \cdots, \tilde{\sigma}_n)$ with interval parameters, the corresponding interval uncertainty-based matrices $\tilde{\mathbb{P}}^{\mathrm{I}}$ and $\tilde{\mathbb{G}}^{\mathrm{I}}$ are calculated to estimate the singular value intervals $\tilde{\sigma}_i^{\mathrm{I}}$. Here, $\tilde{\mathbb{P}}^{\mathrm{I}}$ and $\tilde{\mathbb{G}}^{\mathrm{I}}$ is constructed, and represented as a Lyapunov equation form:

$$\begin{cases} \tilde{\mathbf{A}}^{\mathrm{I}}\tilde{\mathbb{P}}^{\mathrm{I}}+\tilde{\mathbb{P}}^{\mathrm{I}}\left(\tilde{\mathbf{A}}^{\mathrm{I}}\right)^{\mathrm{T}} + \tilde{\mathbf{B}}^{\mathrm{I}}\left(\tilde{\mathbf{B}}^{\mathrm{I}}\right)^{\mathrm{T}}=\mathbf{0} \\ \left(\tilde{\mathbf{A}}^{\mathrm{I}}\right)^{\mathrm{T}}\tilde{\mathbb{G}}^{\mathrm{I}}+\tilde{\mathbb{G}}^{\mathrm{I}}\tilde{\mathbf{A}}^{\mathrm{I}} + \left(\tilde{\mathbf{C}}^{\mathrm{I}}\right)^{\mathrm{T}}\tilde{\mathbf{C}}^{\mathrm{I}}=\mathbf{0} \end{cases} \tag{2.29}$$

By maintaining the linearized term and disregarding the higher-order terms, the deterministic part is obtained as

$$\begin{cases} \tilde{\mathbf{A}}^{c}\tilde{\mathbb{P}}^{c}+\tilde{\mathbb{P}}^{c}\left(\tilde{\mathbf{A}}^{c}\right)^{T}+\tilde{\mathbf{B}}^{c}\left(\tilde{\mathbf{B}}^{c}\right)^{T}=\mathbf{0} \\ \left(\tilde{\mathbf{A}}^{c}\right)^{T}\tilde{\mathbb{G}}^{I}+\tilde{\mathbb{G}}^{c}\tilde{\mathbf{A}}^{c}+\left(\tilde{\mathbf{C}}^{c}\right)^{T}\tilde{\mathbf{C}}^{c}=\mathbf{0} \end{cases} \tag{2.30}$$

and the uncertain part is obtained as

$$\begin{cases} \tilde{\mathbf{A}}^{c}\Delta\tilde{\mathbb{P}}^{I}+\Delta\tilde{\mathbb{P}}^{I}\left(\tilde{\mathbf{A}}^{c}\right)^{T}+\left[\tilde{\mathbf{B}}^{c}\left(\Delta\tilde{\mathbf{B}}^{I}\right)^{T}+\Delta\tilde{\mathbf{B}}^{I}\left(\tilde{\mathbf{B}}^{c}\right)^{T}+\Delta\tilde{\mathbf{A}}^{I}\tilde{\mathbb{P}}^{c}+\tilde{\mathbb{P}}^{c}\left(\Delta\tilde{\mathbf{A}}^{I}\right)^{T}\right]=\mathbf{0} \\ \left(\tilde{\mathbf{A}}^{c}\right)^{T}\Delta\tilde{\mathbb{G}}^{I}+\Delta\tilde{\mathbb{G}}^{I}\tilde{\mathbf{A}}^{c}+\left[\left(\tilde{\mathbf{C}}^{c}\right)^{T}\Delta\tilde{\mathbf{C}}^{I}+\left(\Delta\tilde{\mathbf{C}}^{I}\right)^{T}\tilde{\mathbf{C}}^{c}+\left(\Delta\tilde{\mathbf{A}}^{I}\right)^{T}\tilde{\mathbb{G}}^{c}+\tilde{\mathbb{G}}^{c}\Delta\tilde{\mathbf{A}}^{I}\right]=\mathbf{0} \end{cases} \tag{2.31}$$

Equation (2.30) consists of nominal Lyapunov equations and is straightforward to solve. Since the two equations in Eq. (2.31) also take the form of Lyapunov equations, $\Delta\tilde{\mathbb{P}}$ and $\Delta\tilde{\mathbb{G}}$ can therefore be obtained. Consequently, the intervals can be calculated:

$$\begin{cases} \tilde{\mathbb{P}} \in \tilde{\mathbb{P}}^{I}=\tilde{\mathbb{P}}^{c}+\Delta\tilde{\mathbb{P}}^{I}=\tilde{\mathbb{P}}^{c}+\Delta\tilde{\mathbb{P}}\xi=\left[\tilde{\mathbb{P}}^{c}-\Delta\tilde{\mathbb{P}},\tilde{\mathbb{P}}^{c}+\Delta\tilde{\mathbb{P}}\right] \\ \tilde{\mathbb{G}} \in \tilde{\mathbb{G}}^{I}=\tilde{\mathbb{G}}^{c}+\Delta\tilde{\mathbb{G}}^{I}=\tilde{\mathbb{G}}^{c}+\Delta\tilde{\mathbb{G}}\xi=\left[\tilde{\mathbb{G}}^{c}-\Delta\tilde{\mathbb{G}},\tilde{\mathbb{G}}^{c}+\Delta\tilde{\mathbb{G}}\right] \end{cases} \tag{2.32}$$

According to *Lemma 2.3* and the interval propagation for $\tilde{\mathbb{P}}^{I}$ and $\tilde{\mathbb{G}}^{I}$, the corresponding interval HSVs read:

$$\boldsymbol{\sigma}^{I}=\boldsymbol{\sigma}\left(\tilde{\mathbb{P}}^{I}\right)=\boldsymbol{\sigma}\left(\tilde{\mathbb{G}}^{I}\right)=\tilde{\Sigma}^{I}=\tilde{\Sigma}^{c}+\Delta\tilde{\Sigma}^{I}=\operatorname{diag}\left(\sigma_{1}^{I},\cdots,\sigma_{n}^{I}\right) \tag{2.33}$$

These singular-value intervals are then used for model reduction. According to the equivalence in Eq. (2.33), either $\tilde{\mathbb{P}}^{I}$ and $\tilde{\mathbb{G}}^{I}$ is used to estimate the interval singular values. In what follows, $\tilde{\mathbb{P}}^{I}$ is selected.

2.4 Interval Perturbation-Based SVD Method

From the classical SVD definition, the interval SVD is

$$\tilde{\mathbb{P}}^{I}=\mathbf{U}^{I}\boldsymbol{\Sigma}^{I}\left(\mathbf{V}^{I}\right)^{T} \tag{2.34}$$

where $\mathbf{U}^{I}$ and $\mathbf{V}^{I}$ serve as the left/right singular vector matrices of the balanced system. The nominal part of Eq. (2.34) is

$$\tilde{\mathbb{P}}^{c}=\mathbf{U}^{c}\boldsymbol{\Sigma}^{c}\left(\mathbf{V}^{c}\right)^{T} \tag{2.35}$$

with $\boldsymbol{\Sigma}^{c}$ denoting the non-random singular value, while $\mathbf{U}^{c}$ and $\mathbf{V}^{c}$ contain its nominal left/right singular vectors.

Equation (2.35) can be solved via the conventional SVD, while the uncertain singular values are estimated using the introduced interval perturbation-based SVD approach.

In the presence of a minor perturbation $\Delta\tilde{\mathbb{P}}^{\mathrm{I}}$, keeping only first-order terms yields

$$\begin{cases} \sigma_i^{\mathrm{I}} = \sigma_i^{\mathrm{c}} + \Delta\sigma_i^{\mathrm{I}} + \mathrm{o}\left(\Delta^{(2)}\sigma_i^{\mathrm{I}}\right) \\ \mathbf{u}_i^{\mathrm{I}} = \mathbf{u}_i^{\mathrm{c}} + \Delta\mathbf{u}_i^{\mathrm{I}} + \mathrm{o}\left(\Delta^{(2)}\mathbf{u}_i^{\mathrm{I}}\right) \\ \mathbf{v}_i^{\mathrm{I}} = \mathbf{v}_i^{\mathrm{c}} + \Delta\mathbf{v}_i^{\mathrm{I}} + \mathrm{o}\left(\Delta^{(2)}\mathbf{v}_i^{\mathrm{I}}\right) \end{cases} \tag{2.36}$$

where σ_i^{c} and $\Delta\sigma_i^{\mathrm{I}}$ denote the ith nominal and uncertain singular values, $\mathrm{o}\left(\Delta^{(2)}\sigma_i^{\mathrm{I}}\right)$, $\mathrm{o}\left(\Delta^{(2)}\mathbf{u}_i^{\mathrm{I}}\right)$ and $\mathrm{o}\left(\Delta^{(2)}\mathbf{v}_i^{\mathrm{I}}\right)$ represent 2nd- and higher-order terms. $\mathbf{u}_i^{\mathrm{I}}$, $\mathbf{v}_i^{\mathrm{I}}$, $\mathbf{u}_i^{\mathrm{c}}$, $\mathbf{v}_i^{\mathrm{c}}$, $\Delta\mathbf{u}_i^{\mathrm{I}}$, and $\Delta\mathbf{v}_i^{\mathrm{I}}$ are all vectors in $\mathbf{U}^{\mathrm{I}}$ and $\mathbf{V}^{\mathrm{I}}$, where $\mathbf{u}_i^{\mathrm{I}}$ and $\mathbf{v}_i^{\mathrm{I}}$ denote the ith interval left/right singular vectors; $\mathbf{u}_i^{\mathrm{c}}$ and $\mathbf{v}_i^{\mathrm{c}}$ denote the ith nominal left/right singular vectors; $\Delta\mathbf{u}_i^{\mathrm{I}}$ and $\Delta\mathbf{v}_i^{\mathrm{I}}$ denote the ith radii of uncertain left/right singular vectors, respectively.

From the classical SVD, we have

$$\begin{cases} \left(\mathbf{u}_i^{\mathrm{c}}\right)^{\mathrm{T}}\mathbf{u}_j^{\mathrm{c}} = \delta_{ij} \\ \left(\mathbf{v}_i^{\mathrm{c}}\right)^{\mathrm{T}}\mathbf{v}_j^{\mathrm{c}} = \delta_{ij} \\ \left(\mathbf{u}_i^{\mathrm{c}}\right)^{\mathrm{T}}\tilde{\mathbb{P}}^{\mathrm{c}}\mathbf{v}_j^{\mathrm{c}} = \sigma_i^{\mathrm{c}}\delta_{ij} \end{cases} \tag{2.37}$$

and

$$\begin{cases} \tilde{\mathbb{P}}^{\mathrm{c}}\mathbf{v}_i^{\mathrm{c}} = \sigma_i^{\mathrm{c}}\mathbf{u}_i^{\mathrm{c}}, \\ \left(\mathbf{u}_i^{\mathrm{c}}\right)^{\mathrm{T}}\tilde{\mathbb{P}}^{\mathrm{c}} = \sigma_i^{\mathrm{c}}\left(\mathbf{v}_i^{\mathrm{c}}\right)^{\mathrm{T}} \end{cases} \tag{2.38}$$

where

$$\delta_{ij} = \begin{cases} 0, & i \neq j \\ 1, & i = j \end{cases} \tag{2.39}$$

Uncertainty propagation via interval analysis gives

$$\begin{cases} \Delta\mathbf{u}_i^{\mathrm{I}} = \sum\limits_{k=1}^{m} \left(\partial\mathbf{u}_i\left(\tilde{\mathbb{P}}^{\mathrm{c}}\right) / \partial b_k\right)\Delta b_k^{\mathrm{I}} \\ \Delta\mathbf{v}_i^{\mathrm{I}} = \sum\limits_{k=1}^{m} \left(\partial\mathbf{v}_i\left(\tilde{\mathbb{P}}^{\mathrm{c}}\right) / \partial b_k\right)\Delta b_k^{\mathrm{I}} \end{cases} \tag{2.40}$$

Substitute Eqs. (2.35) and (2.36) into Eq. (2.38) using Eq. (2.37) leads to the first-order relations as

$$\begin{cases} \tilde{\mathbb{P}}^{\mathrm{c}} \Delta \mathbf{v}_i^{\mathrm{I}} + \Delta \tilde{\mathbb{P}}^{\mathrm{I}} \mathbf{v}_i^{\mathrm{c}} = \sigma_i^{\mathrm{c}} \Delta \mathbf{u}_i^{\mathrm{I}} + \Delta \sigma_i^{\mathrm{I}} \mathbf{u}_i^{\mathrm{c}} \\ \left(\tilde{\mathbb{P}}^{\mathrm{c}}\right)^{\mathrm{T}} \Delta \mathbf{u}_i^{\mathrm{I}} + \left(\Delta \tilde{\mathbb{P}}^{\mathrm{I}}\right)^{\mathrm{T}} \mathbf{u}_i^{\mathrm{c}} = \sigma_i^{\mathrm{c}} \Delta \mathbf{v}_i^{\mathrm{I}} + \Delta \sigma_i^{\mathrm{I}} \mathbf{v}_i^{\mathrm{c}} \end{cases} \tag{2.41}$$

and

$$\begin{cases} \mathbf{u}_i^{\mathrm{c}} \Delta \mathbf{u}_i^{\mathrm{I}} = 0 \\ \mathbf{v}_i^{\mathrm{c}} \Delta \mathbf{v}_i^{\mathrm{I}} = 0 \end{cases} \tag{2.42}$$

Expand $\partial \mathbf{u}_i(\tilde{\mathbb{P}}^{\mathrm{c}}) / \partial b_k$ and $\partial \mathbf{v}_i(\tilde{\mathbb{P}}^{\mathrm{c}}) / \partial b_k$ on the basis of $\mathbf{U}^{\mathrm{I}}$ and $\mathbf{V}^{\mathrm{I}}$ as

$$\begin{cases} \partial \mathbf{u}_i(\tilde{\mathbb{P}}^{\mathrm{c}}) / \partial b_k = \sum_{l=1}^{n} \partial \alpha_{li} / \partial b_k \mathbf{u}_i(\tilde{\mathbb{P}}^{\mathrm{c}}) \\ \partial \mathbf{v}_i(\tilde{\mathbb{P}}^{\mathrm{c}}) / \partial b_k = \sum_{l=1}^{n} \partial \beta_{li} / \partial b_k \mathbf{v}_i(\tilde{\mathbb{P}}^{\mathrm{c}}) \end{cases} \tag{2.43}$$

Substituting Eq. (2.43) into (2.41) and multiplying $\left(\mathbf{u}_l^{\mathrm{c}}\right)^{\mathrm{T}}$ and $\left(\mathbf{v}_l^{\mathrm{c}}\right)^{\mathrm{T}}$ can obtain

$$\begin{cases} \sigma_l^{\mathrm{c}} \partial \beta_{li} / \partial b_k + \left(\mathbf{u}_l^{\mathrm{c}}\right)^{\mathrm{T}} \Delta \tilde{\mathbb{P}}^{\mathrm{I}} \mathbf{v}_i^{\mathrm{c}} = \sigma_i^{\mathrm{c}} \partial \alpha_{li} / \partial b_k + \Delta \sigma_i^{\mathrm{I}} \delta_{li} \\ \sigma_l^{\mathrm{c}} \partial \alpha_{li} / \partial b_k + \left(\mathbf{v}_l^{\mathrm{c}}\right)^{\mathrm{T}} \left(\Delta \tilde{\mathbb{P}}^{\mathrm{I}}\right)^{\mathrm{T}} \mathbf{u}_i^{\mathrm{c}} = \sigma_i^{\mathrm{c}} \partial \beta_{li} / \partial b_k + \Delta \sigma_i^{\mathrm{I}} \delta_{li} \end{cases} \tag{2.44}$$

Setting $i = j$ in Eq. (2.44) and using the Eqs. (2.43) and (2.37) yields the first-order singular-value perturbation:

$$\Delta \sigma_i^{\mathrm{I}} = \left(\mathbf{u}_i^{\mathrm{c}}\right)^{\mathrm{T}} \Delta \tilde{\mathbb{P}}^{\mathrm{I}} \mathbf{v}_i^{\mathrm{c}} \tag{2.45}$$

The corresponding radius is

$$\Delta \sigma_i = \left| \left(\mathbf{u}_i^{\mathrm{c}}\right)^{\mathrm{T}} \right| \Delta \tilde{\mathbb{P}} \left| \mathbf{v}_i^{\mathrm{c}} \right| \tag{2.46}$$

Thus, the LB and UB of the ith singular value are

$$\begin{cases} \underline{\sigma_i} = \sigma_i^{\mathrm{c}} - \Delta \sigma_i \\ \overline{\sigma_i} = \sigma_i^{\mathrm{c}} + \Delta \sigma_i \end{cases} \tag{2.47}$$

Consequently, the interval extremes of uncertain singular values are estimated.

2.5 Interval Order-Truncated Criterion

For a nominal balanced system without uncertainty, the reduced order can be defined directly from the singular values of the controllability or observability matrices. In contrast, when uncertainties are present, these singular values become interval quantities, making it impossible to apply the standard truncation rules. The challenge becomes even greater when adjacent interval singular values overlap, leading to an unclear truncated order. To address this issue, a new truncation criterion and an accompanying order-selection strategy are developed, providing a more appropriate solution for systems with densely distributed singular-value intervals.

2.5.1 Interval-Based Possibility of Order-Truncated Criterion

According to the classical SVD, the singular values are ordered in decreasing magnitude as:

$$\sigma_1 \geq \cdots \geq \sigma_i \geq \cdots \sigma_n \tag{2.48}$$

On the basis of the interval analysis method, the deterministic components of the singular values are also arranged in descending order:

$$\sigma_1^{\mathrm{c}} \geq \cdots \geq \sigma_i^{\mathrm{c}} \geq \cdots \sigma_n^{\mathrm{c}} \tag{2.49}$$

All singular values lie within corresponding intervals. Consequently, three possible interval relations of σ_i^{I} and $\sigma_{i+1}^{\mathrm{I}}$ may occur, as shown in Fig. 2.1.

In the first situation, $\overline{\sigma_{i+1}} < \underline{\sigma_i}$, two neighboring uncertain intervals are with $\sigma_{i+1}^{\mathrm{I}} < \sigma_i^{\mathrm{I}}$, which indicates that every element of $\sigma_{i+1}^{\mathrm{I}}$ is strictly smaller than every element of σ_i^{I}. In the second situation, $\overline{\sigma_{i+1}} = \underline{\sigma_i}$, two neighboring uncertain intervals are with $\sigma_{i+1}^{\mathrm{I}} \leq \sigma_i^{\mathrm{I}}$, The equality occurs only at the bound where the two intervals meet, while all other values in $\sigma_{i+1}^{\mathrm{I}}$ remain less than those in σ_i^{I}. However, in the third situation, $\overline{\sigma_{i+1}} > \underline{\sigma_i}$, the interval disturbances arise, which make the truncation order difficult to determine.

To address this issue, an interval possibility is introduced for truncating order in control systems with interval uncertainties. For better illustration, the 1D representation in Fig. 2.1 is reformulated into a 2D representation in Fig. 2.2. By treating two neighboring singular-value intervals as reference axes, the three scenarios are redrawn in this new layout. Since the first two cases are straightforward, the focus here is on the third case, where $\overline{\sigma_{i+1}} > \underline{\sigma_i}$.

To identify whether the truncation order of an uncertain system should be i or $i+1$ when the two adjacent singular value intervals are densely distributed, a threshold value σ_{cr} is introduced, as illustrated in Fig. 2.3. The interval-based probabilities $P(\sigma_i^{\mathrm{I}} > \sigma_{\mathrm{cr}})$ and $P(\sigma_{i+1}^{\mathrm{I}} > \sigma_{\mathrm{cr}})$ are then defined as

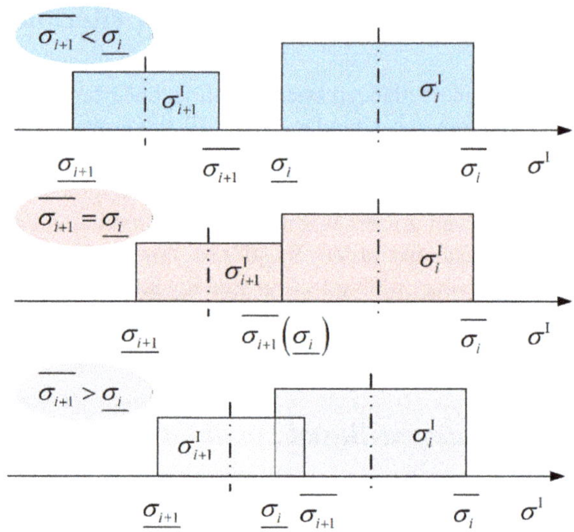

Fig. 2.1 Three relationships of 1D interval between neighboring singular-value intervals

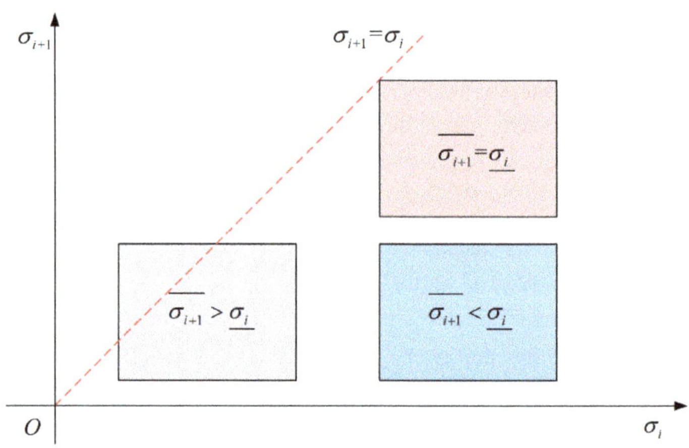

Fig. 2.2 Three relationships of 2D interval between neighboring singular-value intervals

$$\begin{cases} P\left(\sigma_i^{\mathrm{I}} > \sigma_{\mathrm{cr}}\right) = \left\{\sigma_i \middle| \exists \sigma_i \in \sigma_i^{\mathrm{I}}, \text{s.t.} \sigma_i > \sigma_{\mathrm{cr}}\right\} \\ P\left(\sigma_{i+1}^{\mathrm{I}} > \sigma_{\mathrm{cr}}\right) = \left\{\sigma_{i+1} \middle| \exists \sigma_{i+1} \in \sigma_{i+1}^{\mathrm{I}}, \text{s.t.} \sigma_{i+1} > \sigma_{\mathrm{cr}}\right\} \end{cases} \tag{2.50}$$

Since $\sigma_{i+1}^{\mathrm{I}} < \sigma_i^{\mathrm{I}}$, the coordinates $\left(\sigma_i^{\mathrm{c}}, \sigma_{i+1}^{\mathrm{c}}\right)$ always lie right beside the line $\sigma_{i+1}=\sigma_i$. Thus, the interval possibility $P\left(\sigma_i^{\mathrm{I}} > \sigma_{i+1}^{\mathrm{I}}\right)$ is given by

$$P\left(\sigma_i^{\mathrm{I}} > \sigma_{i+1}^{\mathrm{I}}\right) = \left\{\sigma_i, \sigma_{i+1} \middle| \exists \sigma_i \in \sigma_i^{\mathrm{I}}, \exists \sigma_{i+1} \in \sigma_{i+1}^{\mathrm{I}}, \text{s.t.} \sigma_i > \sigma_{i+1}\right\} = 100\% \tag{2.51}$$

Hence, three possibilities in Eq. (2.50) and (2.51) is expressed as

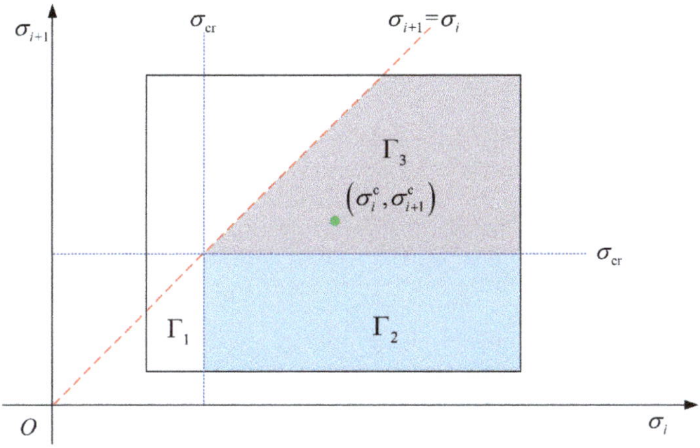

Fig. 2.3 2D interval-wise truncated singular value

$$\begin{cases} P(\sigma_i^{\mathrm{I}} > \sigma_{\mathrm{cr}}) = S_{\Gamma_i} / S_\Gamma \\ P(\sigma_{i+1}^{\mathrm{I}} > \sigma_{\mathrm{cr}}) = S_{\Gamma_{i+1}} / S_\Gamma \end{cases} \tag{2.52}$$

Here S_Γ denotes the interval rectangle area located on the diagonal-line right side. S_{Γ_i} and $S_{\Gamma_{i+1}}$ denote the range of $\sigma_i^{\mathrm{I}} > \sigma_{\mathrm{cr}}$ and $\sigma_{i+1}^{\mathrm{I}} > \sigma_{\mathrm{cr}}$, and can be computed as

$$\begin{cases} S_{\Gamma_{i+1}} = 1/2(2\overline{\sigma_i} - \overline{\sigma_{i+1}} - \sigma_{\mathrm{cr}})(\overline{\sigma_{i+1}} - \sigma_{\mathrm{cr}}) \\ S_{\Gamma_i} = S_{\Gamma_{i+1}} + (\overline{\sigma_i} - \sigma_{\mathrm{cr}})\left(\sigma_{\mathrm{cr}} - \underline{\sigma_{i+1}}\right) \end{cases} \tag{2.53}$$

Lemma 2.4 *To establish the interval-based truncation criterion in IROM, the interval possibility ratio* η *is introduced as*

$$\eta = S_{\Gamma_{i+1}} / S_{\Gamma_i} \tag{2.54}$$

The order-truncated interval metric w is then formulated as

$$w = \begin{cases} i, & \text{if } \eta \le 0.5 \\ i+1, & \text{if } \eta > 0.5 \end{cases}, \quad 1 \le w \le n-1 \tag{2.55}$$

namely

$$\{\sigma_w^{\mathrm{I}}\} = \begin{cases} \{\sigma_1^{\mathrm{I}}, \cdots, \sigma_i^{\mathrm{I}}\}, & \text{if } \eta \le 0.5 \\ \{\sigma_1^{\mathrm{I}}, \cdots, \sigma_i^{\mathrm{I}}, \sigma_{i+1}^{\mathrm{I}}\}, & \text{if } \eta > 0.5 \end{cases} \tag{2.56}$$

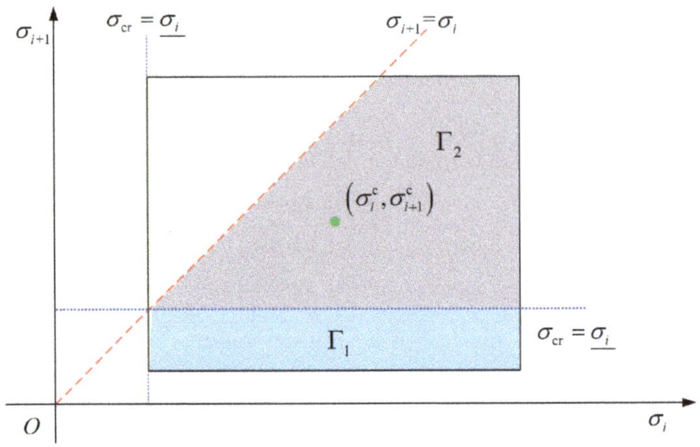

Fig. 2.4 Unique interval truncation strategy for the 2D singular-value set

From the above derivation process, it is clear that both the singular value intervals and the value that has been truncated σ_{cr} play decisive roles in determining the truncation order. While the singular value intervals can be obtained through the proposed estimation method, identifying an appropriate σ_{cr} remains the key challenge. Here we provide a direct approach.

Lemma 2.5 *Without loss of generality, set* $\sigma_{\mathrm{cr}} = \underline{\sigma_i}$ *as shown in* Fig. 2.4 *as*

2.5.2 Order-Truncated Error and Possibility

The error and its possibility of IROM can be stated as follows.

Theorem 2.1 *When* $\underline{\sigma_w} < \sigma_{\mathrm{cr}} < \overline{\sigma_{w+1}}$ *and* $\left[\hat{\sigma}_1^{\mathrm{I}}, \cdots, \hat{\sigma}_w^{\mathrm{I}}\right]$ *are the truncated singular value intervals, the possibility of the order-truncated error* $P(\varepsilon)$ *is*

$$\eta^* = S^*_{\Gamma_{i+1}} \big/ S^*_{\Gamma_i} \tag{2.57}$$

where

$$\begin{cases} S^*_{\Gamma_{i+1}} = S_{\Gamma_{i+1}}\big|_{\sigma_{\mathrm{cr}}=\underline{\sigma_i}} = 1/2\left(2\overline{\sigma_i} - \overline{\sigma_{i+1}} - \sigma_{\mathrm{cr}}\right)\left(\overline{\sigma_{i+1}} - \underline{\sigma_i}\right) \\ S^*_{\Gamma_i} = S_{\Gamma_i}\big|_{\sigma_{\mathrm{cr}}=\underline{\sigma_i}} = S_{\Omega_{i+1}} + 2\Delta\sigma_i\left(\underline{\sigma_i} - \underline{\sigma_{i+1}}\right) \end{cases} \tag{2.58}$$

The areas of Γ and Γ_i coincide in Fig. 2.4, namely $S^*_{\Gamma_i} = S^*_{\Gamma}$. Substituting Eq. (2.57) into Eqs. (2.55) and (2.56) can determine the truncated order.

$$P\left(\varepsilon \middle| \varepsilon \le \sigma_{\mathrm{cr}} + \sum_{i=w+2}^{n} \overline{\sigma_i}\right) = 1 - S_{\Gamma_{i+1}} / S_{\Gamma} \tag{2.59}$$

where $\varepsilon = \left\| \tilde{\Sigma}^{\mathrm{I}} - \hat{\Sigma}^{\mathrm{I}} \right\|_{H_\infty}$ denotes the truncation error.

Proof The rank-cut residual of the deterministic reduced-order model (Antoulas 2005).

$$\varepsilon \le \sum_{i=w+1}^{n} \sigma_i \tag{2.60}$$

Applying interval analysis, since $\sigma_i \le \overline{\sigma_i}$, Eq. (2.60) becomes

$$\varepsilon \le \sum_{i=w+1}^{n} \overline{\sigma_i} = \sigma_{w+1}^{\mathrm{I}} + \sum_{i=w+2}^{n} \overline{\sigma_i} \tag{2.61}$$

Utilizing the second term in Eq. (2.52), and considering the probability of 2 events, we have

$$\begin{aligned} P\left(\sigma_{w+1}^{\mathrm{I}} > \sigma_{\mathrm{cr}}\right) &= P\left(\left[\hat{\sigma}_1^{\mathrm{I}}, \cdots, \hat{\sigma}_w^{\mathrm{I}}\right] \bigcap \sigma_{w+1}^{\mathrm{I}} > \sigma_{\mathrm{cr}}\right) \\ &= P\left(\left[\hat{\sigma}_1^{\mathrm{I}}, \cdots, \hat{\sigma}_w^{\mathrm{I}}\right]\right) \bigcap P\left(\sigma_{w+1}^{\mathrm{I}} > \sigma_{\mathrm{cr}}\right) = S_{\Gamma_{i+1}} / S_{\Gamma} \end{aligned} \tag{2.62}$$

Therefore,

$$P\left(\sigma_{w+1}^{\mathrm{I}} \le \sigma_{\mathrm{cr}}\right) = 1 - S_{\Gamma_{i+1}} / S_{\Gamma} \tag{2.63}$$

Substituting Eq. (2.63) into Eq. (2.61) completes the proof. 

Corollary 2.1 As $\sigma_{\mathrm{cr}} = \underline{\sigma_w}$ in *Theorem 2.1*, the corresponding $P(\varepsilon^*)$ becomes.

$$P\left(\varepsilon \middle| \varepsilon \le \underline{\sigma_w} + \sum_{i=w+2}^{n} \overline{\sigma_i}\right) = 1 - \eta^* \tag{2.64}$$

Proof For $\sigma_{\mathrm{cr}} = \underline{\sigma_w}$ as shown in Fig. 2.4, it follows that

$$\begin{cases} \Gamma_i \triangleq \Gamma \\ S_{\Gamma_i} \equiv S_{\Gamma} \end{cases} \tag{2.65}$$

Substituting Eq. (2.65) into Eq. (2.59) and applying Eq. (2.57) yields the result. □

2.5.3 Optimization of Order Selection

Considering the optimization framework, the optimal solution of states can be formulated to be an optimization problem:

$$\begin{aligned} &\text{find} \quad \hat{\mathbf{x}}^{\mathrm{T}}=[\hat{x}_1,\cdots,\hat{x}_w]^{\mathrm{T}} \\ &\min \quad \left\|\left[\hat{\sigma}_1^{\mathrm{c}},\cdots,\hat{\sigma}_w^{\mathrm{c}}\right]-\left[\sigma_1^{\mathrm{c}},\cdots,\sigma_w^{\mathrm{c}}\right]\right\| \\ &\text{s.t.} \quad \hat{\mathbf{x}}^{\mathrm{T}}=[\hat{x}_1,\cdots,\hat{x}_w]^{\mathrm{T}}=\mathbf{x}_{\mathrm{sub}}^{\mathrm{T}}\subset\mathbf{x}^{\mathrm{T}}=[x_1,\cdots,x_n]^{\mathrm{T}} \\ &\qquad 1\le w\le n-1 \end{aligned} \tag{2.66}$$

Here $\hat{\mathbf{x}}^{\mathrm{T}}$ denotes the total w selected states associated with the w retained orders. $\mathbf{x}_{\mathrm{sub}}^{\mathrm{T}}$ is the corresponding subvector of the full state vector $\mathbf{x}^{\mathrm{T}}$. $\left[\hat{\sigma}_1^{\mathrm{c}},\cdots,\hat{\sigma}_w^{\mathrm{c}}\right]$ and $\left[\sigma_1^{\mathrm{c}},\cdots,\sigma_w^{\mathrm{c}}\right]$ capture the leading w singular values of the balanced and full systems, respectively.

Once the states are chosen, the reduced interval state-space model is expressed as

$$\hat{\Sigma}^{\mathrm{c}}+\Delta\hat{\Sigma}^{\mathrm{I}}=\hat{\Sigma}^{\mathrm{I}}=\left(\begin{array}{c|c}\hat{\mathbf{A}}^{\mathrm{I}} & \hat{\mathbf{B}}^{\mathrm{I}}\\ \hline \hat{\mathbf{C}}^{\mathrm{I}} & \end{array}\right)=\left(\begin{array}{c|c}\hat{\mathbf{A}}^{\mathrm{c}} & \hat{\mathbf{B}}^{\mathrm{c}}\\ \hline \hat{\mathbf{C}}^{\mathrm{c}} & \end{array}\right)+\left(\begin{array}{c|c}\Delta\hat{\mathbf{A}}^{\mathrm{I}} & \Delta\hat{\mathbf{B}}^{\mathrm{I}}\\ \hline \Delta\hat{\mathbf{C}}^{\mathrm{I}} & \end{array}\right) \tag{2.67}$$

where the superscript "$\wedge$" indicates the reduced model, which can be reconstructed as submatrices from Eq. (2.27) through the corresponding selected positions $\hat{\mathbf{x}}^{\mathrm{T}}$ obtained by the optimal selection of Eq. (2.66).

After reconstruction of the reduced model as Eq. (2.67), the radius of the reduced output $\Delta\hat{\mathbf{y}}$ can be solved utilizing Eq. (2.14) with $\Delta\mathbf{x}$. Thus, the uncertain output $\hat{\mathbf{y}}^{\mathrm{I}}$ in the interval form is given by

$$\hat{\mathbf{y}}^{\mathrm{I}}=\left[\underline{\hat{\mathbf{y}}},\overline{\hat{\mathbf{y}}}\right]=\left[\hat{\mathbf{y}}^{\mathrm{c}}-\Delta\hat{\mathbf{y}},\hat{\mathbf{y}}^{\mathrm{c}}+\Delta\hat{\mathbf{y}}\right] \tag{2.68}$$

Similarly, the state variables interval radii can also be derived following Eq. (2.68).

2.6 Numerical Example

2.6.1 Spring-Damper-Mass System

A spring-damper-mass system with three degree-of-freedom (DoFs) is used to demonstrate the performance of the IROM, as shown in Fig. 2.5. The nominal variables are given by $[m_1, m_2, m_3] = [1, 3, 10]$, $[c_1, c_2, c_3] = [3, 7.5, 13.5]$, and $[k_1, k_2, k_3] = [3000, 1800, 1200]$. The uncertainty radii of the intervals are specified as $\Delta m_i = 0.02$, $\Delta c_i = 0.2$, and $\Delta k_i = 10$ for $i = 1, 2, 3$. The input is $u_3 = 20\cos(10\pi t)$ with the sampling time of 0.004 s, and the orientations of input and output are identical. All simulations are performed on a computer with a CPU: 2.20 GHz (16 cores) and 16GB RAM.

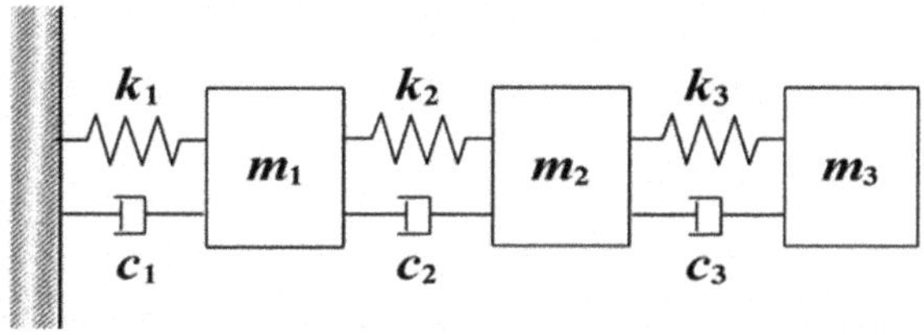

Fig. 2.5 The three DoFs spring-damper-mass system

The system state-space equations are expressed in interval form rather than deterministic values because of the propagation of parameter uncertainty. The two-norm of the initial full-order system, the balanced full-order system, and the balanced reduced-order system obtained by IROM are listed in Table 2.1. It is noted that in Table 2.1, the two-norm of the LB exceeds the two-norm of UB, which results from negative matrix elements. To assess the accuracy of IROM, an MCS with 100,000 runs of the deterministic ROM is carried out for comparison. The results show that the state-space equations obtained by IROM are in close agreement with the outcomes from MCS. Furthermore, the initial 4 HSV uncertainty segments of the uncertain controllability matrix $\tilde{\mathbb{P}}^{\mathrm{I}}$ are calculated as shown in Table 2.2. HSV segments from IROM closely match the results derived via MCS.

Since the proximity of the nominal HSVs of the third and fourth orders (σ_3^c=0.0260 and σ_4^c=0.0256), the third- and fourth-order HSV segments computed by IROM overlap. This interval interference is also reflected in MCS, which confirms the necessity of applying the proposed interval possibility method for order truncation.

Table 2.1 Two-norm of state equations in the spring-damper-mass system

	$\lVert\overline{\mathbf{A}}\rVert_2$		$\lVert\underline{\mathbf{A}}\rVert_2$		$\lVert\overline{\mathbf{B}}\rVert_2$		$\lVert\underline{\mathbf{B}}\rVert_2$	
	IROM	MCS	IROM	MCS	IROM	MCS	IROM	MCS
Original system	5121.61	5126.09	5298.57	5298.69	1.02	1.02	0.98	0.98
Balanced system	77.48	77.51	77.75	77.78	0.58	0.58	0.59	0.59
Reduced order system	32.22	32.26	32.89	32.86	0.55	0.55	0.56	0.56

Table 2.2 Reduced-order Hankel singular gains for the mass-spring-damper chain

		IROM		MCS	
Singular value intervals	Order	LB	UB	LB	UB
	1	0.0841	0.0920	0.0857	0.0905
	2	0.0841	0.0917	0.0855	0.0903
	3	0.0247	0.0272	0.0253	0.0266
	4	0.0241	0.0271	0.0250	0.0263
Computation time (s)		0.261		2379	

For this purpose, the truncated value is chosen to be $\sigma_{cr} = \sigma_3$=0.0247. Using the interval possibility-based measure, the ratio between the 3rd and 4th orders is calculated as $\eta^* = 0.6575$. Since $\eta^* > 0.5$, the 4th-order state of the full system should be preserved. The comparative results of the 3rd and 4th order HSVs calculated from both MCS and IROM are shown in Fig. 2.6.

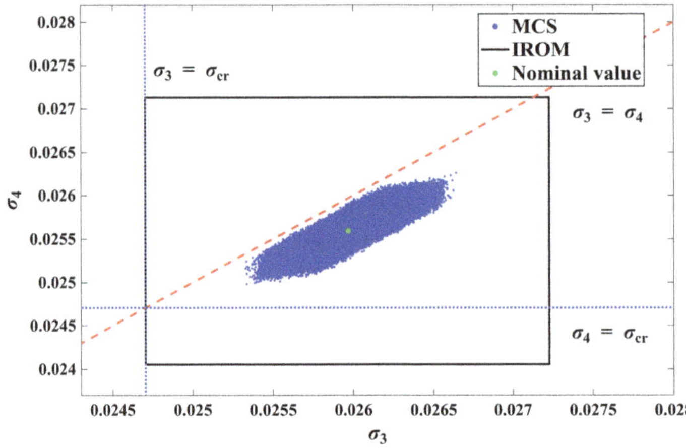

Fig. 2.6 HSV benchmarking at the third and fourth orders across IROM and MCS outputs

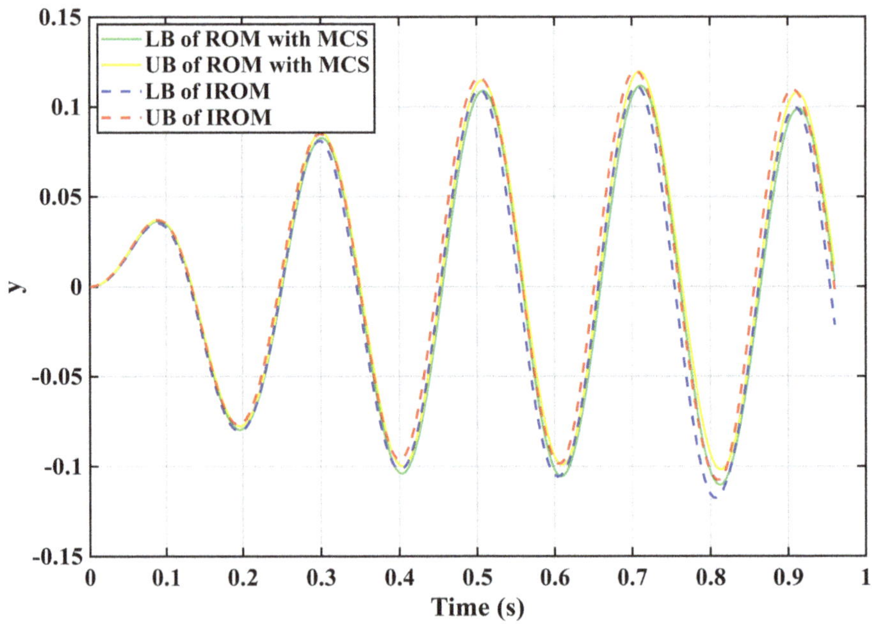

Fig. 2.7 Comparison of output between IROM and MCS

2.6.2 *Truss System*

A five-span truss assembly is simulated and illustrated in Fig. 2.8. The nominal span or height of every horizontal or vertical member is $l = 1\text{m}$, the structural density is $\rho = 7675\text{kg/m}^3$, Young's modulus is $E = 2 \times 10^{11}\text{Pa}$, and the section area is $S = 1 \times 10^{-4}\text{m}^2$. The uncertainty in E is set to one percent of its deterministic values. The system actuation is defined as $u_{12} = 100 \times \delta(0)$ (N), where $\delta(0)$ is a Dirac delta input applied at the 12th degree of freedom. The system output is the vertical velocity at the 10th degree of freedom. The actuation and measurement axes align.

The two-norm of the initial full-order system, the balanced full-order system, and the balanced reduced-order system are listed in Table 2.3. Since the uncertainty of Young's modulus does not propagate into the input matrix **B**, no interval uncertainty is observed in **B**. Therefore, Table 2.3 only presents the LB and UB of the matrix **A**. The leading 18 HSV interval bands under parameter uncertainty of $\widetilde{\mathbb{P}}^{\mathrm{I}}$ are shown in Table 2.4. The UB and LB calculated utilizing IROM appear very near to those obtained utilizing MCS.

Because the deterministic HSV at orders 17 and 18 are nearly coincident (σ_{17}^{c}=2.749×10^{-7} and σ_{18}^{c}=2.738×10^{-7}), the interval-possibility index serves as a rank-cut rule. For this case, the truncated value is determined to be $\sigma_{\mathrm{cr}} = \underline{\sigma}_{17}$=$2.673 \times 10^{-7}$. The ratio between the 17th and 18th orders is calculated as $\eta^* = 0.9645$ utilizing the interval possibility-based criterion. Since $\eta^* > 0.5$, the

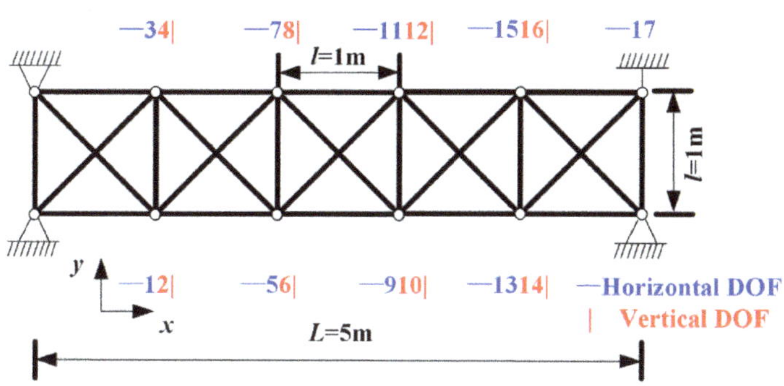

Fig. 2.8 The five-span truss system

Table 2.3 Two-norm of state equations in the truss system

	$\Vert\overline{\mathbf{A}}\Vert_2$		$\Vert\underline{\mathbf{A}}\Vert_2$	
	IROM	MCS	IROM	MCS
Original system (1×10^7)	9.335	9.354	9.354	9.337
Balanced system (1×10^3)	9.651	9.648	9.990	9.705
Reduced order system (1×10^3)	9.555	9.557	9.766	9.557

Table 2.4 HSV set of the condensed truss model (1×10^{-7})

		IROM		MCS	
Singular value intervals	Order	LB	UB	LB	UB
	1	98.95	101.06	99.41	100.60
	2	97.29	99.32	97.65	98.84
	3	28.39	29.10	28.59	28.97
	4	28.15	28.81	28.31	28.65
	5	16.13	16.56	16.22	16.50
	6	16.02	16.39	16.07	16.32
	7	10.20	10.55	10.29	10.49
	8	10.12	10.46	10.18	10.38
	9	8.80	9.05	8.86	9.00
	10	8.73	8.99	8.78	8.92
	11	5.13	5.39	5.20	5.32
	12	5.13	5.33	5.18	5.27
	13	3.93	4.07	3.97	4.03
	14	3.89	4.04	3.93	4.00
	15	3.22	3.35	3.25	3.33
	16	3.19	3.32	3.22	3.29
	17	2.67	2.82	2.71	2.80
	18	2.67	2.81	2.70	2.78
Computation time (s)		0.356		11,986	

18th order of the system can be maintained. The H_∞-norm error estimate between the reduced-order system and the original system is 0.1161. The comparison of the 17th- and 18th-order HSV calculated by MCS and IROM is shown in Fig. 2.9.

In addition, Figs. 2.7 and 2.10 present the time-domain output trajectory over time intervals for the reduced-order system obtained through the IROM, respectively, contrasted with the full-model output obtained via Monte Carlo sampling.

Regarding precision, the LB and UB generated by IROM coincide exactly with Monte Carlo reference data. Regarding computational performance, the computational times listed in Tables 2.2 and 2.4 show that IROM requires only about 0.1% of the time consumed by MCS. This demonstrates that when all uncertain variables vary within their prescribed intervals, IROM can predict the system measurement both precisely and efficiently.

Therefore, the results confirm that the interval approach is both accurate and reliable. Beyond the spring-mass-damper and truss examples, the IROM method is also applicable to further higher-order linear plants, exemplified by spacecraft dynamics, etc.

Fig. 2.9 HSV cross-check at ranks 17–18 between interval ROM and Monte Carlo samples

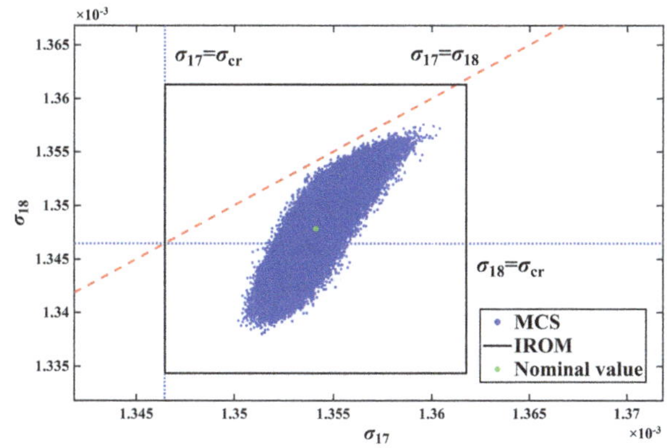

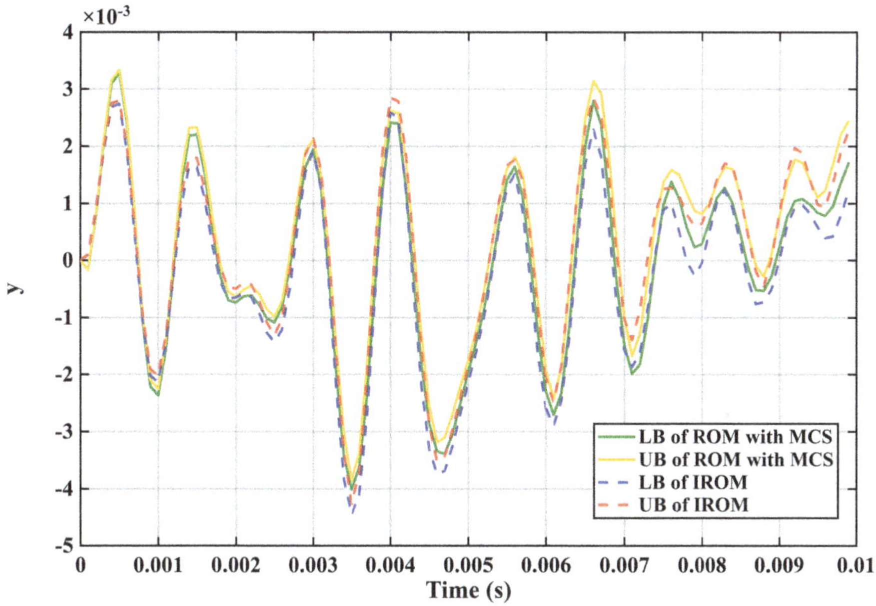

Fig. 2.10 Contrast between MCS and IROM of output in the truss system

2.7 Summary

Over the past decades, deterministic balanced ROM methods have been thoroughly developed and applied to a variety of control systems, leading to notable improvements in both accuracy and computational efficiency. In addition, ROM approaches based on probability and statistics can extend the conventional nominal framework to handle uncertainty to some degree. However, their reliable application requires a large number of samples; directly applying probabilistic ROM methods with limited samples can introduce errors, potentially causing the reduced-order system to diverge from the initial and even resulting in system failure. To address these challenges, this work introduces a new IROM for linear systems. The chapter investigates an interval-based ROM framework, including the interval form of the state-space model, coefficient matrices, order-truncated criterion, the interval perturbation-based SVD, and the upper bound of the order-truncation error. These components are presented through definitions, formula derivations, theorem proofs, and schematic illustrations, highlighting the method's novelty and completeness. Numerical examples confirm the accuracy of the proposed IROM by estimating uncertain singular values. Both the interval order-truncated criterion and the system output demonstrate superior accuracy and computational speed in comparison to MCS. Consequently, the introduced IROM provides a practical complement to probabilistic ROM methods, particularly in small-sample scenarios.

References

Antoulas AC (2005) Approximation of large-scale dynamical systems. Soc Ind Appl Math

Chen L, Cheng R, Li S, Lian H, Zheng C, Bordas SP (2022b) A sample-efficient deep learning method for multivariate uncertainty qualification of acoustic–vibration interaction problems. Comput Methods Appl Mech Eng 393:114784

Chen K, Lin J, Qiu Y, Liu F, Song Y (2022a) Model predictive control for wind farm power tracking with deep learning-based reduced order modeling. IEEE Trans Ind Inform 18:7484–7493

Farlessyost W, Singh S (2022) Reduced order dynamical models for complex dynamics in manufacturing and natural systems using machine learning. Nonlinear Dyn 110:1613–1631

Ivagnes A, Stabile G, Mola A, Iliescu T, Rozza G (2023) Hybrid data-driven closure strategies for reduced order modeling. Appl Math Comput 448:127920

Lee HS, Park C, Park P (2024) A search algorithm for the compact uncertainty region of system elements with interval and affine constraint properties. Appl Math Comput 477:128823

Mierlo C, Burmberger L, Daub M, Duddeck F, Faes MG, Moens D (2022) Interval methods for lack-of-knowledge uncertainty in crash analysis. Mech Syst Signal Process 168:108574

Moore B (1981) Principal component analysis in linear systems: Controllability, observability, and model reduction. IEEE Trans Autom Control 26:17–32

Mullis C, Roberts R (1976) Synthesis of minimum roundoff noise fixed point digital filters. IEEE Trans Circuits Syst 23:551–562

Pu H, Guangwei W, Yingkai C, Zhaokui W (2023) Reduction of space debris collision prediction uncertainty based on Q-Sat precise orbit. Space Sci Technol 3:0005

Shen Y, Wu Z-G, Meng D (2021) Nonsynchronous model reduction for uncertain 2-D Markov jump systems. IEEE Trans. Cybern. 52:10177–10186

Strazzullo M, Vicini F (2023) POD-Based reduced order methods for optimal control problems governed by parametric partial differential equation with varying boundary control. Appl Math Comput 457:128191

Wang X, Wang X, Su H, Lam J (2022) Reduced-order interval observer based consensus for MASs with time-varying interval uncertainties. Automatica 135:109989

Xu Y, Jiang M, Zhu M, Hu N, Zhao C, Wang J (2024) Reduced-complexity decoding of 3D product codes for satellite communications. Space Sci Technol 4:0096

Zhang K, Wu S, Wu Z (2020) Robust enhanced control strategy of a solar power satellite using multiple sensors. J Guid Control Dyn 43:338–346. https://doi.org/10.2514/1.G004733

Zhang S, Devriendt H, Belle L, Desmet W (2024) Substructuring-based parametric reduced-order modelling for structural dynamic predictions of bolted assemblies. Mech Syst Signal Process 218:111513

Zhu X, Pedrycz W, Li Z (2022) A granular approach to interval output estimation for rule-based fuzzy models. IEEE Trans Cybern 52:7029–7038. https://doi.org/10.1109/TCYB.2020.3025668

Chapter 3
Uncertain Reduced-Order Model Based on Convex Set

Abstract This chapter addresses the challenge of model order reduction for linear systems subject to non-probabilistic uncertainties. A novel convex set-based reduced-order model (CSBROM) framework is established to effectively handle parameter uncertainties with unknown-but-bounded (UBB) and parameter correlation of uncertainties using convex set theory rather than large-sample statistical methods. The approach begins by reconstructing the uncertain system into a convex set-based (CSB) state-space equation via an order-extended matrix technique, ensuring both controllability and stability. Subsequently, the classical balanced truncation method is generalized to the convex domain. A convex set perturbation-based singular value decomposition (SVD) algorithm is derived to accurately estimate the bounds of uncertain Hankel singular values (HSVs) by solving the CSB Lyapunov equation. To determine the optimal dimension of the reduced system, a novel truncation criterion is proposed based on the geometric possibility of uncertain singular value distributions. Numerical examples confirm that the proposed CSBROM achieves accuracy comparable to Monte Carlo simulations (MCSs) with significantly higher computational efficiency.

3.1 Introduction

As introduced in Sect. 2.2, ROM offers a streamlined approach that balances accuracy and computational efficiency (Su et al. 2023). ROM techniques have found proven use in areas including manipulator control (Wang et al. 2020), flexible multibody systems (Peng and Kan 2022), nonlinear system modeling (Imtiaz and Akhtar 2020), spacecraft thermal analysis (Qian et al. 2015), and spacecraft deployment, effectively reducing computational costs while preserving key system dynamics.

While the previous chapter focused on interval perturbation methods for model order reduction under uncertainty, this chapter shifts the attention to convex set perturbation techniques. Both methods tackle the inherent uncertainties in control systems, yet they offer distinct methodologies for managing these uncertainties (Maulik et al.

C. Yang and Y. Xia, *Set Theory-Based Spacecraft Dynamics and Control*, Intelligent Control and Learning Systems 26, https://doi.org/10.1007/978-981-95-9179-4_3

2021; Meng and Zhou 2020), each with its own advantages in different scenarios (Zhu and Li 2022; Zhu et al. 2022a).

Convex set-based perturbation has emerged as a promising alternative to traditional probabilistic methods, especially in cases where limited data makes probabilistic analysis impractical (Cao et al. 2021; Time-variant Reliability Modeling Based on Hybrid Non-probability Method n.d.). Compared with interval uncertainty, convex sets take into account the correlations among uncertain variables, which can reduce the conservatism of uncertainty analysis in some scenarios, including controller design (Lü et al. 2022), output estimation (Zhu et al. 2022b), and decision making (Liu et al. 2022). This chapter explores the theoretical foundations and practical applications of convex set-based reduced-order models (CSBROM), with a focus on their efficiency and robustness in uncertain environments.

We begin by formulating a convex set-based state-space equation, extending the order of the system, and solving it based on the proofs of controllability and stability. A perturbation-based SVD is then proposed to estimate the uncertain singular values, a key step in understanding system behavior under uncertainty. Additionally, we propose a novel order-truncation criterion specific to convex sets, offering a balance between reduced computational complexity and model precision. An analytical worst-case truncation error limit is established and verified, ensuring robust performance in reduced-order systems.

This chapter offers a comprehensive view of the tools available for addressing uncertainty in control systems. The proposed CSB method provides a complement to the IROM framework in Chap. 2, expanding the applicability of ROM techniques in practical engineering contexts.

3.2 Convex Set-Based (CSB) Space-State Equation

3.2.1 Second Type of Set Theory: CSB Uncertainties

When the correlation features among multi-dimensional uncertainties are considered, the convex set can be formulated as

$$\mathbf{b} = [b_1, \ldots, b_k \ldots, b_m]^{\mathrm{T}} \in \mathbf{b}^{\mathrm{CS}} = \left\{ \mathbf{b} | (\mathbf{b} - \mathbf{b}^{\mathrm{c}})^{\mathrm{T}} \mathbf{M}\big(\mathbf{b}^{\mathrm{r}}, \boldsymbol{\rho}_{\mathbf{b}}\big)(\mathbf{b} - \mathbf{b}^{\mathrm{c}}) \leq 1 \right\} \tag{3.1}$$

where

$$\mathbf{b}^{\mathrm{c}} = \{b_k^{\mathrm{c}}\} = \big(\overline{\mathbf{b}} + \underline{\mathbf{b}}\big)/2, \quad \mathbf{b}^{\mathrm{r}} = \{b_k^{\mathrm{r}}\} = \big(\overline{\mathbf{b}} - \underline{\mathbf{b}}\big)/2 \tag{3.2}$$

where $\mathbf{b}$ represents an m-dimensional convex vector and b_k denotes its k th component. $\mathbf{b}^{\mathrm{CS}}$ refers to the convex set. The vectors $\overline{\mathbf{b}}$ and $\underline{\mathbf{b}}$ respectively define the UB and LB of $\mathbf{b}$. $\mathbf{b}^{\mathrm{c}}$, $\mathbf{b}^{\mathrm{r}}$, and $\boldsymbol{\rho}_{\mathbf{b}} = \big[\rho\big(b_i, b_j\big)\big]_{m \times m}$ are the nominal/radius vector, and the correlation coefficient matrix of $\mathbf{b}$. b_k^{c} and b_k^{r} represent the nominal and radius the k

th convex elements b_k^{CS}, while b_k^{CS} is the k th element of $\mathbf{b}^{\mathrm{CS}}$. The symmetric positive-definite matrix $\mathbf{M}(\mathbf{b}^{\mathrm{r}}, \boldsymbol{\rho}_{\mathbf{b}})$ characterizes the degree of dispersion and the dependency among all convex variables, and it depends on both $\mathbf{b}^{\mathrm{r}}$ and $\boldsymbol{\rho}_{\mathbf{b}}$.

According to Jiang et al. (2011), Wang et al. (2018), it can be obtained as

$$\mathbf{M}(\mathbf{b}^{\mathrm{r}}, \boldsymbol{\rho}_{\mathbf{b}}) = \mathbf{C}_{\mathbf{b}}^{-1}(\mathbf{b}^{\mathrm{r}}, \boldsymbol{\rho}_{\mathbf{b}}) = \left(\mathbf{b}^{\mathrm{r}}(\mathbf{b}^{\mathrm{r}})^{\mathrm{T}} \circ \boldsymbol{\rho}_{\mathbf{b}}\right)^{-1} = \begin{bmatrix} \mathrm{Cov}(b_1, b_1) & \cdots & \mathrm{Cov}(b_1, b_m) \\ \vdots & \ddots & \vdots \\ \mathrm{Cov}(b_m, b_1) & \cdots & \mathrm{Cov}(b_m, b_m) \end{bmatrix}^{-1} \tag{3.3}$$

and

$$\mathrm{Cov}(b_i, b_j) = \rho(b_i, b_j) b_i^{\mathrm{r}} b_j^{\mathrm{r}}, \qquad 1 \le i, j \le m \tag{3.4}$$

where $\mathbf{C}_{\mathbf{b}}(\mathbf{b}^{\mathrm{r}}, \boldsymbol{\rho}_{\mathbf{b}}) = \left[\mathrm{Cov}(b_i, b_j)\right]_{m \times m}$ denotes the covariance matrix, and the Hadamard operator "$\circ$" indicates element-wise multiplication between corresponding entries of two vectors or matrices. Using the Cholesky decomposition, the symmetric matrix $\mathbf{M}$ which satisfies the positive-definiteness condition can be diagonalized by an orthogonal matrix $\mathbf{H}$ as

$$\mathbf{H}^{\mathrm{T}}\mathbf{M}\mathbf{H} = \boldsymbol{\Lambda}, \mathbf{H}^{\mathrm{T}}\mathbf{H} = \mathbf{I} \tag{3.5}$$

where $\boldsymbol{\Lambda}$ is a diagonal matrix. The normalized vector is then given by

$$\boldsymbol{\xi}_{\mathbf{b}} = \boldsymbol{\Lambda}^{1/2}\mathbf{H}^{\mathrm{T}}(\mathbf{b} - \mathbf{b}^{\mathrm{c}}) \tag{3.6}$$

Therefore, Eq. (3.1) can be reconstructed in the equivalent representation:

$$\mathbf{b} = \mathbf{b}^{\mathrm{c}} + \mathbf{H}\Lambda^{-1/2}\boldsymbol{\xi}_{\mathbf{b}} = \mathbf{b}^{\mathrm{c}} + \boldsymbol{\delta}_{\mathbf{b}} \tag{3.7}$$

3.2.2 Construction of CSB State-Space Equation

Similar to the interval uncertainty-based state-space equation in Sect. 2.3, considering the nominal state-space equation defined as Eq. (2.5), to account for uncertainty, the CSB analysis method mentioned above is applied to transform Eq. (2.5) into

$$\dot{\mathbf{x}}(t) = \mathbf{A}(\mathbf{b})\mathbf{x}(t) + \mathbf{B}(\mathbf{b})\mathbf{u}(\mathbf{b}, t), \qquad \mathbf{x}(t) \in \mathbb{R}^n, \mathbf{u}(\mathbf{b}, t) \in \mathbb{R}^p \tag{3.8}$$

To address Eq. (3.8), the convex set perturbation method derived from the natural set expansion is used. Focusing on $\mathbf{A}(\mathbf{b})$ as an example, the CSB expression can be expressed as

$$\mathbf{A}^{\mathrm{CS}}(\mathbf{b}) = \left[A_{ij}(\mathbf{b})\right] = \mathbf{A}^{\mathrm{c}}(\mathbf{b}) + \mathbf{A}^{\mathrm{r}}(\mathbf{b}) \circ \boldsymbol{\xi}_{\mathbf{A}} = \mathbf{A}^{\mathrm{c}}(\mathbf{b}) + \boldsymbol{\delta}_{\mathbf{A}} \tag{3.9}$$

and $\mathbf{A}^{\mathrm{c}}(\mathbf{b}) = \left[A^{\mathrm{c}}_{ij}(\mathbf{b})\right] = \mathbf{A}(\mathbf{b}^{\mathrm{c}})$

$$A^{\mathrm{CS}}_{ij}(\mathbf{b}) = A_{ij}(\mathbf{b}^{\mathrm{c}}) + \mathbf{h}^{\mathrm{T}}_{A_{ij}}\boldsymbol{\delta}_{\mathbf{x}} = A^{\mathrm{c}}_{ij}(\mathbf{b}) + \mathbf{h}^{\mathrm{T}}_{A_{ij}}\mathbf{H}\Lambda^{-1/2}\boldsymbol{\xi}_{\mathbf{b}} = A^{\mathrm{c}}_{ij}(\mathbf{b}) + \delta_{A_{ij}} \tag{3.10}$$

where $\boldsymbol{\xi}_{\mathbf{A}}$ represents a unit hypersphere region, and $\boldsymbol{\delta}_{\mathbf{A}} = \left[\delta_{A_{ij}}\right]$ is the convex perturbation norm. $\mathbf{h}_{A_{ij}} = \left[\partial A_{ij}(\mathbf{b}^{\mathrm{c}})/\partial b_1, \ldots, \partial A_{ij}(\mathbf{b}^{\mathrm{c}})/\partial b_k, \ldots \partial A_{ij}(\mathbf{b}^{\mathrm{c}})/\partial b_m\right]^{\mathrm{T}}$ is the gradient vector.

Since the intrinsic correlation of $\mathbf{b}$, the Lagrangian function L_{ij} can be defined as

$$L_{ij} = A_{ij}(\mathbf{b}^{\mathrm{c}}) + \mathbf{h}^{\mathrm{T}}_{A_{ij}}\boldsymbol{\delta}_{\mathbf{b}} + \lambda_{A_{ij}}\left(\boldsymbol{\delta}^{\mathrm{T}}_{\mathbf{b}}\mathbf{M}\left(\mathbf{b}^{\mathrm{r}}, \boldsymbol{\rho}_{\mathbf{b}}\right)\boldsymbol{\delta}_{\mathbf{b}} - 1\right) \tag{3.11}$$

where $\lambda_{A_{ij}}$ represents the Lagrange multiplier.

Differentiating Eq. (3.11) gives

$$\partial L_{ij}/\partial\tilde{\delta}_{\mathbf{x}} = \mathbf{h}^{\mathrm{T}}_{A_{ij}} + 2\lambda_{A_{ij}}\mathbf{M}\left(\mathbf{b}^{\mathrm{r}}, \boldsymbol{\rho}_{\mathbf{b}}\right)\tilde{\delta}_{\mathbf{b}} = 0 \tag{3.12}$$

for which the solution is

$$\tilde{\delta}_{\mathbf{b}} = -1/\left(2\lambda_{A_{ij}}\right)\mathbf{M}^{-1}\left(\mathbf{b}^{\mathrm{r}}, \boldsymbol{\rho}_{\mathbf{b}}\right)\mathbf{h}_{A_{ij}} \tag{3.13}$$

By substituting Eq. (3.13) into the bounds of Eq. (3.1), $\lambda_{A_{ij}}$ can be determined as

$$\lambda_{A_{ij}} = \pm 1/2\sqrt{\mathbf{h}^{\mathrm{T}}_{A_{ij}}\mathbf{M}^{-1}\left(\mathbf{b}^{\mathrm{r}}, \boldsymbol{\rho}_{\mathbf{b}}\right)\mathbf{h}_{A_{ij}}} \tag{3.14}$$

Substituting Eqs. (3.13) and (3.14) into Eq. (3.10) results in

$$\begin{aligned} A^{\mathrm{CS}}_{ij}(\mathbf{b}) &= A_{ij}(\mathbf{b}^{\mathrm{c}}) + \mathbf{h}^{\mathrm{T}}_{A_{ij}}\delta_{\mathbf{b}} \\ &\in \left[A_{ij}(\mathbf{b}^{\mathrm{c}}) - \sqrt{\mathbf{h}^{\mathrm{T}}_{A_{ij}}\mathbf{M}^{-1}(\mathbf{b}^{\mathrm{r}}, \rho_{\mathbf{b}})\mathbf{h}_{A_{ij}}}, A_{ij}(\mathbf{b}^{\mathrm{c}}) + \sqrt{\mathbf{h}^{\mathrm{T}}_{A_{ij}}\mathbf{M}^{-1}(\mathbf{b}^{\mathrm{r}}, \rho_{\mathbf{b}})\mathbf{h}_{A_{ij}}}\right] \end{aligned} \tag{3.15}$$

In the same way, the input matrices and vectors can be expressed as

$$\begin{cases} \mathbf{B}^{\mathrm{CS}}(\mathbf{b}) = \mathbf{B}^{\mathrm{c}}(\mathbf{b}) + \mathbf{B}^{\mathrm{r}}(\mathbf{b}) \circ \boldsymbol{\xi}_{\mathbf{B}} = \mathbf{B}^{\mathrm{c}}(\mathbf{b}) + \boldsymbol{\delta}_{\mathbf{B}} \\ \mathbf{u}^{\mathrm{CS}}(\mathbf{b}) = \mathbf{u}^{\mathrm{c}}(\mathbf{b}) + \mathbf{u}^{\mathrm{r}}(\mathbf{b}) \circ \boldsymbol{\xi}_{\mathbf{u}} = \mathbf{u}^{\mathrm{c}}(\mathbf{b}) + \boldsymbol{\delta}_{\mathbf{u}} \end{cases} \tag{3.16}$$

Consequently, Eq. (3.8) can be decomposed into deterministic and uncertain components. The deterministic part is expressed as

$$\dot{\mathbf{x}}^{\mathrm{c}}(t) = \mathbf{A}^{\mathrm{c}}(\mathbf{b})\mathbf{x}^{\mathrm{c}}(t) + \mathbf{B}^{\mathrm{c}}(\mathbf{b})\mathbf{u}^{\mathrm{c}}(\mathbf{b}, t)\mathbf{x}^{\mathrm{c}}(t) \tag{3.17}$$

For the uncertain part, since the partial derivative $\partial \mathbf{x}^{\mathrm{c}}(t)/\partial \mathbf{b}$ in relation to the uncertain state response $\mathbf{x}^{\mathrm{c}}(t)$ in terms of the uncertainty $\mathbf{b}$ is undetermined, the extended-order matrix approach is applied to solve it. Taking the derivative operator to Eq. (3.8) and ignoring the second-order derivative terms, we can obtain:

For the uncertain part, since the derivative $\partial \mathbf{x}^{\mathrm{c}}(t)/\partial \mathbf{b}$ of $\mathbf{x}^{\mathrm{c}}(t)$ with respect to $\mathbf{b}$ is undetermined, the extended-order matrix approach is introduced. Taking the derivative operator in Eq. (3.8) with respect to b_k and neglecting second-order derivatives gives

$$\begin{aligned}\partial \dot{\mathbf{x}}^{\mathrm{c}}(t)/\partial b_k =& \partial \mathbf{A}^{\mathrm{c}}(\mathbf{b})/\partial b_k \mathbf{x}^{\mathrm{c}}(t) + \mathbf{A}^{\mathrm{c}}(\mathbf{b})\partial \mathbf{x}^{\mathrm{c}}(t)/\partial b_k \\ &+ \partial \mathbf{B}^{\mathrm{c}}(\mathbf{b})/\partial b_k \mathbf{u}^{\mathrm{c}}(\mathbf{b}, t) + \mathbf{B}^{\mathrm{c}}(\mathbf{b})\partial \mathbf{u}^{\mathrm{c}}(\mathbf{b}, t)/\partial b_k\end{aligned} \tag{3.18}$$

Remark 3.1 Equations (3.17) and (3.18) can be reorganized into an extended-order matrix form as

$$\begin{aligned}\begin{bmatrix}\dot{\mathbf{x}}^{\mathrm{c}}(t) \\ \partial \dot{\mathbf{x}}^{\mathrm{c}}(t)/\partial b_k\end{bmatrix} =& \begin{bmatrix}\mathbf{A}^{\mathrm{c}}(\mathbf{b}) & \mathbf{0} \\ \partial \mathbf{A}^{\mathrm{c}}(\mathbf{b})/\partial b_k & \mathbf{A}^{\mathrm{c}}(\mathbf{b})\end{bmatrix}\begin{bmatrix}\mathbf{x}^{\mathrm{c}}(t) \\ \partial \mathbf{x}^{\mathrm{c}}(t)/\partial b_k\end{bmatrix} \\ &+ \begin{bmatrix}\mathbf{B}^{\mathrm{c}}(\mathbf{b}) & \mathbf{0} \\ \partial \mathbf{B}^{\mathrm{c}}(\mathbf{b})/\partial b_k & \mathbf{B}^{\mathrm{c}}(\mathbf{b})\end{bmatrix}\begin{bmatrix}\mathbf{u}^{\mathrm{c}}(\mathbf{b}, t) \\ \partial \mathbf{u}^{\mathrm{c}}(\mathbf{b}, t)/\partial b_k\end{bmatrix}\end{aligned} \tag{3.19}$$

Equation (3.19) can be reformulated as

$$\dot{\mathbb{X}}\ominus(t) = \mathbb{A}\ominus(\mathbf{b})\mathbb{X}(t) + \mathbb{B}\ominus(\mathbf{b})\mathbb{U}(\mathbf{b}, t), \qquad \mathbb{X} \in \mathbb{R}^{2n} \tag{3.20}$$

where $\dot{\mathbb{X}}(t) = \left[(\dot{\mathbf{x}}^{\mathrm{c}}(t))^{\mathrm{T}}, \left(\partial \dot{\mathbf{x}}^{\mathrm{c}}(t)/\partial b_k\right)^{\mathrm{T}}\right]^{\mathrm{T}}$, $\mathbb{X}(t) = \left[(\mathbf{x}^{\mathrm{c}}(t))^{\mathrm{T}}, \left(\partial \mathbf{x}^{\mathrm{c}}(t)/\partial b_k\right)^{\mathrm{T}}\right]^{\mathrm{T}}$, $\mathbb{A}\ominus(\mathbf{b}) = \begin{bmatrix}\mathbf{A}^{\mathrm{c}}(\mathbf{b}) & \mathbf{0} \\ \partial \mathbf{A}^{\mathrm{c}}(\mathbf{b})/\partial b_k & \mathbf{A}^{\mathrm{c}}(\mathbf{b})\end{bmatrix}$, $\mathbb{B}\ominus(\mathbf{b}) = \begin{bmatrix}\mathbf{B}^{\mathrm{c}}(\mathbf{b}) & \mathbf{0} \\ \partial \mathbf{B}^{\mathrm{c}}(\mathbf{b})/\partial b_k & \mathbf{B}^{\mathrm{c}}(\mathbf{b})\end{bmatrix}$, and $\mathbb{U}(\mathbf{b}, t) = \left[(\mathbf{u}^{\mathrm{c}}(\mathbf{b}, t))^{\mathrm{T}}, \left(\partial \mathbf{u}^{\mathrm{c}}(\mathbf{b}, t)/\partial b_k\right)^{\mathrm{T}}\right]^{\mathrm{T}}$.

The solution method of the CSB state-space equation has been researched in Sect. 2.3 after $\partial \mathbf{x}^{\mathrm{c}}(t)/\partial b_k$ was obtained:

$$\begin{aligned}\left[\partial \boldsymbol{x}^{\mathrm{c}}(t)/\partial b_k\right]_{1\times m} =& \left[\partial x_i^{\mathrm{c}}(t)/\partial b_k\right]_{n\times m} = \begin{bmatrix}\partial x_1^{\mathrm{c}}(t)/\partial b_1 & \cdots & \partial x_1^{\mathrm{c}}(t)/\partial b_m \\ \vdots & \ddots & \vdots \\ \partial x_n^{\mathrm{c}}(t)/\partial b_1 & \cdots & \partial x_n^{\mathrm{c}}(t)/\partial b_m\end{bmatrix} \\ =& \left[\boldsymbol{h}_{x_1}, \ldots, \boldsymbol{h}_{x_i}, \ldots, \boldsymbol{h}_{x_n}\right]^{\mathrm{T}}\end{aligned} \tag{3.21}$$

where $\mathbf{x} = [x_1, \ldots, x_i, \ldots, x_n]^{\mathrm{T}}$. Thus, the partial derivative vector $\mathbf{h}_{x_i}$ is

$$\boldsymbol{h}_{x_i} = \left[\partial x_i^{\mathrm{c}}(t)/\partial b_1, \ldots, \partial x_i^{\mathrm{c}}(t)/\partial b_k, \ldots \partial x_i^{\mathrm{c}}(t)/\partial b_m\right]^{\mathrm{T}} \tag{3.22}$$

Finally, the uncertain state bound can be derived as

$$x_i^{\mathrm{CS}} = x_i^{\mathrm{c}} + \mathbf{h}_{x_i}^{\mathrm{T}}\boldsymbol{\delta}_{\mathbf{b}} \in \left[x_i^{\mathrm{c}} - \sqrt{\mathbf{h}_{x_i}^{\mathrm{T}}\mathbf{M}^{-1}(\mathbf{b}^{\mathrm{r}}, \boldsymbol{\rho}_{\mathbf{b}})\mathbf{h}_{x_i}}, x_i^{\mathrm{c}} + \sqrt{\mathbf{h}_{x_i}^{\mathrm{T}}\mathbf{M}^{-1}(\mathbf{b}^{\mathrm{r}}, \boldsymbol{\rho}_{\mathbf{b}})\mathbf{h}_{x_i}}\right] \tag{3.23}$$

3.2.3 Solution Method of the CSB State-Space Equation

Similar to the solution of the interval uncertainty-based state-space equation introduced in Sect. 2.3, for the deterministic LTI equation with non-zero forcing given in Eq. (2.5), the corresponding solution can be represented as Eq. (2.16). Based on the deterministic solution method, a novel CSB-based LTI equation with non-zero forcing described in Eq. (3.19) can be formulated and solved accordingly. Using Taylor expansion, the deterministic component and its partial derivatives are computed as follows:

$$\mathbf{x}^{\mathrm{c}}(t) = \exp(\mathbf{A}^{\mathrm{c}}(\mathbf{b})t)\mathbf{x}^{\mathrm{c}}(0) + \int_0^t \exp(\mathbf{A}^{\mathrm{c}}(\mathbf{b})(t-\tau))\mathbf{B}^{\mathrm{c}}(\mathbf{b})\mathbf{u}^{\mathrm{c}}(\mathbf{b}, \tau)\mathrm{d}\tau \tag{3.24}$$

and

$$\begin{aligned}\frac{\partial \boldsymbol{x}^{\mathrm{c}}(t)}{\partial b_k} =& \exp\left[\boldsymbol{A}^{\mathrm{c}}(\boldsymbol{b})t\right]\frac{\partial \boldsymbol{x}^{\mathrm{c}}(t)}{\partial b_k}\bigg|_{t=0} \\ &+ \int_0^t \exp\left[\boldsymbol{A}^{\mathrm{c}}(\boldsymbol{b})(t-\tau)\right]\left[\boldsymbol{B}^{\mathrm{c}}(\boldsymbol{b})\frac{\partial \boldsymbol{u}^{\mathrm{c}}(\boldsymbol{b}, \tau)}{\partial b_k} + \frac{\partial \boldsymbol{B}^{\mathrm{c}}(\boldsymbol{b})}{\partial b_k}\boldsymbol{u}^{\mathrm{c}}(\boldsymbol{b}, \tau)\right]\mathrm{d}\tau \\ &+ \exp\left[\boldsymbol{A}^{\mathrm{c}}(\boldsymbol{b})t\right]\frac{\partial \boldsymbol{A}^{\mathrm{c}}(\boldsymbol{b})}{\partial b_k}\boldsymbol{x}^{\mathrm{c}}(0)t \\ &+ \int_0^t (t-\tau)\exp\left[\boldsymbol{A}^{\mathrm{c}}(\boldsymbol{b})(t-\tau)\right]\frac{\partial \boldsymbol{A}^{\mathrm{c}}(\boldsymbol{b})}{\partial b_k}\boldsymbol{B}^{\mathrm{c}}(\boldsymbol{b})\boldsymbol{u}^{\mathrm{c}}(\boldsymbol{b}, \tau)\mathrm{d}\tau\end{aligned} \tag{3.25}$$

where $\mathbf{x}^{\mathrm{c}}(0)$ and $\partial \dot{x}^{\mathrm{c}}(t)/\partial b_k|_{t=0}$ denote the original deterministic state and its corresponding partial derivative, respectively.

Consequently, Eqs. (3.24) and (3.25) can be combined as

$$\begin{bmatrix} \boldsymbol{x}^{\mathrm{c}}(t) \\ \partial \boldsymbol{x}^{\mathrm{c}}(t)/\partial b_k \end{bmatrix} = \begin{bmatrix} \exp[\boldsymbol{A}^{\mathrm{c}}(\boldsymbol{b})t] & \boldsymbol{0} \\ \exp[\boldsymbol{A}^{\mathrm{c}}(\boldsymbol{b})t]\frac{\partial \boldsymbol{A}^{\mathrm{c}}(\boldsymbol{b})}{\partial b_k}t & \exp[\boldsymbol{A}^{\mathrm{c}}(\boldsymbol{b})t] \end{bmatrix} \begin{bmatrix} \boldsymbol{x}^{\mathrm{c}}(0) \\ \left.\frac{\partial \boldsymbol{x}^{\mathrm{c}}(t)}{\partial b_k}\right|_{t=0} \end{bmatrix}$$
$$+ \int_0^t \begin{bmatrix} \exp[\boldsymbol{A}^{\mathrm{c}}(\boldsymbol{b})(t-\tau)]\boldsymbol{B}^{\mathrm{c}}(\boldsymbol{b}) & \boldsymbol{0} \\ \exp[\boldsymbol{A}^{\mathrm{c}}(\boldsymbol{b})(t-\tau)]\left[(t-\tau)\frac{\partial \boldsymbol{A}^{\mathrm{c}}(\boldsymbol{b})}{\partial b_k}\boldsymbol{B}^{\mathrm{c}}(\boldsymbol{b}) + \frac{\partial \boldsymbol{B}^{\mathrm{c}}(\boldsymbol{b})}{\partial b_k}\right] & \exp[\boldsymbol{A}^{\mathrm{c}}(\boldsymbol{b})(t-\tau)]\boldsymbol{B}^{\mathrm{c}}(\boldsymbol{b}) \end{bmatrix} \begin{bmatrix} \boldsymbol{u}^{\mathrm{c}}(\boldsymbol{b},\tau) \\ \frac{\partial \boldsymbol{u}^{\mathrm{c}}(\boldsymbol{b},\tau)}{\partial b_k} \end{bmatrix} \mathrm{d}\tau \tag{3.26}$$

According to the previously defined quantities $\mathbb{X}(t)$, $\mathbb{A}\ominus(\mathbf{b})$, $\mathbb{B}\ominus(\mathbf{b})$, and $\mathbb{U}(\mathbf{b}, t)$, Eq. (3.26) can be compactly rewritten as

$$\mathbb{X}(t) = \hat{\mathbb{A}}(\mathbf{b}, t)\mathbb{X}(0) + \int_0^t \left[\hat{\mathbb{B}}(\mathbf{b}, t, \tau)\mathbb{U}(\mathbf{b}, \tau)\right]\mathrm{d}\tau, \ \mathbb{X} \in \mathbb{R}^{2n} \tag{3.27}$$

where

$$\hat{\mathbb{A}}(\mathbf{b}, t) = \begin{bmatrix} \exp[\mathbf{A}^{\mathrm{c}}(\mathbf{b})t] & \mathbf{0} \\ \exp[\mathbf{A}^{\mathrm{c}}(\mathbf{b})t]\frac{\partial \mathbf{A}^{\mathrm{c}}(\mathbf{b})}{\partial b_k}t & \exp[\mathbf{A}^{\mathrm{c}}(\mathbf{b})t] \end{bmatrix} \tag{3.28}$$

and

$$\hat{\mathbb{B}}(\mathbf{b}, t, \tau) = \begin{bmatrix} \exp[\mathbf{A}^{\mathrm{c}}(\mathbf{b})(t-\tau)]\mathbf{B}^{\mathrm{c}}(\mathbf{b}) & \mathbf{0} \\ \exp[\mathbf{A}^{\mathrm{c}}(\mathbf{b})(t-\tau)]\left[(t-\tau)\frac{\partial \mathbf{A}^{\mathrm{c}}(\mathbf{b})}{\partial b_k}\mathbf{B}^{\mathrm{c}}(\mathbf{b}) + \frac{\partial \mathbf{B}^{\mathrm{c}}(\mathbf{b})}{\partial b_k}\right] & \exp[\mathbf{A}^{\mathrm{c}}(\mathbf{b})(t-\tau)]\mathbf{B}^{\mathrm{c}}(\mathbf{b}) \end{bmatrix} \tag{3.29}$$

Therefore, the partial derivative term $\partial \mathbf{x}^{\mathrm{c}}(t)/\partial b_k$ can be directly determined using the proposed method. By iteratively computing the partial derivatives for all k from 1 to m, the complete derivative matrix $\left[\partial x_i^{\mathrm{c}}(t)/\partial b_k\right]_{n\times m}$ in Eq. (3.21) can be obtained, from which the partial derivative $\mathbf{h}_{x_i}$ is derived. Consequently, the state response bounds $x_i^{\mathrm{c}}(\mathbf{b})$ can be accurately estimated based on the convex set.

3.3 CSB Reduction Framework

3.3.1 CSB Balance System

As discussed in Sect. 2.4, directly reducing the model may change the intrinsic characteristics of the primary system. To ensure that the essential dynamic features are preserved during the reduction process, the system should first be transformed into a balanced realization before performing ROM. Balanced realizations maintain the key invariance properties of the full system throughout the ROM procedure, thus preventing performance degradation.

According to the balanced realization theory introduced in Eq. (2.23) to Eq. (2.26), the reachability-energy and observation-energy matrices of a balanced system are equal and diagonalizable by the HSVs. These properties will be employed to construct the CSB balanced system.

Using Lemmas 2.1 and 2.2 presented in Sect. 2.4, and based on the convex set, the partial derivatives for the balanced system parameters in relation to the uncertainty $\partial \tilde{\Xi}^{c} / \partial b_k$ can be expressed as

$$\partial \tilde{\Xi}^{c} / \partial b_k = \left(\partial \tilde{\mathbf{A}}^{c} / \partial b_k \,\middle|\, \partial \tilde{\mathbf{B}}^{c} / \partial b_k \right) \tag{3.30}$$

where

$$\begin{cases} \partial \tilde{\mathbf{A}}^{c} / \partial b_k = \mathbf{T} \partial \mathbf{A}^{c}(\mathbf{b}) / \partial b_k \mathbf{T}^{-1} \\ \partial \tilde{\mathbf{B}}^{c} / \partial b_k = \mathbf{T} \partial \mathbf{B}^{c}(\mathbf{b}) / \partial b_k \end{cases} \tag{3.31}$$

To perform reduction for the CSB balanced system, the corresponding controllability matrix $\tilde{\mathbb{P}}^{CS}$ and observability matrix $\tilde{\mathbb{G}}^{CS}$ are constructed to estimate the bounds of the singular values $\tilde{\sigma}_i^{CS}$. Considering the equality of singular values established in Eq. (2.33), it is sufficient to employ only the controllability matrix $\tilde{\mathbb{P}}$ of the balanced system in the following derivations.

The controllability matrix $\tilde{\mathbf{P}}$ can be obtained by solving the Lyapunov equation:

$$\tilde{\mathbf{A}} \tilde{\mathbb{P}} + \tilde{\mathbb{P}} \tilde{\mathbf{A}}^{\mathrm{T}} + \tilde{\mathbf{B}} \tilde{\mathbf{B}}^{\mathrm{T}} = 0 \tag{3.32}$$

Taking the partial derivatives of Eq. (3.32) to $\mathbf{b}$, the nominal and derivative forms of the Lyapunov equation are, respectively, written as

$$\tilde{\mathbf{A}}^{c} \tilde{\mathbb{P}}^{c} + \tilde{\mathbb{P}}^{c} \left(\tilde{\mathbf{A}}^{c} \right)^{\mathrm{T}} + \tilde{\mathbf{B}}^{c} \left(\tilde{\mathbf{B}}^{c} \right)^{\mathrm{T}} = 0 \tag{3.33}$$

and

$$\begin{aligned} &\tilde{\mathbf{A}}^{c} \partial \tilde{\mathbb{P}}^{c} / \partial b_k + \partial \tilde{\mathbb{P}}^{c} / \partial b_k \left(\tilde{\mathbf{A}}^{c} \right)^{\mathrm{T}} + \tilde{\mathbf{B}}^{c} \left(\partial \tilde{\mathbf{B}}^{c} \right)^{\mathrm{T}} \Big/ \partial b_k \\ &+ \partial \tilde{\mathbf{B}}^{c} / \partial b_k \left(\tilde{\mathbf{B}}^{c} \right)^{\mathrm{T}} + \partial \tilde{\mathbf{A}}^{c} / \partial b_k \tilde{\mathbb{P}}^{c} + \tilde{\mathbb{P}}^{c} \left(\partial \tilde{\mathbf{A}}^{c} \right)^{\mathrm{T}} \Big/ \partial b_k = \mathbf{0} \end{aligned} \tag{3.34}$$

Equation (3.33) represents a deterministic Lyapunov equation which can be directly solved, while Eq. (3.34) is a Lyapunov equation in relation to $\partial \tilde{\mathbb{P}}^{c}(\mathbf{b}) / \partial b_k$, which can also be solved analogously. Therefore, the uncertain controllability matrix $\tilde{\mathbb{P}}^{CS}$ under the convex set can be obtained as

$$\tilde{P}_{ij}^{CS} = \tilde{P}_{ij}^{c} + \mathbf{h}_{\tilde{P}_{ij}}^{\mathrm{T}} \boldsymbol{\delta}_{\mathbf{b}} \in \left[\tilde{P}_{ij}^{c} - \sqrt{\mathbf{h}_{\tilde{P}_{ij}}^{\mathrm{T}} \mathbf{M}^{-1} \left(\mathbf{b}^{r}, \boldsymbol{\rho}_{\mathbf{b}} \right) \mathbf{h}_{\tilde{P}_{ij}}}, \tilde{P}_{ij}^{c} + \sqrt{\mathbf{h}_{\tilde{P}_{ij}}^{\mathrm{T}} \mathbf{M}^{-1} \left(\mathbf{b}^{r}, \boldsymbol{\rho}_{\mathbf{b}} \right) \mathbf{h}_{\tilde{P}_{ij}}} \right] \tag{3.35}$$

where

$$\mathbf{h}_{P_{ij}} = \left[\partial \tilde{P}_{ij}^{c} \big/ \partial b_1 \cdots \partial \tilde{P}_{ij}^{c} \big/ \partial b_k \cdots \partial \tilde{P}_{ij}^{c} \big/ \partial b_m \right]^{\mathrm{T}} \tag{3.36}$$

According to Lemma 2.1 in Sect. 2.4 and the CSB uncertainty propagation of the controllability matrix $\tilde{\mathbb{P}}^{\mathrm{CM}}$ derived in this section, the corresponding convex set HSVs are

$$\tilde{\boldsymbol{\sigma}}^{\mathrm{CS}} = \boldsymbol{\sigma}\left(\tilde{\mathbb{P}}^{\mathrm{CS}}\right) = \tilde{\Sigma}^{\mathrm{CS}} = \tilde{\Sigma}^{\mathrm{c}} + \tilde{\Sigma}^{\mathrm{r}} = \mathrm{diag}\left(\tilde{\sigma}_1^{\mathrm{CS}}, \ldots, \tilde{\sigma}_n^{\mathrm{CS}}\right) \tag{3.37}$$

where $\tilde{\boldsymbol{\sigma}}^{\mathrm{CS}} = \left\{\tilde{\boldsymbol{\sigma}}_i^{\mathrm{CS}}\right\}, i = 1, \ldots, n$ denotes the CSB HSV of the balanced system. These singular values will be applied for model reduction in the subsequent section.

Based on the interval balanced system equation discussed above, a convex set perturbation–based SVD method is now proposed to estimate a tighter bound for the uncertain singular values.

The deterministic SVD plays a crucial part in obtaining the nominal singular values and vectors of the system matrix as Eq. (2.35). Considering Eq. (2.36) to Eq. (2.39) and the convex set uncertainty propagation, substituting Eqs. (2.35) and (2.36) into Eq. (2.38) and applying Eq. (2.37) yield the following first-order relationships:

$$\begin{cases} \tilde{\mathbb{P}}^{\mathrm{c}} \dfrac{\partial \mathbf{v}_i^{\mathrm{c}}}{\partial b_k} + \dfrac{\partial \tilde{\mathbb{P}}^{\mathrm{c}}}{\partial b_k} \mathbf{v}_i^{\mathrm{c}} = \sigma_i^{\mathrm{c}} \dfrac{\partial \mathbf{u}_i^{\mathrm{c}}}{\partial b_k} + \dfrac{\partial \sigma_i^{\mathrm{c}}}{\partial b_k} \mathbf{u}_i^{\mathrm{c}} \\ \left(\mathbf{u}_i^{\mathrm{c}}\right)^{\mathrm{T}} \dfrac{\partial \tilde{\mathbb{P}}^{\mathrm{c}}}{\partial b_k} + \dfrac{\partial \left(\mathbf{u}_i^{\mathrm{c}}\right)^{\mathrm{T}}}{\partial b_k} \tilde{\mathbb{P}}^{\mathrm{c}} = \sigma_i^{\mathrm{c}} \dfrac{\partial (\mathbf{v}_i^{\mathrm{c}})^{\mathrm{T}}}{\partial b_k} + \dfrac{\partial \sigma_i^{\mathrm{c}}}{\partial b_k} \left(\mathbf{v}_i^{\mathrm{c}}\right)^{\mathrm{T}} \end{cases} \tag{3.38}$$

and

$$\mathbf{u}_i^{\mathrm{c}} \partial \mathbf{u}_i^{\mathrm{c}} \big/ \partial b_k = 0, \ \mathbf{v}_i^{\mathrm{c}} \partial \mathbf{v}_i^{\mathrm{c}} \big/ \partial b_k = 0 \tag{3.39}$$

The derivatives of the singular vectors to the uncertain parameters can be expanded in the basis of deterministic singular vectors:

$$\partial \mathbf{u}_i^{\mathrm{c}} \big/ \partial b_k = \sum\nolimits_{l=1}^{n} \partial \alpha_{li} \big/ \partial b_k \mathbf{u}_i^{\mathrm{c}}, \ \partial \mathbf{v}_i^{\mathrm{c}} \big/ \partial b_k = \sum\nolimits_{l=1}^{n} \partial \beta_{li} \big/ \partial b_k \mathbf{v}_i^{\mathrm{c}} \tag{3.40}$$

where $\partial \alpha_{li} \big/ \partial b_k$ represent the projection length of $\partial \mathbf{u}_i^{\mathrm{c}} \big/ \partial b_k$ onto the $\mathbf{u}_i^{\mathrm{c}}$-based space in the l th dimension, and $\partial \beta_{li} \big/ \partial b_k$ represent the projection length of $\partial \mathbf{v}_i^{\mathrm{c}} \big/ \partial b_k$ onto the $\mathbf{v}_i^{\mathrm{c}}$-based space in the l th dimension, respectively. Substitute Eq. (3.40) into (3.38) and left-multiply both sides by $\left(\mathbf{u}_l^{\mathrm{c}}\right)^{\mathrm{T}}$ and $\left(\mathbf{v}_l^{\mathrm{c}}\right)^{\mathrm{T}}$ gives

$$\begin{cases} \sigma_l^{\mathrm{c}} \dfrac{\partial \beta_{li}}{\partial b_k} + \left(\mathbf{u}_l^{\mathrm{c}}\right)^{\mathrm{T}} \dfrac{\partial \tilde{\mathbb{P}}^{\mathrm{c}}}{\partial b_k} \mathbf{v}_i^{\mathrm{c}} = \sigma_i^{\mathrm{c}} \dfrac{\partial \alpha_{li}}{\partial b_k} + \dfrac{\partial \sigma_i^{\mathrm{c}}}{\partial b_k} \delta_{li} \\ \sigma_l^{\mathrm{c}} \dfrac{\partial \alpha_{li}}{\partial b_k} + \left(\mathbf{v}_l^{\mathrm{c}}\right)^{\mathrm{T}} \dfrac{\partial \left(\tilde{\mathbb{P}}^{\mathrm{c}}\right)^{\mathrm{T}}}{\partial b_k} \mathbf{u}_i^{\mathrm{c}} = \sigma_i^{\mathrm{c}} \dfrac{\partial \beta_{li}}{\partial b_k} + \dfrac{\partial \sigma_i^{\mathrm{c}}}{\partial b_k} \delta_{li} \end{cases} \tag{3.41}$$

Combining Eq. (3.41) where $i = l$ is set with the results of substituting Eq. (3.40) into Eq. (2.37), the partial derivative of the singular value to the uncertainties can be written as

$$\partial\sigma_i^{\mathrm{c}} / \partial b_k = \left(\mathbf{u}_i^{\mathrm{c}}\right)^{\mathrm{T}} \partial\tilde{\mathbb{P}}_i^{\mathrm{c}} / \partial b_k \mathbf{v}_i^{\mathrm{c}} \tag{3.42}$$

Consequently, the uncertain singular value σ_i^{CS} under the convex set can be obtained as

$$\sigma_i^{\mathrm{CS}} = \sigma_i^{\mathrm{c}} + \mathbf{h}_{\sigma_i}^{\mathrm{T}} \boldsymbol{\delta}_{\mathrm{b}} \in \left[\sigma_i^{\mathrm{c}} - \sqrt{\mathbf{h}_{\sigma_i}^{\mathrm{T}} \mathbf{M}^{-1}\left(\mathbf{b}^{\mathrm{r}}, \boldsymbol{\rho}_{\mathrm{b}}\right)\mathbf{h}_{\sigma_i}}, \sigma_i^{\mathrm{c}} + \sqrt{\mathbf{h}_{\sigma_i}^{\mathrm{T}} \mathbf{M}^{-1}\left(\mathbf{b}^{\mathrm{r}}, \boldsymbol{\rho}_{\mathrm{b}}\right)\mathbf{h}_{\sigma_i}}\right] \tag{3.43}$$

where

$$\mathbf{h}_{\sigma_i} = \left[\partial\sigma_i^{\mathrm{c}} / \partial b_1 \cdots \partial\sigma_i^{\mathrm{c}} / \partial b_k \cdots \partial\sigma_i^{\mathrm{c}} / \partial b_m\right]^{\mathrm{T}} \tag{3.44}$$

Thus, the convex set bounds of each uncertain singular value are determined as

$$\underline{\sigma_i} = \sigma_i^{\mathrm{c}} - \sigma_i^{\mathrm{r}}, \overline{\sigma_i} = \sigma_i^{\mathrm{c}} + \sigma_i^{\mathrm{r}} \tag{3.45}$$

where

$$\sigma_i^{\mathrm{r}} = \sqrt{\mathbf{h}_{\sigma_i}^{\mathrm{T}} \mathbf{M}^{-1}\left(\mathbf{b}^{\mathrm{r}}, \boldsymbol{\rho}_{\mathrm{b}}\right)\mathbf{h}_{\sigma_i}} \tag{3.46}$$

where $\underline{\sigma_i}$ and $\overline{\sigma_i}$ denote the LB and UB of the i th CSB uncertain singular value, respectively.

Consequently, by applying this convex set perturbation–based SVD, the bounds of the uncertain singular values are effectively estimated. These results will serve as the foundation for defining the order-truncation criterion and the subsequent optimization of model order selection.

3.3.2 CSB Order-Truncated Criterion

As discussed in Chap. 2, for deterministic cases, the singular values are arranged as Eq. (2.49). Each singular value lies in the corresponding convex set derived through the proposed perturbation-based estimation. Considering two adjacent uncertain singular values, σ_i^{CM} and $\sigma_{i+1}^{\mathrm{CM}}$, three possible two-dimensional relationships (far away, adjacent, and cross) may occur, as illustrated in Fig. 3.1. The bisector of the first quadrant corresponds to the equality $\sigma_{i+1} = \sigma_i$. Since the first two cases (far away and adjacent) are straightforward, the discussion focuses on the third case, where the upper bound of the $i+1$ th singular value exceeds the lower bound of the i-th singular value, i.e., $\overline{\sigma_{i+1}} > \underline{\sigma_i}$.

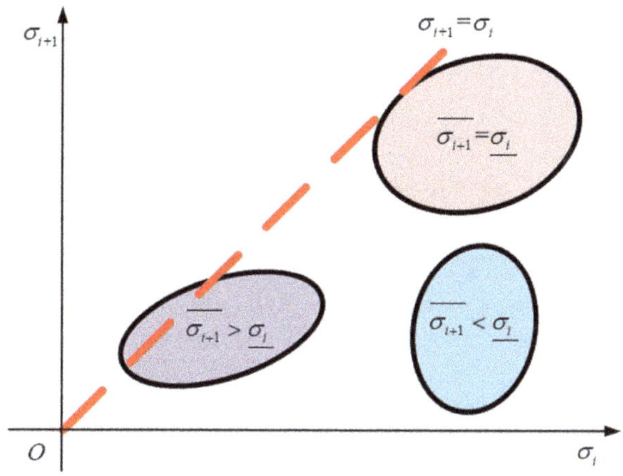

Fig. 3.1 X 3 CSB relationships for the two uncertain singular values

To determine the truncation order for uncertain systems with closely distributed singular values, a truncation criterion σ_{cr} is proposed, as shown in Fig. 3.2.

In this two-dimensional convex set space, three lines are defined: the bisector of the first quadrant $\sigma_{i+1} = \sigma_i$, the horizontal line $\sigma_{i+1} = \sigma_{\text{cr}}$, and the vertical line $\sigma_i = \sigma_{\text{cr}}$. These lines intersect the oblique elliptical region representing the convex uncertainty of the two singular values. The CSB possibilities that the uncertain singular values exceed the criterion are defined as

$$\begin{cases} P\left(\sigma_i^{\text{CS}} > \sigma_{\text{cr}}\right) = \left\{\sigma_i \middle| \exists \sigma_i \in \sigma_i^{\text{CS}}, \text{ s.t. } \sigma_i > \sigma_{\text{cr}}\right\} \\ P\left(\sigma_{i+1}^{\text{CS}} > \sigma_{\text{cr}}\right) = \left\{\sigma_{i+1} \middle| \exists \sigma_{i+1} \in \sigma_{i+1}^{\text{CS}}, \text{ s.t. } \sigma_{i+1} > \sigma_{\text{cr}}\right\} \end{cases} \tag{3.47}$$

Because of the inherent ordering $\sigma_{i+1}^{\text{CS}} < \sigma_i^{\text{CS}}$, the coordinates $\left(\sigma_i^{\text{c}}, \sigma_{i+1}^{\text{c}}\right)$ must be located right beside $\sigma_{i+1} = \sigma_i$. Therefore, the CSB possibility of the event $P\left(\sigma_i^{\text{CS}} \geq \sigma_{i+1}^{\text{CS}}\right)$ is

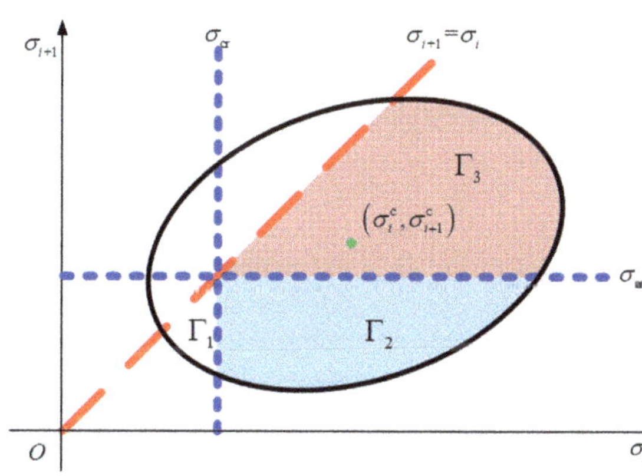

Fig. 3.2 Truncated value across the two-dimensional CSB singular values

$$P\left(\sigma_i^{\mathrm{CS}} \geq \sigma_{i+1}^{\mathrm{CS}}\right) = \left\{\sigma_i, \sigma_{i+1} \middle| \exists \sigma_i \in \sigma_i^{\mathrm{CM}}, \exists \sigma_{i+1} \in \sigma_{i+1}^{\mathrm{CM}}, \text{s.t.} \sigma_i \geq \sigma_{i+1}\right\} = 100\% \tag{3.48}$$

Thus, the possibilities in Eq. (3.47) are solved as

$$P\left(\sigma_i^{\mathrm{CS}} > \sigma_{\mathrm{cr}}\right) = S_{\Gamma_i} / S_{\Gamma}, P\left(\sigma_{i+1}^{\mathrm{CS}} > \sigma_{\mathrm{cr}}\right) = S_{\Gamma_{i+1}} / S_{\Gamma} \tag{3.49}$$

where $\Gamma = \Gamma_1 \cup \Gamma_2 \cup \Gamma_3$ denotes the area of the CSB oblique ellipse right beside the bisector $\sigma_{i+1} = \sigma_i$. $\Gamma_i = \Gamma_2 \cup \Gamma_3$ and $\Gamma_{i+1} = \Gamma_3$ represent the domains satisfying $\sigma_i^{\mathrm{CM}} > \sigma_{\mathrm{cr}}$ and $\sigma_{i+1}^{\mathrm{CM}} > \sigma_{\mathrm{cr}}$. S_Γ, S_{Γ_i} and $S_{\Gamma_{i+1}}$ are the corresponding regions of these domains.

Lemma 3.1 The CSB possibility ratio η is calculated to formulate the CSB order-truncated criterion as

$$\eta = S_{\Gamma_{i+1}} / S_{\Gamma_i} \tag{3.50}$$

Accordingly, the CSB truncation order w is determined by

$$w = \begin{cases} i, & \text{if } \eta \leq 0.5 \\ i+1, & \text{if } \eta > 0.5 \end{cases}, \quad 1 \leq w \leq n-1 \tag{3.51}$$

That is,

$$\left\{\sigma_w^{\mathrm{CS}}\right\} = \begin{cases} \left\{\sigma_1^{\mathrm{CS}}, \ldots, \sigma_i^{\mathrm{CS}}\right\}, & \text{if } \eta \leq 0.5 \\ \left\{\sigma_1^{\mathrm{CS}}, \ldots, \sigma_i^{\mathrm{CS}}, \sigma_{i+1}^{\mathrm{CS}}\right\}, & \text{if } \eta > 0.5 \end{cases} \tag{3.52}$$

The determination of σ_{cr}, along with the bisector $\sigma_{i+1} = \sigma_i$ and the geometric characteristics of the oblique ellipse, plays a crucial role in defining the truncation boundary. Since the uncertain singular values have been estimated via the proposed method, the next step is to determine an appropriate value for σ_{cr}. Selecting σ_{cr} is challenging, yet essential for defining the truncated order. A direct geometric approach is adopted here.

Lemma 3.2 Without loss of generality, set $\sigma_{\mathrm{cr}} = \underline{\sigma_i}$ as illustrated in Fig. 3.3 as

$$\eta^* = S_{\Gamma_{i+1}^*} / S_{\Gamma_i^*} \tag{3.53}$$

As shown in Fig. 3.3, the regions Γ^* and Ω_i^* are equivalent, i.e., $\Gamma^* = \Gamma_i^* = \Gamma_1^* \cup \Gamma_2^*$, and $S_{\Gamma^*} = S_{\Gamma_i^*}$. Meanwhile, $\Gamma_{i+1}^* = \Gamma_2^*$ as indicated in Fig. 3.3. Substituting Eq. (3.53) into the truncation criterion of Eqs. (3.51) and (3.52) enables the determination of the truncation order.

Theorem 3.1 When $\underline{\sigma_w} < \sigma_{\mathrm{cr}} < \overline{\sigma_{w+1}}$ and $\left[\hat{\sigma}_1^{\mathrm{CS}}, \ldots, \hat{\sigma}_w^{\mathrm{CS}}\right]$ denote the truncated CSB singular values, the possibility $P(\varepsilon)$ satisfies

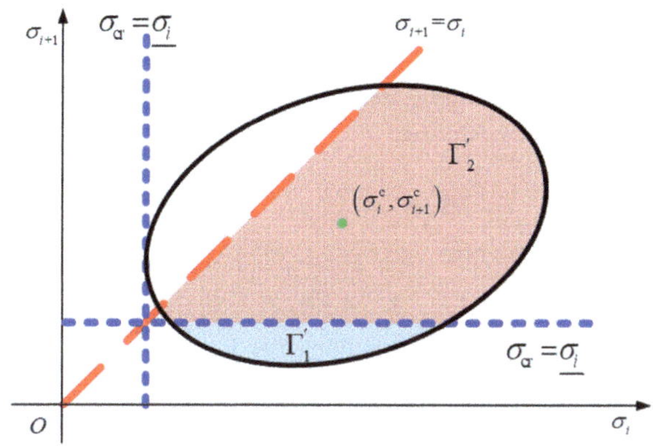

Fig. 3.3 Special truncation across the two-dimensional CSB singular values

$$P\left(\varepsilon\middle|\varepsilon \leq \sigma_{\mathrm{cr}} + \sum\nolimits_{i=w+2}^{n} \overline{\sigma_i}\right) = 1 - S_{\Gamma_{i+1}} / S_{\Gamma} \tag{3.54}$$

Here $\varepsilon = \left\| \tilde{\Xi}^{\mathrm{CS}} - \hat{\Xi}^{\mathrm{CS}} \right\|_{H_\infty}$ represents the truncation error.

Proof For deterministic ROM (Wang et al. 2018), the truncation error satisfies Eq. ?. Considering the convex set relation $\sigma_i \leq \overline{\sigma_i}$, Eq. ? becomes:

$$\varepsilon \leq \sum\nolimits_{i=w+1}^{n} \overline{\sigma_i} = \sigma_{w+1}^{\mathrm{CM}} + \sum\nolimits_{i=w+2}^{n} \overline{\sigma_i} \tag{3.55}$$

Based on the second expression in Eq. (3.49), the possibility of the two CSB events yields

$$\begin{aligned} P\left(\sigma_{w+1}^{\mathrm{CS}} > \sigma_{\mathrm{cr}}\right) &= P\left(\left[\hat{\sigma}_1^{\mathrm{CS}}, \cdots, \hat{\sigma}_w^{\mathrm{CS}}\right] \bigcap \sigma_{w+1}^{\mathrm{CS}} > \sigma_{\mathrm{cr}}\right) \\ &= P\left(\left[\hat{\sigma}_1^{\mathrm{CS}}, \cdots, \hat{\sigma}_w^{\mathrm{CS}}\right]\right) \bigcap P\left(\sigma_{w+1}^{\mathrm{CS}} > \sigma_{\mathrm{cr}}\right) = S_{\Gamma_{i+1}} / S_{\Gamma} \end{aligned} \tag{3.56}$$

Thus, the CSB possibility becomes

$$P\left(\sigma_{w+1}^{\mathrm{CS}} \leq \sigma_{\mathrm{cr}}\right) = 1 - S_{\Gamma_{i+1}} / S_{\Gamma} \tag{3.57}$$

Substituting Eq. (3.57) into Eq. (3.55) completes the proof. □

Corollary 3.1 When $\sigma_{\mathrm{cr}} = \underline{\sigma_w}$ in *Theorem 3.1*, the corresponding CSB possibility $P(\varepsilon^*)$ is.

$$P\left(\varepsilon\middle|\varepsilon \leq \underline{\sigma_w} + \sum\nolimits_{i=w+2}^{n} \overline{\sigma_i}\right) = 1 - \eta^* \tag{3.58}$$

Proof As shown in Fig. 3.3, when $\sigma_{\mathrm{cr}} = \underline{\sigma_w}$,

$$\Gamma_i \triangleq \Gamma, \quad S_{\Gamma_i} \equiv S_{\Gamma} \tag{3.59}$$

Substituting Eq. (3.59) into Eq. (3.54) and applying Eq. (3.53) completes the proof. □

After determining the truncation order, the optimal reduced state parameters can be derived by solving:

$$\begin{aligned}
&\text{find} \quad \hat{\mathbf{x}}^{\mathrm{T}} = [\hat{x}_1, \ldots, \hat{x}_w]^{\mathrm{T}} \\
&\min \quad \left\| \left[\hat{\sigma}_1^{\mathrm{c}}, \ldots, \hat{\sigma}_w^{\mathrm{c}}\right] - \left[\sigma_1^{\mathrm{c}}, \ldots, \sigma_w^{\mathrm{c}}\right] \right\| \\
&\text{s.t.} \quad \hat{\mathbf{x}}^{\mathrm{T}} = [\hat{x}_1, \ldots, \hat{x}_w]^{\mathrm{T}} = \mathbf{x}_{\mathrm{sub}}^{\mathrm{T}} \subset \mathbf{x}^{\mathrm{T}} = [x_1, \ldots, x_n]^{\mathrm{T}} \\
&\qquad 1 \le w \le n-1
\end{aligned} \tag{3.60}$$

where $\hat{\mathbf{x}}^{\mathrm{T}} = [\hat{x}_1, \ldots, \hat{x}_w]^{\mathrm{T}}$ represents the selected w state parameters corresponding to the retained singular values. $\mathbf{x}_{\mathrm{sub}}^{\mathrm{T}}$ is the subset of the original state vector $\mathbf{x}^{\mathrm{T}}$, and $\left[\hat{\sigma}_1^{\mathrm{c}}, \ldots, \hat{\sigma}_w^{\mathrm{c}}\right]$ $\left[\sigma_1^{\mathrm{c}}, \ldots, \sigma_w^{\mathrm{c}}\right]$ are the leading w values of the reduced and full systems, respectively.

Accordingly, the reduced CSB state-space model can be reconstructed as

$$\hat{\Xi}^{\mathrm{c}} + \hat{\Xi}^{\mathrm{r}} = \hat{\Xi}^{\mathrm{CS}} = \left(\hat{\mathbf{A}}^{\mathrm{CS}} \,\middle|\, \hat{\mathbf{B}}^{\mathrm{CS}}\right) = \left(\hat{\mathbf{A}}^{\mathrm{c}} \,\middle|\, \hat{\mathbf{B}}^{\mathrm{c}}\right) + \left(\hat{\mathbf{A}}^{\mathrm{r}} \,\middle|\, \hat{\mathbf{B}}^{\mathrm{r}}\right) \tag{3.61}$$

where the superscript "^" represents the reduced model reconstructed according to the optimized state subset $\hat{\mathbf{x}}^{\mathrm{T}} = [\hat{x}_1, \ldots, \hat{x}_w]^{\mathrm{T}}$ obtained from Eq. (3.60).

3.4 Numerical Example

A three DoF spring-damper-mass system, illustrated in Fig. 2.5 of Sect. 2.7, is employed to judge the performance of the CSBROM method. The deterministic variables of the system are given as $\left[m_1^{\mathrm{c}}, m_2^{\mathrm{c}}, m_3^{\mathrm{c}}\right] = [1, 2, 10]$, $\left[c_1^{\mathrm{c}}, c_2^{\mathrm{c}}, c_3^{\mathrm{c}}\right] = [3, 9, 15]$, and $\left[k_1^{\mathrm{c}}, k_2^{\mathrm{c}}, k_3^{\mathrm{c}}\right] = [3000, 1500, 1000]$, where m_i^{c}, c_i^{c}, k_i^{c} $(i = 1, 2, 3)$ denote the nominal values of mass/damping/stiffness coefficients. The variances of these variables are defined as $\mathrm{cov}(m_i, m_i) = 0.004$, $\mathrm{cov}(c_i, c_i) = 0.225$, and $\mathrm{cov}(k_i, k_i) = 1000$, $i = 1, 2, 3$, with no covariance between different parameters. The system is subjected to an input excitation of $u_3 = 10\cos(12\pi t)$ using a time step of 0.004 s.

To benchmark the effectiveness of the introduced uncertain SVD methods, the CSBROM results are compared with those obtained from 100,000 MCS conducted on the deterministic ROM. Table 3.1 presents the first four bounds of the uncertain singular values of the controllability matrix $\tilde{\mathbb{P}}$. The UB/LB of the singular values calculated by CSBROM closely match those derived from the MCS results.

Since the third and fourth singular values ($\sigma_3^{\mathrm{c}} = 0.0253$ and $\sigma_4^{\mathrm{c}} = 0.0246$) are nearly identical, the corresponding bounds obtained by CSBROM exhibit an interference relationship. Consequently, the convex set possibility-based order-truncation criterion is adopted to determine the appropriate truncation.

Table 3.1 Results of the proposed method compared with MCS

		CSBROM		MCS	
Singular value bounds	Order	LB	UB	LB	UB
	1	0.0642	0.0686	0.0644	0.0687
	2	0.0636	0.0687	0.0638	0.0684
	3	0.0242	0.0264	0.0243	0.0263
	4	0.0238	0.0255	0.0238	0.0254
Computation time (s)		0.223		2814	

A two-dimensional convex set is established between the third and fourth singular values. Following Lemma 2.4 in Sect. 2.6, the truncation threshold is set to $\sigma_{cr} = \underline{\sigma_3} = 0.0242$. Based on the convex set possibility analysis, the computed ratio η^* between the third and fourth singular values is 0.8119 using the proposed CSBROM method. Since $\eta^* > 0.5$, the fourth-order component is preserved. Figure 3.4 illustrates the third and fourth singular values obtained through both the CMROM and MCS.

Figure 3.5 displays the output range (position of No.3 DoF) obtained via the proposed CSBROM and MCS methods throughout the time history. In terms of accuracy, the upper and lower bound curves (UB and LB) determined by CSBROM coincide precisely with those from MCS across the entire simulation period. Regarding computational efficiency, the CSBROM method achieves significantly faster computation compared to MCS, as summarized in Table 3.1. These results collectively confirm that the proposed CSBROM method achieves both high precision and efficiency.

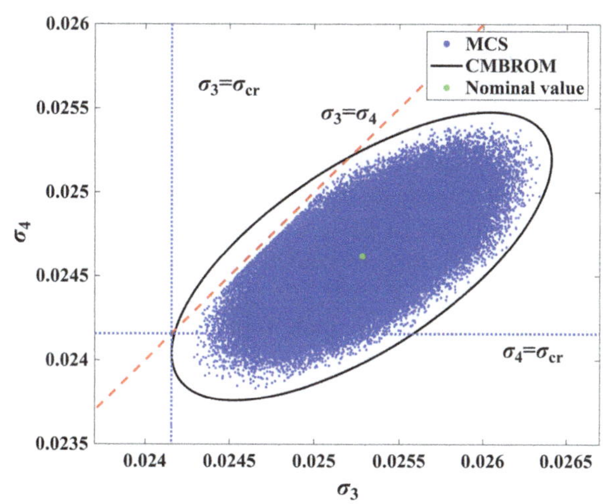

Fig. 3.4 Comparison of the third- and fourth-order singular values between CSBROM and MCS

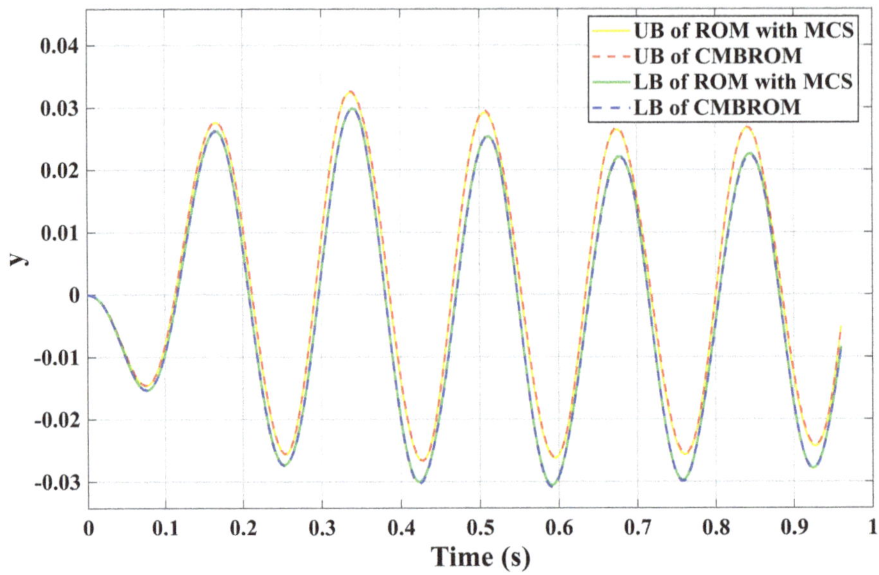

Fig. 3.5 Comparison of output between CSBROM and MCS

3.5 Summary

This chapter introduces the CSBROM framework to overcome the limitations of probabilistic approaches that require extensive sampling data. Grounded in non-probabilistic convex set theory considering both UBB and parameter correlation of uncertainties, the method constructs a CSB state-space equation using an order-extended matrix technique, allowing for the analytical solution of uncertain states. To enable model reduction, the classical balanced truncation is generalized by solving the CSB Lyapunov equation to derive uncertain controllability and observability Gramians. Subsequently, a convex set perturbation-based SVD algorithm is developed to accurately estimate the bounds of uncertain HSVs. A novel truncation criterion, utilizing CSB possibility and the relationships of singular value distributions, is proposed to objectively select the reduced order. Experimental results verify that CSBROM provides approximation accuracy comparable to MCSs while drastically reducing computational time, proving its efficiency for uncertain linear control systems.

References

Cao L, Liu J, Xie L, Jiang C, Bi R (2021) Non-probabilistic polygonal convex set model for structural uncertainty quantification. Appl Math Model 89:504–518. https://doi.org/10.1016/j.apm.2020.07.025

Imtiaz H, Akhtar I (2020) Nonlinear closure modeling in reduced order models for turbulent flows: a dynamical system approach. Nonlinear Dyn 99:479–494. https://doi.org/10.1007/s11071-019-05087-2

Jiang C, Han X, Lu GY, Liu J, Zhang Z, Bai YC (2011) Correlation analysis of non-probabilistic convex model and corresponding structural reliability technique. Comput Methods Appl Mech Eng 200:2528–2546. https://doi.org/10.1016/j.cma.2011.04.007

Liu BH, Li H, Bu X, Zhou X (2022) Optimization-based three-way decisions with interval-valued intuitionistic fuzzy information. IEEE Trans Cybern, 1–15. https://doi.org/10.1109/TCYB.2022.3151899

Lü H, Yang K, Huang X, Shangguan W-B, Zhao K (2022) Uncertainty and correlation propagation analysis of powertrain mounting systems based on multi-ellipsoid convex model. Mech Syst Signal Process 173:109058. https://doi.org/10.1016/j.ymssp.2022.109058

Maulik TB, Ramachandra N, Mason LR, Pan I (2021) Latent-space time evolution of non-intrusive reduced-order models using Gaussian process emulation. Phys Nonlinear Phenom 416:132797. https://doi.org/10.1016/j.physd.2020.132797

Meng ZZ, Zhou H (2020) A novel experimental data-driven exponential convex model for reliability assessment with uncertain-but-bounded parameters. Appl Math Model 77:773–787. https://doi.org/10.1016/j.apm.2019.08.010

Peng NS, Kan Z (2022) Data-driven model order reduction with proper symplectic decomposition for flexible multibody system. Nonlinear Dyn 107:173–203. https://doi.org/10.1007/s11071-021-06990-3

Qian J, Wang Y, Song H, Pant K, Peabody H, Ku J, Butler CD, Projection-based reduced-order modeling for spacecraft thermal analysis. J Spacecr Rocketshttps://doi.org/10.2514/1.A33117

Reduced-Order Modeling for Flexible Spacecraft Deployment and Dynamics, https://doi.org/10.2514/6.2021-1385

Su QC, Sun S, Bing Z, Knoll A (2023) Output feedback control of fuzzy systems via reduced-order approximation technique. IEEE Trans Syst Man Cybern Syst 1–12. https://doi.org/10.1109/TSMC.2023.3270289

Time-variant reliability modeling based on hybrid non-probability method. Arch Appl Mech, https://doi.org/10.1007/s00419-019-01605-1

Wang L, Wang X, Wu D, Zhang X (2018) Non-probabilistic time-variant reliability assessment (NTRA) for the active control of vibration systems with convex uncertainties. ISA Trans 83:276–289. https://doi.org/10.1016/j.isatra.2018.08.018

Wang XL, Zhang P, Wu M (2020) Control strategy based on model reduction and online intelligent calculation for planar n -link underactuated manipulators. IEEE Trans Syst Man Cybern Syst 50:1046–1054. https://doi.org/10.1109/TSMC.2017.2761862

Zhu WP, Li Z (2022) A granular approach to interval output estimation for rule-based fuzzy models. IEEE Trans Cybern 52:7029–7038. https://doi.org/10.1109/TCYB.2020.3025668

Zhu PT, He Z, Zhang C, Fang Q, Chen G (2022a) Nonlinear model predictive control of USC boiler-turbine power units in flexible operations via input convex neural network. Energy 255:124486. https://doi.org/10.1016/j.energy.2022.124486

Zhu X, Pedrycz W, Li Z (2022b) A granular approach to interval output estimation for rule-based fuzzy models. IEEE Trans Cybern 52:7029–7038. https://doi.org/10.1109/TCYB.2020.3025668

Part II
Set Theory-Based Attitude Determination and Control

Chapter 4
Robust Attitude Determination Based on Interval Uncertainty Theory

Abstract This chapter introduces an attitude determination framework designed for satellite missions characterized by limited on-orbit data and uncertain measurements. Unlike traditional probabilistic approaches that require extensive statistical sampling, the methodology adopted here utilizes non-probabilistic interval analysis techniques to model system uncertainties as unknown-but-bounded (UBB) parameters. A key contribution discussed is the development of a narrow bound theory, which employs interval dimension-wise analysis (IDWA) rooted in polynomial chaos expansions. This approach effectively mitigates the overestimation issues often associated with conventional interval perturbation methods, ensuring more precise bound estimation for the attitude determination system. The classical Wahba problem is reformulated into a multi-objective optimization context, aiming to simultaneously minimize the nominal loss function and its uncertainty radius. To solve this complex problem, the strength Pareto evolutionary algorithm based on reference direction (SPEA/R) is utilized, balancing convergence and diversity in the solution set. Numerical simulations confirm that this interval-based strategy achieves high attitude estimation accuracy and computational efficiency, offering a resilient alternative for spacecraft attitude operating in data-sparse environments.

4.1 Introduction

As space missions become increasingly frequent, the number of spacecraft and satellites carrying out tasks such as detection and navigation in orbit continues to grow, making accurate attitude determination a critical requirement (Hasan et al. 2023; Wu et al. 2018; Zhang et al. 2021). Achieving precise satellite attitude determination at a relatively low cost has become one of the key performance indicators for on-orbit operations. In recent years, the integration of spaceborne gravity gradiometers and gyroscopes with advanced algorithms has enabled more accurate and efficient attitude determination (Sun et al. 2022, 2023), thereby enhancing the overall effectiveness of satellite missions in space.

C. Yang and Y. Xia, *Set Theory-Based Spacecraft Dynamics and Control*, Intelligent Control and Learning Systems 26, https://doi.org/10.1007/978-981-95-9179-4_4

The growing complexity of satellite on-orbit missions presents significant challenges to the field of attitude determination. Recent advances indicate a growing focus on integrating advanced algorithms with hardware systems to better address the diverse requirements of real-world space applications (Hsu and Chen 2022; Lee and Gallucci 2023). Nevertheless, system uncertainties and measurement noise continue to undermine the accuracy of attitude determination. To solve this problem, a series of methods have been developed (Cheng et al. 2021; He et al. 2021; Gan et al. 2021). Unlike ground-based environments, the space on-orbit setting is characterized by a scarcity of high-quality sample data, often plagued by incomplete information such as uncertainty and noise. Traditional probabilistic and statistical approaches are often inadequate for quantifying such limited and imprecise data due to their high computational and experimental costs (Jiang et al. 2011).

In response, emerging non-probabilistic interval analysis methods have gained attention in recent years as effective tools for uncertainty quantification when data samples are limited (Qiu and Elishakoff 1998; Ben-Haim 1994). These methods have demonstrated promising results in various aerospace applications. However, existing techniques based solely on interval perturbation theory exhibit limited applicability and are primarily effective for scenarios with small uncertainties. To address these limitations and bridge the existing gaps, this chapter utilizes the IDWA method in spacecraft attitude determination, aiming to enhance the capability of handling multi-source and large-scale uncertainties in attitude determination systems.

4.2 Nominal Model of Attitude Determination

In the Wahba attitude determination problem (Wahba 1965), the aim is to determine the optimal rotation matrix, denoted as $\boldsymbol{\Upsilon}$, by utilizing at least two distinct and independent sets of measured vectors. This rotation matrix is obtained by minimizing an objective function that follows a weighted least-squares approach. $\boldsymbol{\Upsilon}$, a commonly used cost function seeks to minimize, is expressed in the following form:

$$J(\boldsymbol{\Upsilon}) = \sum_{j=1}^{N} \varsigma_i \|\mathbf{m}_i - \boldsymbol{\Upsilon}\boldsymbol{r}_i\|^2 \tag{4.1}$$

where $\mathbf{m}_i$ refers to the ith measurement and $\mathbf{r}_i$ is the matching reference vector, respectively. ς_i represents a strictly positive coefficient corresponding to the ith observation pair $\mathbf{m}_i$ and $\mathbf{r}_i$. The spacecraft attitude can be represented using a quaternion, which is denoted as

$$\mathbf{q} = \left[q_1\, q_2\, q_3\, q_0\right]^{\mathrm{T}} = \left[\boldsymbol{\xi}^{\mathrm{T}}\, q_0\right]^{\mathrm{T}} \tag{4.2}$$

where q_0 and $\boldsymbol{\xi}$ denotes the scalar part and the vector part in the quaternion, respectively. The rotation matrix $\boldsymbol{\Upsilon}$ is obtained from the quaternion $\mathbf{q}$ as

$$\boldsymbol{\Upsilon} = \left(q_0 - \boldsymbol{\xi}^{\mathrm{T}}\boldsymbol{\xi}\right)\mathbf{I}_3 + 2\boldsymbol{\xi}\boldsymbol{\xi}^{\mathrm{T}} - 2q_0\boldsymbol{\xi}^{\times} \tag{4.3}$$

where $\boldsymbol{\xi}^{\times}$ is the skew-symmetric matrix constructed from the vector $\boldsymbol{\xi}$.

The technique has drawn increasing scholarly attention because it achieves globally optimal solutions without depending on local linearization or overly simplified modeling conditions (Horn 1987; Ahmed et al. 2012). Moreover, it is capable of handling an arbitrary number of noisy measurements. However, in practical satellite operations, measurement errors are inevitably introduced by various sensors, such as magnetometers, sun sensors, and star sensors. To ensure reliable attitude determination under such conditions, it becomes essential to incorporate uncertainty modeling into the problem formulation. Accordingly, the development of a robust approach that enables accurate attitude determination in the presence of sensor inaccuracies is of critical importance.

4.3 Davenport Formula

Most commonly used algorithms for addressing the attitude determination problem are based on the q-method originally proposed by Davenport (Keat 1977). Given that $\boldsymbol{\Upsilon}$ is a rotation matrix, which indicates its orthogonal property, while $\mathbf{m}_i$ and $\mathbf{r}_i$ are normalized vectors, the following relationship holds:

$$\begin{aligned} J(\boldsymbol{\Upsilon}) &= \frac{1}{2}\sum_{j=1}^{N}\|\mathbf{m}_i - \boldsymbol{\Upsilon}\mathbf{r}_i\|^2 = \frac{1}{2}\sum_{j=1}^{N}\left(\mathbf{m}_i^{\mathrm{T}}\mathbf{m}_i - 2\mathbf{m}_i^{\mathrm{T}}\boldsymbol{\Upsilon}\mathbf{r}_i + \mathbf{r}_i^{\mathrm{T}}\boldsymbol{\Upsilon}^{\mathrm{T}}\boldsymbol{\Upsilon}\mathbf{r}_i\right) \\ &= N - \sum_{j=1}^{N}\mathbf{m}_i^{\mathrm{T}}\boldsymbol{\Upsilon}\mathbf{r}_i = N - \mathrm{tr}\left(\mathbf{M}^{\mathrm{T}}\boldsymbol{\Upsilon}\mathbf{R}\right) \end{aligned} \tag{4.4}$$

where $\mathbf{R} = [\mathbf{r}_1\ \mathbf{r}_2 \ldots \mathbf{r}_n]$ and $\mathbf{M} = [\mathbf{m}_1\ \ \mathbf{m}_2 \ldots \mathbf{m}_n]$. By using Eq. (4.3), $\mathrm{tr}\left(\mathbf{M}^{\mathrm{T}}\boldsymbol{\Upsilon}\mathbf{R}\right)$ can be rewritten as

$$\begin{aligned} \mathrm{tr}\left(\mathbf{M}^{\mathrm{T}}\boldsymbol{\Upsilon}\mathbf{R}\right) &= \mathrm{tr}\left(\mathbf{M}^{\mathrm{T}}\left[\left(q_0^2 - \boldsymbol{\xi}^{\mathrm{T}}\boldsymbol{\xi}\right)\mathbf{I}_3 + 2\boldsymbol{\xi}\boldsymbol{\xi}^{\mathrm{T}} - 2q_0\boldsymbol{\xi}^{\times}\right]\mathbf{R}\right) \\ &= \left(q_0^2 - \boldsymbol{\xi}^{\mathrm{T}}\boldsymbol{\xi}\right)\mathrm{tr}\left(\mathbf{M}^{\mathrm{T}}\mathbf{R}\right) + 2\mathrm{tr}\left(\boldsymbol{\xi}\boldsymbol{\xi}^{\mathrm{T}}\mathbf{R}\mathbf{M}^{\mathrm{T}}\right) - 2q_0\mathrm{tr}\left(\mathbf{M}^{\mathrm{T}}\boldsymbol{\xi}^{\times}\mathbf{R}\right) \end{aligned} \tag{4.5}$$

Defining $\boldsymbol{\Psi} = \mathbf{M}^{\mathrm{T}}\mathbf{R}$, $\sigma = \mathrm{tr}(\boldsymbol{\Psi})$, $\mathbf{Q} = \boldsymbol{\Psi} + \boldsymbol{\Psi}^{\mathrm{T}}$ and $\mathbf{z}^{\mathrm{T}} = \left[\, \Psi_{23} - \Psi_{32}\ \Psi_{31} - \Psi_{13}\ \Psi_{12} - \Psi_{21} \,\right]$, then one can get

$$2\mathrm{tr}\left(\boldsymbol{\xi}\boldsymbol{\xi}^{\mathrm{T}}\mathbf{R}\mathbf{M}^{\mathrm{T}}\right) = 2\boldsymbol{\xi}^{\mathrm{T}}\mathbf{R}\mathbf{M}^{\mathrm{T}}\boldsymbol{\xi} = \boldsymbol{\xi}^{\mathrm{T}}\left(\mathbf{R}\mathbf{M}^{\mathrm{T}} + \mathbf{R}^{\mathrm{T}}\mathbf{M}\right)\boldsymbol{\xi} = \boldsymbol{\xi}^{\mathrm{T}}\mathbf{Q}\boldsymbol{\xi} \tag{4.6}$$

Given that $\mathbf{z}^{\times} = \boldsymbol{\Psi}^{\mathrm{T}} - \boldsymbol{\Psi}$, $\boldsymbol{\xi}^{\times^{\mathrm{T}}} = -\boldsymbol{\xi}^{\times}$ and $\mathrm{tr}\left(\boldsymbol{\xi}^{\times}\mathbf{z}^{\times}\right) = -2\boldsymbol{\xi}^{\mathrm{T}}\mathbf{z}$, one can get

$$
\begin{aligned}
2q_0\mathrm{tr}\left(\mathbf{M}^{\mathrm{T}}\boldsymbol{\xi}^{\times}\mathbf{R}\right) &= q_0\mathrm{tr}\left(\boldsymbol{\xi}^{\times}\boldsymbol{\Psi}^{\mathrm{T}} + \boldsymbol{\Psi}\boldsymbol{\xi}^{\times^{\mathrm{T}}}\right) = q_0\mathrm{tr}\left(\boldsymbol{\xi}^{\times}\boldsymbol{\Psi}^{\mathrm{T}} + \boldsymbol{\xi}^{\times^{\mathrm{T}}}\boldsymbol{\Psi}\right) \\
&= q_0\mathrm{tr}\left(\boldsymbol{\xi}^{\times}\left(\boldsymbol{\Psi}^{\mathrm{T}} - \boldsymbol{\Psi}\right)\right) = q_0\mathrm{tr}\left(\boldsymbol{\xi}^{\times}\mathbf{z}^{\times}\right) = -2q_0\boldsymbol{\xi}^{\mathrm{T}}\mathbf{z}
\end{aligned}
\tag{4.7}
$$

Substitute Eqs. (4.6) and (4.7) into Eq. (4.5), one can get

$$
\begin{aligned}
\mathrm{tr}\left(\mathbf{M}^{\mathrm{T}}\boldsymbol{\Upsilon}\mathbf{R}\right) &= (q_0^2 - \boldsymbol{\xi}^{\mathrm{T}}\boldsymbol{\xi})\sigma + \boldsymbol{\xi}^{\mathrm{T}}\mathbf{H}\boldsymbol{\xi} + 2q_0\boldsymbol{\xi}^{\mathrm{T}}\mathbf{z} \\
&= \begin{bmatrix} q_0 & \xi^{\mathrm{T}} \end{bmatrix}\begin{bmatrix} \mathbf{Q} - \sigma\mathbf{I} & \mathbf{z}^{\mathrm{T}} \\ \mathbf{z} & \sigma \end{bmatrix}\begin{bmatrix} q_0 \\ \xi \end{bmatrix} = \mathbf{q}^{\mathrm{T}}\mathbf{K}\mathbf{q}
\end{aligned}
\tag{4.8}
$$

where $\mathbf{K} = \begin{bmatrix} \mathbf{Q} - \sigma\mathbf{I} & \mathbf{z}^{\mathrm{T}} \\ \mathbf{z} & \sigma \end{bmatrix}$ is the Davenport matrix. Thus,

$$
\min J(\boldsymbol{\Upsilon}) = \left(N - \max \mathrm{tr}\left(\mathbf{M}^{\mathrm{T}}\boldsymbol{\Upsilon}\mathbf{R}\right)\right) = N - \max \mathbf{q}^{\mathrm{T}}\mathbf{K}\mathbf{q}
\tag{4.9}
$$

4.4 Interval Uncertainty-Based Attitude Determination Problem

4.4.1 Uncertain Attitude Determination Problem

The interval uncertainty analysis approach is capable of addressing the attitude determination problem. The cost function given in Eq. (4.1) can be reformulated into two separate components as follows:

$$
J(\boldsymbol{\Upsilon}) = J_{\mathrm{m}} + \varepsilon\Delta J
\tag{4.10}
$$

where J_{m} and ΔJ represent the center and the radius of $J(\boldsymbol{\Upsilon})$, respectively. ε denotes the interval $[-1, 1]$. The expressions for computing J_{m} and ΔJ are given by the following two equations, respectively (Ghadiri et al. 2022).

$$
\begin{cases}
J_{\mathrm{m}} = 1 - \mathbf{q}^{\mathrm{T}}\mathbf{K}_{\mathrm{m}}\mathbf{q} \\
\Delta J = -\mathbf{q}^{\mathrm{T}}\mathbf{K}_{\Delta\mathrm{r}}\mathbf{q} - \mathbf{q}^{\mathrm{T}}\mathbf{K}_{\Delta\mathrm{b}}\mathbf{q} - \mathbf{q}^{\mathrm{T}}\mathbf{K}_{\Delta\mathrm{r},\Delta\mathrm{b}}\mathbf{q}
\end{cases}
\tag{4.11}
$$

where $\mathbf{K}_{\mathrm{m}}$, $\mathbf{K}_{\Delta\mathrm{b}}$, $\mathbf{K}_{\Delta\mathrm{r}}$, $\mathbf{K}_{\Delta\mathrm{b},\Delta\mathrm{r}}$ are the Davenport matrices, which are determined as follows:

$$\begin{cases} \mathbf{K}_{\mathrm{m}} = \begin{bmatrix} \boldsymbol{\Psi}_{\mathrm{m}} + \boldsymbol{\Psi}_{\mathrm{m}}^{\mathrm{T}} - \mathrm{tr}(\boldsymbol{\Psi}_{\mathrm{m}})\mathbf{I}_3 & \mathbf{Z}_{\mathrm{m}} \\ \mathbf{Z}_{\mathrm{m}}^{\mathrm{T}} & \mathrm{tr}(\Psi_{\mathrm{m}}) \end{bmatrix} \\ \mathbf{K}_{\Delta\mathrm{m}} = \begin{bmatrix} \boldsymbol{\Psi}_{\Delta\mathrm{m}} + \boldsymbol{\Psi}_{\Delta\mathrm{m}}^{\mathrm{T}} - \mathrm{tr}(\boldsymbol{\Psi}_{\Delta\mathrm{m}})\mathbf{I}_3 & \mathbf{Z}_{\Delta\mathrm{m}} \\ \mathbf{Z}_{\Delta\mathrm{m}}^{\mathrm{T}} & \mathrm{tr}(\boldsymbol{\Psi}_{\Delta\mathrm{m}}) \end{bmatrix} \\ \mathbf{K}_{\Delta\mathrm{b}} = \begin{bmatrix} \boldsymbol{\Psi}_{\Delta\mathrm{b}} + \boldsymbol{\Psi}_{\Delta\mathrm{b}}^{\mathrm{T}} - \mathrm{tr}(\boldsymbol{\Psi}_{\Delta\mathrm{b}})\mathbf{I}_3 & \mathbf{Z}_{\Delta\mathrm{b}} \\ \mathbf{Z}_{\Delta\mathrm{b}}^{\mathrm{T}} & \mathrm{tr}(\boldsymbol{\Psi}_{\Delta\mathrm{b}}) \end{bmatrix} \\ \mathbf{K}_{\Delta\mathrm{m}\Delta\mathrm{b}} = \begin{bmatrix} \boldsymbol{\Psi}_{\Delta\mathrm{m}\Delta\mathrm{b}} + \boldsymbol{\Psi}_{\Delta\mathrm{m}\Delta\mathrm{b}}^{\mathrm{T}} - \mathrm{tr}(\boldsymbol{\Psi}_{\Delta\mathrm{m}\Delta\mathrm{b}})\mathbf{I}_3 & \mathbf{Z}_{\Delta\mathrm{m}\Delta\mathrm{b}} \\ \mathbf{Z}_{\Delta\mathrm{m}\Delta\mathrm{b}}^{\mathrm{T}} & \mathrm{tr}(\boldsymbol{\Psi}_{\Delta\mathrm{m}\Delta\mathrm{b}}) \end{bmatrix} \end{cases} \tag{4.12}$$

$\boldsymbol{\Psi}_i$ and $\mathbf{Z}_i$, where $i = \mathrm{m}, \Delta\mathrm{m}, \Delta\mathrm{b}, \Delta\mathrm{m}\Delta\mathrm{b}$, are given as

$$\begin{cases} \boldsymbol{\Psi}_{\mathrm{m}} = \mathbf{M}^{\mathrm{m}}\mathbf{W}\mathbf{R}^{\mathrm{m}^{\mathrm{T}}} \\ \mathbf{Z}_{\mathrm{m}} = (\mathbf{M}^{\mathrm{m}} \times \mathbf{R}^{\mathrm{m}})\mathbf{W} \\ \boldsymbol{\Psi}_{\Delta\mathrm{m}} = \mathbf{GaM}^{\mathrm{r}}\mathbf{W}\mathbf{R}^{\mathrm{m}^{\mathrm{T}}} \\ \mathbf{Z}_{\Delta\mathrm{m}} = (\mathbf{GaM}^{\mathrm{r}} \times \mathbf{R}^{\mathrm{m}})\mathbf{W} \\ \boldsymbol{\Psi}_{\Delta\mathrm{m}} = \mathbf{M}^{\mathrm{m}}\mathbf{W}\mathbf{R}^{\mathrm{r}^{\mathrm{T}}}\mathbf{b}^{\mathrm{T}}\mathbf{H}^{\mathrm{T}} \\ \mathbf{Z}_{\Delta\mathrm{m}} = (\mathbf{M}^{\mathrm{m}} \times \mathbf{HbR}^{\mathrm{r}})\mathbf{W} \\ \boldsymbol{\Psi}_{\Delta\mathrm{m}\ \Delta\mathrm{b}} = \mathbf{GaM}^{\mathrm{r}}\mathbf{W}\mathbf{R}^{\mathrm{r}^{\mathrm{T}}}\mathbf{b}^{\mathrm{T}}\mathbf{H}^{\mathrm{T}} \\ \mathbf{Z}_{\Delta\mathrm{m}\ \Delta\mathrm{b}} = (\mathbf{GaM}^{\mathrm{r}} \times \mathbf{HbR}^{\mathrm{r}})\mathbf{W} \end{cases} \tag{4.13}$$

where $\mathbf{R}^{\mathrm{m}} = [\mathbf{r}_1\ \mathbf{r}_2 \ldots \mathbf{r}_n]$ and $\mathbf{M}^{\mathrm{m}} = [\mathbf{m}_1\ \mathbf{m}_2 \ldots \mathbf{m}_n]$. $\mathbf{R}^{\mathrm{r}}$ and $\mathbf{M}^{\mathrm{r}}$ denote the uncertainty bound corresponding to the measurement and reference vectors, respectively. The elements in the diagonal weight matrix $\mathbf{W}$ are all non-negative. Matrices $\mathbf{H}$ and $\mathbf{G}$ are shape matrices, with their columns representing generating vectors that define the uncertainty geometry. Diagonal matrices $\mathbf{a}$ and $\mathbf{b}$ contain interval components within standard bounds. The interval-based framework facilitates the calculation of ΔJ, representing the set of cost functions under bounded uncertainties. Equations (4.11)–(4.13) effectively construct a multi-objective optimization task to tackle the attitude determination problem. By solving this task, measurement errors and system uncertainties can both be systematically accounted for, thereby enhancing the robustness of attitude determination under real on-orbit conditions.

4.4.2 Attitude Determination with IDWA Method

Considering the proposed attitude determination task, due to the presence of squared norms, classical interval analysis techniques, such as interval perturbation theory, are not adequate for accurately capturing the inherent nonlinearities of the system. To effectively address this limitation, we introduce a novel methodology, known as the IDWA method (Xu et al. 2018; Wang et al. 2018). Drawing inspiration from polynomial chaos expansion, the IDWA method combines Bernstein polynomial approximation with extremal response vectors that represent the upper and lower bounds of system behavior. This framework is specifically designed to overcome the challenges posed by nonlinear uncertainty propagation, offering a more adaptive and accurate solution tailored to the attitude determination problem characteristics.

The minimum n_{P} for Bernstein integration can be computed with the roots of the n_{P}-order Bernstein polynomial $P_{n_{\mathrm{P}}}(\mathbf{w})$, as expressed below:

$$n_{\mathrm{P}} \geq \left[\frac{n_{\mathrm{T}} + n_{\mathrm{s}}}{2}\right] + 1 \tag{4.14}$$

where n_{T} denotes the order of nonlinearity present in the attitude determination problem based on interval uncertainty, and n_{s} represents the order of the optimal least-squares fit. Based on these parameters, the solutions to the o_{P}-degree $P_{n_{\mathrm{P}}}(w)$ can be derived as

$$\hat{\mathbf{w}} = \left[\hat{w}_1, \hat{w}_2, \ldots, \hat{w}_{n_{\mathrm{P}}}\right]^{\mathrm{T}} \tag{4.15}$$

The Gaussian integration points are selected as the solutions to $P_{n_{\mathrm{P}}}(w)$, which are developed to effectively characterize the interval-valued parameter space. A mapping from points in Eq. (4.15) to the kth interval variable x_k in the interval parameter space x^{I} is established by forming a sample matrix, which is used to compute the estimated coefficients of the response surface:

$$\mathcal{X}_k = \left[\hat{\mathbf{x}}_k^{(1)}, \ldots, \hat{\mathbf{x}}_k^{(m_{\mathrm{P}})}\right]^{\mathrm{T}} = \begin{bmatrix} x_1^{\mathrm{m}} & \cdots & x_{k-1}^{\mathrm{m}} & \hat{x}_k^{(1)} & x_{k+1}^{\mathrm{m}} & \cdots & x_{n_{\mathrm{u}}}^{\mathrm{m}} \\ \vdots & & \vdots & \vdots & \vdots & & \vdots \\ x_1^{\mathrm{m}} & \cdots & x_{k-1}^{\mathrm{m}} & \hat{x}_k^{(n_{\mathrm{P}})} & x_{k+1}^{\mathrm{m}} & \cdots & x_{n_{\mathrm{u}}}^{\mathrm{m}} \end{bmatrix}, \quad k = 1, 2, \ldots, n_{\mathrm{u}} \tag{4.16}$$

$$\hat{x}_k^{(i)} = x_k^{\mathrm{m}} + \alpha_k \cdot \hat{w}_i\,,\ i = 1, 2, \ldots, n_{\mathrm{P}} \tag{4.17}$$

As defined in Eq. (4.16), every row in $\mathcal{X}_k$ corresponds to a unique parameter combination of input. These parameter sets are derived from the interval vector product introduced in Eq. (4.13). Based on this configuration, the interval factors can be incorporated into the computation of the cost function radius ΔJ, which is estimated as follows:

$$\Delta J\left(\hat{\mathbf{x}}_k^{(i)}\right) = -\mathbf{q}^{\mathrm{T}}\mathbf{K}_{\Delta\mathrm{r}}\left(\hat{\mathbf{x}}_k^{(i)}\right)\mathbf{q} - \mathbf{q}^{\mathrm{T}}\mathbf{K}_{\Delta\mathrm{b}}\left(\hat{\mathbf{x}}_k^{(i)}\right)\mathbf{q} - \mathbf{q}^{\mathrm{T}}\mathbf{K}_{\Delta\mathrm{r},\Delta\mathrm{b}}\left(\hat{\mathbf{x}}_k^{(i)}\right)\mathbf{q},\ k = 1, 2, \ldots, n_{\mathrm{u}} \tag{4.18}$$

By performing n_{u} computation of ΔJ, every interval term in Eq. (4.18) yields n_{r} components. Hence, ΔJ can be effectively addressed as the objective function in a structured way without the loss of generality. By applying the IDWA method to the sampled domain, the integrated cost function $\mathbf{j}_s$, which incorporates all ΔJ, is formulated as follows:

$$\mathbf{j}_s = \left[\hat{\mathbf{j}}_1, \ldots, \hat{\mathbf{j}}_k, \ldots, \hat{\mathbf{j}}_{n_u}\right],\ k = 1, 2, \ldots, n_u \tag{4.19}$$

where the vector $\hat{\mathbf{j}}_k$ corresponds to the kth interval variable, which is derived as

$$\hat{\mathbf{j}}_k = \left[\mathbf{j}_k^{(1)}, \mathbf{j}_k^{(2)}, \ldots, \mathbf{j}_k^{(n_P)}\right] = \begin{bmatrix} j_k^{(1,1)} & j_k^{(2,1)} & \cdots j_k^{(n_P,1)} \\ \vdots & \vdots & \vdots \\ j_k^{(1,j)} & j_k^{(2,j)} & \cdots j_k^{(n_P,j)} \\ \vdots & \vdots & \vdots \\ j_k^{(1,n_r)} & j_k^{(2,n_r)} & \cdots j_k^{(n_P,n_r)} \end{bmatrix},\ j = 1, 2, \ldots, n_r \tag{4.20}$$

where $\mathbf{j}_k^{(i)}$ denotes the target function associated with the ith sampling points within the kth uncertain variable. The jth row corresponds to the jth element of the target function vector, which can be computed by aggregating the results across all sampling points in this formulation.

Hence, the minimum and maximum of each element in $\mathbf{j}_s$ determine the LB $\underline{j_s^{(j)}}$ and UB $\overline{j_s^{(j)}}$, respectively. This can be written as follows:

$$\begin{cases} \underline{j_s^{(j)}} = \min\{\mathbf{j}_s(1), \ldots, \mathbf{j}_s(j), \ldots, \mathbf{j}_s(n_r)\} \\ \overline{j_s^{(j)}} = \max\{\mathbf{j}_s(1), \ldots, \mathbf{j}_s(j), \ldots, \mathbf{j}_s(n_r)\} \end{cases},\ j = 1, 2, \ldots, n_r \tag{4.21}$$

where $\mathbf{j}_s(j)$ denotes the set of all components in the jth row of $\mathbf{j}_s$, representing the jth element of the target function across all sample points. Based on this, the LB $\underline{\mathbf{j}_s}$ and UB $\overline{\mathbf{j}_s}$ in the objective matrix can be denoted as follows:

$$\begin{cases} \underline{\mathbf{j}_s} = \left[\underline{j_s^{(1)}}, \underline{j_s^{(2)}}, \ldots, \underline{j_s^{(n_r)}}\right]^T \\ \overline{\mathbf{j}_s} = \left[\overline{j_s^{(1)}}, \overline{j_s^{(2)}}, \ldots, \overline{j_s^{(n_r)}}\right]^T \end{cases} \tag{4.22}$$

Consequently, one can determine the bounds of the target function as follows:

$$\mathbf{j}_s^I = \left[\underline{\mathbf{j}_s}, \overline{\mathbf{j}_s}\right] \tag{4.23}$$

Any continuous function defined on $[-1, 1]$ can be approximated by employing the Bernstein polynomial basis function as

$$L_k^{(j)}(w) = \sum_{D=0}^{n_s}\left[\sum_{i}^{n_P}\frac{(2D+1)t_k^{(i,j)}P_H(\hat{w}_i)}{(1-\hat{w}_i^2)\left[P'_{n_P}(\hat{w}_i)\right]^2}\right]P_H(w)\,,\ j = 1, 2, \ldots, n_r \tag{4.24}$$

where $w \in [-1, 1]$. The term $P_H(\hat{w}_i)$ denotes the value of the D-order Bernstein polynomial evaluated at $\hat{w}_i$, while $P'_{n_P}(\hat{w}_i)$ denotes the value of n_P-order Bernstein polynomial with $\hat{w}_i$. The D th order Bernstein polynomial, denoted as $P_D(w)$, is expressed as

$$P_D(t) = \sum_{k=0}^{m}\omega_{s_k}\cdot\frac{m!}{k!(m-k)!}\left(\frac{w+1}{2}\right)^k\left(\frac{-w+1}{2}\right)^{m-k} \tag{4.25}$$

The coefficients ω_{sk} in Eq. (4.25) are obtained as

$$\boldsymbol{\omega}_s = \left[\omega_{s_0}\ \omega_{s_1}\ \ldots\ \omega_{s_m}\right]^T = \left(\mathbf{B}^T\mathbf{B}\right)^{-1}\mathbf{B}^T\boldsymbol{\lambda}_{ri,s} \tag{4.26}$$

In this equation, the Bernstein-Vandermonde matrix is denoted as

$$\mathbf{B} = \begin{bmatrix} b_{m_0}(t_0) & b_{m_1}(t_0) & \ldots & b_{m_m}(t_0) \\ b_{m_0}(t_1) & b_{m_1}(t_1) & \ldots & b_{m_m}(t_1) \\ \vdots & \vdots & \ddots & \vdots \\ b_{m_0}(t_m) & b_{m_1}(t_m) & \ldots & b_{m_m}(t_m) \end{bmatrix} \tag{4.27}$$

and $\boldsymbol{\lambda}_{ri,s}$ can be represented as

$$\boldsymbol{\lambda}_{ri,s} = \left[\lambda_{ri,s_0}, \ldots, \lambda_{ri,s_k}, \ldots, \lambda_{ri,s_m}\right]^T \tag{4.28}$$

In Eq. (4.28), λ_{ri,s_k} is the real part of $\hat{\mathbf{x}}_k$. By applying Eq. (4.24), one can approximate a continuous function defined over $[-1, 1]$. Consequently, this enables the derivation of an estimate for Eq. (4.18). To determine the corresponding interval bounds, it is necessary to compute the extreme points of Eq. (4.24). Considering classical extreme value theory for continuous functions, an extremum occurs either at the variable interval bound or at the zero point of the derivative. Accordingly, the resulting extreme points can be denoted as

$$\mathbf{w}_{root} = \left[w_k^{(j,1)}, w_k^{(j,2)}, \ldots, w_k^{(j,n_{root})}\right] \tag{4.29}$$

Equation (4.29) shows the extremum points corresponding to the jth element of the target function, with respect to the kth interval variables, as defined in Eq. (4.19).

Taking into account the variable domain of the polynomial approximation model $L_k^{(j)}$, the values obtained from Eq. (4.29) require further adjustment. Specifically, if an extremum point yields an imaginary value, it is assigned a value of 1. For real-valued extremum points that lie outside the defined variable range, it is assigned a value of -1. In all other cases, no modification is applied. Accordingly, the revised extreme points of $L_k^{(j)}$ can be transformed to $\{\mathbf{w}_{\text{root}}, -1, 1\}$. Based on these revised points, the minimum and maximum values of the kth element of the target function with respect to the jth interval parameter can be computed as

$$\begin{cases} L_k^{(j)}\left(w_{k,\min}^{(j)}\right) = \min\limits_{w \in \{\mathbf{w}_{\text{root}}, -1, 1\}} L_k^{(j)}(w) \\ L_k^{(j)}\left(w_{k,\max}^{(j)}\right) = \max\limits_{w \in \{\mathbf{w}_{\text{root}}, -1, 1\}} L_k^{(j)}(w) \end{cases}, k = 1, 2, \ldots, n_{\text{u}} \tag{4.30}$$

By applying a consistent methodology across all interval parameters to compute extremal bounds for each element of the target function, the corresponding LB and UB of jth element of the target function over $[-1, 1]$ can be obtained. As a result, the associated input $\mathbf{w}_{\min}^{(j)}$ and $\mathbf{w}_{\max}^{(j)}$, can be determined accordingly:

$$\begin{cases} \mathbf{w}_{\min}^{(j)} = \left[w_{1,\min}^{(j)}, w_{2,\min}^{(j)}, \ldots, w_{n_{\text{u}},\min}^{(j)}\right]^{\text{T}} \\ \mathbf{w}_{\max}^{(j)} = \left[w_{1,\max}^{(j)}, w_{2,\max}^{(j)}, \ldots, w_{n_{\text{u}},\max}^{(j)}\right]^{\text{T}} \end{cases}, j = 1, 2, \ldots, n_{\text{r}} \tag{4.31}$$

According to Eq. (4.17), the mapping of the results from Eq. (4.14) into the interval parameter space can be performed as

$$\begin{cases} \mathbf{x}_{\min}^{(j)} = \left[x_{1,\min}^{(j)}, x_{2,\min}^{(j)}, \ldots, x_{n_{\text{u}},\min}^{(j)}\right]^{\text{T}} \\ \mathbf{x}_{\max}^{(j)} = \left[x_{1,\max}^{(j)}, x_{2,\max}^{(j)}, \ldots, x_{n_{\text{u}},\max}^{(j)}\right]^{\text{T}} \end{cases}, j = 1, 2, \ldots, n_{\text{r}} \tag{4.32}$$

Therefore, substituting Eq. (4.32) into Eq. (4.18) yields the following results:

$$\begin{cases} (\varepsilon\Delta J)_{\min}^{(j)} = -\mathbf{q}^{\text{T}}\mathbf{K}_{\Delta\text{r}}\left(\mathbf{x}_{\min}^{(j)}\right)\mathbf{q} - \mathbf{q}^{\text{T}}\mathbf{K}_{\Delta\text{b}}\left(\mathbf{x}_{\min}^{(j)}\right)\mathbf{q} - \mathbf{q}^{\text{T}}\mathbf{K}_{\Delta\text{r},\Delta\text{b}}\left(\mathbf{x}_{\min}^{(j)}\right)\mathbf{q} \\ (\varepsilon\Delta J)_{\max}^{(j)} = -\mathbf{q}^{\text{T}}\mathbf{K}_{\Delta\text{r}}\left(\mathbf{x}_{\max}^{(j)}\right)\mathbf{q} - \mathbf{q}^{\text{T}}\mathbf{K}_{\Delta\text{b}}\left(\mathbf{x}_{\max}^{(j)}\right)\mathbf{q} - \mathbf{q}^{\text{T}}\mathbf{K}_{\Delta\text{r},\Delta\text{b}}\left(\mathbf{x}_{\max}^{(j)}\right)\mathbf{q} \end{cases} \tag{4.33}$$

where $j = 1, 2, \ldots, n_{\text{r}}$, $(\varepsilon\Delta J)_{\min}^{(j)}$ and $(\varepsilon\Delta J)_{\max}^{(j)}$ represent the approximation of the LB and UB of the j th component of the target function computed by the IDWA method. A comprehensive estimation of the objective function $\varepsilon\Delta J_{\text{a}}$ can be obtained by iteratively executing the IDWA procedure n_{r} times, leading to the following expression:

$$\varepsilon\Delta J_{\text{a}} = \left[(\varepsilon\Delta J_{\text{a}})_{\min}^{(1)}, \ldots, (\varepsilon\Delta J_{\text{a}})_{\min}^{(n_{\text{r}})}, (\varepsilon\Delta J_{\text{a}})_{\max}^{(1)}, \ldots, (\varepsilon\Delta J_{\text{a}})_{\max}^{(n_{\text{r}})}\right] \tag{4.34}$$

Given that the extreme bounds of each element are interpreted as the bounds of the estimated response, considering that the LB of ε is the additive inverse of its UB, one can obtain the following formula:

$$\overline{\Delta J_{\text{a}}^{(j)}} = \max\{\Delta J_{\text{a}}(1), \ldots, \Delta J_{\text{a}}(j), \ldots, \Delta J_{\text{a}}(n_{\text{r}})\} \quad , \quad j = 1, 2, \ldots, n_{\text{r}} \tag{4.35}$$

where $\Delta J_{\mathrm{a}}(j)$ denotes all components within the jth row of $\Delta \mathbf{J}_{\mathrm{a}}$, representing the jth component in all estimated target functions. Consequently, the LB and UB of the approximated functions, represented as $\underline{\Delta \mathbf{J}_{\mathrm{a}}}$ and $\overline{\Delta \mathbf{J}_{\mathrm{a}}}$, are derived as

$$\begin{cases} \underline{\Delta \mathbf{J}_{\mathrm{a}}} = \left[\underline{\Delta J_{\mathrm{a}}^{(1)}}, \underline{\Delta J_{\mathrm{a}}^{(2)}}, \ldots, \underline{\Delta J_{\mathrm{a}}^{(n_{\mathrm{r}})}}\right]^{\mathrm{T}} \\ \overline{\Delta \mathbf{J}_{\mathrm{a}}} = \left[\overline{\Delta J_{\mathrm{a}}^{(1)}}, \overline{\Delta J_{\mathrm{a}}^{(2)}}, \ldots, \overline{\Delta J_{\mathrm{a}}^{(n_{\mathrm{r}})}}\right]^{\mathrm{T}} \end{cases} \tag{4.36}$$

Hence, the LB and UB of the approximated target function are formally derived as

$$\varepsilon \Delta \mathbf{J}_{\mathrm{a}}^{\mathrm{I}} = \left[\underline{\Delta \mathbf{J}_{\mathrm{a}}}, \overline{\Delta \mathbf{J}_{\mathrm{a}}}\right] \tag{4.37}$$

Therefore, an accurate interval estimation for the attitude determination problem is obtained.

4.5 A Bi-objective Optimization Framework-Based Robust Attitude Determination

4.5.1 Optimization Problem Formation

To achieve robust attitude determination, a non-constrained bi-objective optimization problem is formulated, with the objective of simultaneously minimizing both the J_{m} and ΔJ. This optimization problem is mathematically expressed as

$$\begin{aligned} &\text{find} \quad \mathbf{q} \\ &\min \quad J_{\mathrm{m}} \\ &\min \quad \Delta J \end{aligned} \tag{4.38}$$

The solution for this problem described by Eq. (4.38) results in a Pareto optimal set. Moreover, an additional selection criterion is required to identify the most suitable solution $\mathbf{q}$. In this chapter, the optimal solution $\mathbf{q}$ is selected from the Pareto-optimal set by selecting the element associated with the smallest attitude determination error $\delta \upsilon$. The attitude determination error is defined as

$$\delta \upsilon = \arccos\left\{\frac{1}{2}[\mathrm{trace}(\Delta \boldsymbol{\Upsilon}) - 1]\right\} \tag{4.39}$$

where $\Delta \boldsymbol{\Upsilon}$ is defined as

$$\Delta \boldsymbol{\Upsilon} = \boldsymbol{\Upsilon}_{\mathrm{est}}^{\mathrm{T}} \boldsymbol{\Upsilon}_{\mathrm{exact}} \tag{4.40}$$

where $\boldsymbol{\Upsilon}_{\text{est}}$ is defined in Eq. (4.3).

4.5.2 A Developed MOOA: SPEA/R

The Strength Pareto Evolutionary Algorithm based on Reference Directions (SPEA/R) (Jiang and Yang 2017) is a sophisticated multi-objective optimization algorithm (MOOA) particularly well-suited for addressing the robust attitude determination problem described above. SPEA/R demonstrates strong capabilities in effectively balancing convergence, diversity, and feasibility within the solution set. The algorithm emphasizes diversity preservation as a primary objective, while maintaining adequate convergence towards the Pareto front. It incorporates a reference direction-based density estimator, which significantly enhances computational efficiency and reduces overall runtime.

The optimization process begins with the generation of a diverse reference direction set, which is strategically employed to partition the objective space. The algorithm's effectiveness hinges critically on this reference direction set, as it is used to guide the association and distribution of solutions within the population. Based on this guidance, the algorithm proceeds to evolve the population and generate subsequent generations. To select the most promising individuals, a fitness evaluation mechanism is implemented that considers both global and local convergence properties, enabling effective differentiation and selection of all candidate solutions.

Each population member is characterized by a corresponding point in the solution space, determined through the optimization objectives. Evolution of the population is realized through repeated iterations that incorporate both the set of reference directions and a defined fitness assignment scheme, aiming to drive the solutions toward the Pareto front. This iterative process enables the SPEA/R-based optimization framework to efficiently converge to high-quality trade-off solutions. As a result, the application of SPEA/R demonstrates superior effectiveness in solving the multi-objective robust attitude determination problem, highlighting its capability in handling complex, real-world optimization scenarios.

4.6 Numerical Example

Equations (4.10)–(4.13) effectively reformulate the attitude determination problem as a bi-objective optimization task. Then, the SPEA/R algorithm is utilized to solve the optimization task, enabling an evaluation of how input accuracy influences the performance of the solution. This part presents two distinct series of measurements and corresponding reference vectors to verify the robustness and stability of the introduced approach. These pairs are defined as

$$\begin{cases} \mathbf{r}_1 = [0.2673\,0.5345 0.8018]^{\mathrm{T}} \\ \mathbf{r}_2 = [-0.3124\,0.937\,-0.1562]^{\mathrm{T}} \\ \mathbf{m}_1 = [0.7749 0.3448\,0.5297]^{\mathrm{T}} \\ \mathbf{m}_2 = [0.6296 0.6944\quad -0.3486]^{\mathrm{T}} \end{cases} \tag{4.41}$$

In practical spacecraft operations, sensor errors inevitably lead to deviations of both reference and measurement vectors from true values. To reflect the sensor errors, a 5% error margin is introduced for each vector element in the numerical example. Accordingly, the LB and UB of each component are set at 95% and 105% of their nominal values, respectively.

An initial population size of 80 is adopted in the optimization algorithm, and the algorithm is run for 10,000 iterations. The solution corresponding to the attitude minimum estimation error is shown in Table 4.1.

To further evaluate the performance and robustness of the proposed approach, MCS, a fully multiplicative unscented Kalman filter (FMUKF) (Fully Multiplicative Unscented Kalman Filter for Attitude Estimation), and a particle filter (Djuric et al. 2003) are applied to address the uncertainty in the system. Comparative results are summarized in Table 4.1. The proposed method yields determinations that are closely aligned with those obtained by MCS and particle filter, thereby confirming its validity and reliability. In comparison with the FMUKF, the proposed approach yields reduced attitude determination errors, indicating improved accuracy and greater robustness in managing bounded uncertainty.

Furthermore, to provide a more in-depth demonstration of the proposed method's robustness, a case study involving increased complexity is introduced. For this example, the measurement and reference vectors are represented by two distinct pairs of unit vectors, which are specified as follows:

Table 4.1 The result achieves the minimum δa with different methods

Method	Approximated q	Attitude deviation
MCS	$[0.2489\quad 0\quad 0.4778\quad 0.8425]^{\mathrm{T}}$	1.41°
FMUKF (Fully Multiplicative Unscented Kalman Filter for Attitude Estimation)	$[0.2654\quad -0.024\quad 0.4700\quad 0.8415]^{\mathrm{T}}$	1.53°
Particle filter (Djuric et al. 2003)	$[0.2487\quad 0.002\quad 0.4776\quad 0.8426]^{\mathrm{T}}$	1.41°
IDWA	$[0.2463\quad -0.0243\quad 0.4780\quad 0.8428]^{\mathrm{T}}$	1.41°

$$\begin{cases} \mathbf{r}_1 = [-0.1517 \quad -0.9669 \quad 0.2050]^{\mathrm{T}} \\ \mathbf{r}_2 = [-0.8393 \quad 0.4494 \quad -0.3044]^{\mathrm{T}} \\ \mathbf{r}_3 = [-0.0886 \quad -0.5856 \quad -0.8000]^{\mathrm{T}} \\ \mathbf{r}_4 = [0.8814 \quad -0.0303 \quad 0.5202]^{\mathrm{T}} \\ \mathbf{m}_1 = [0.8273 \quad 0.5541 \quad -0.0920]^{\mathrm{T}} \\ \mathbf{m}_2 = [-0.8285 \quad 0.5522 \quad -0.0955]^{\mathrm{T}} \\ \mathbf{m}_3 = [0.2155 \quad 0.5522 \quad 0.8022]^{\mathrm{T}} \\ \mathbf{m}_4 = [0.5570 \quad -0.7442 \quad -0.2884]^{\mathrm{T}} \end{cases} \tag{4.42}$$

By applying the aforementioned method to optimize Eq. (4.10), and considering a 5% component-wise measurement error, the resulting Pareto front for the robust bi-objective optimization is presented in Fig. 4.1. The result shown corresponds to 10,000 iterations with an initial population size of 80.

As shown in Fig. 4.1 and based on previous analysis, it can be observed that both J_{m} and ΔJ yield nonnegative values, which is consistent with the distribution seen in Fig. 4.1. The Pareto front allows for the selection of the best compromise solution using the criterion defined in Eq. (4.39). Three representative solutions (No. 1, No. 40, and No. 80) are drawn in purple in Fig. 4.1. The δa of each solution are shown in Fig. 4.2. δa converges to approximately 1.26° (No. 1), 1.25° (No. 40), and 1.247° (No. 80) after 10,000 iterations, respectively. These results indicate that as J_{m} increases, the δa decreases. As J_{m} stands for the upper bound of the objective function's center, a greater value indicates a higher degree of loss. Consequently, one can conclude that an increase in J_{m} generally leads to a relatively larger δa.

Furthermore, to assess the computational precision and computational speed of the proposed method, the same optimization problem was also solved using MCS over

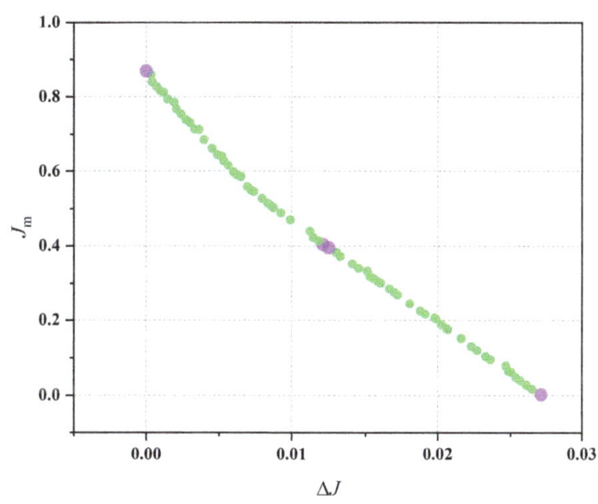

Fig. 4.1 The resulting Pareto front

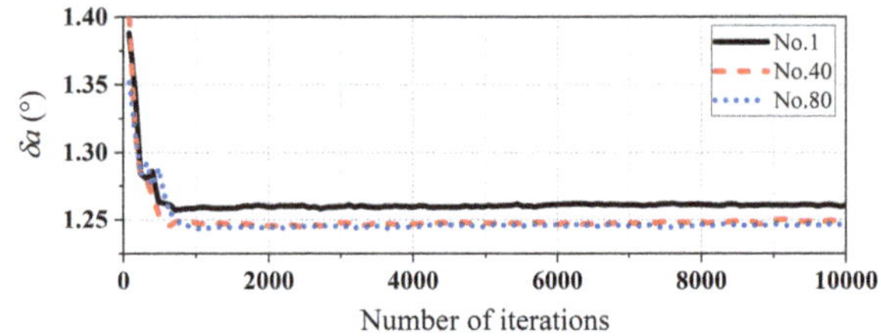

Fig. 4.2 No. 1, 40, and 80 solutions in the Pareto set

10,000 iterations. While the MCS-based method demands 643 s for computation, the proposed approach completes in just 3 s, demonstrating a significant improvement in computational efficiency.

$\mathbf{q}$ represents the estimated and corresponding attitude errors for each method. Despite the differences in the estimated $\mathbf{q}$ values, the proposed approach demonstrates comparable attitude determination accuracy to that of the Particle Filter and MCS methods, indicating that it effectively captures the full parameter space and maintains comparable generalization performance while offering a substantial reduction in computational cost. Moreover, the introduced approach demonstrates enhanced outcomes relative to the FMUKF, yielding an attitude determination error of only 1.24°, which indicates a high level of accuracy. This result highlights the robustness and effectiveness of the proposed approach even under uncertainties (Table 4.2).

The demonstrated capability of the introduced approach is further assessed by analyzing the statistical properties of the attitude determination errors, specifically their mean and variance. Following 100 independent optimization trials, the mean attitude error was determined to be 1.256°, with an associated variance of approximately 0.003, demonstrating the stability and consistency of the approach. The root mean square (RMS) of the attitude error is also used as a supplementary performance indicator. The RMS of δa over the entire optimization process is illustrated in Fig. 4.3. As observed, after around 1000 iterations, the RMS approximately converges to 1.25°, indicating that the method not only converges efficiently but also maintains a tightly bounded error range throughout the optimization process. This behavior highlights the fast convergence rate and the ability to achieve robust performance under interval uncertainties of the introduced approach.

Table 4.2 The result achieves the minimum δa with different methods

Approach	Approximated $\mathbf{q}$	Attitude deviation
MCS	$[0.3468\ \ -0.0123\ \ 0.5021\ \ 0.7921]^T$	1.24°
FMUKF (Fully Multiplicative Unscented Kalman Filter for Attitude Estimation)	$[0.3523\ \ 0.0240\ \ 0.5304\ \ 0.7707]^T$	1.36°
Particle filter (Djuric et al. 2003)	$[0.3469\ \ -0.0125\ \ 0.5020\ \ 0.7921]^T$	1.24°
IDWA	$[0.3471\ \ -0.0132\ \ 0.5033\ \ 0.7912]^T$	1.24°

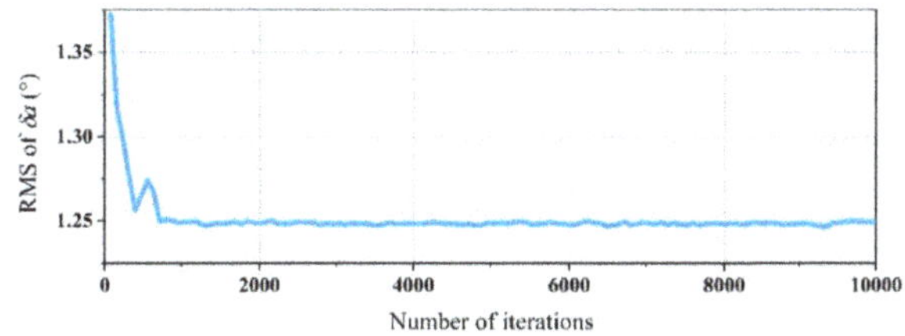

Fig. 4.3 The attitude error RMS of each iteration

Table 4.3 Attitude error δa under different error levels

Percentage of error (%)	δa (°)
1	1.194
3	1.213
5	1.234
7	1.25
9	1.272
11	1.288
13	**1.33**

Table 4.3 illustrates the variation in δa under different error percentages applied to the input vectors. From the results, it can be observed that as the error level increases, the attitude error δa also increases; however, the rate of growth remains relatively small. Specifically, δa increases by approximately 0.02° for every 2% rise in input error, with the exception of the 13% error case, where the increase reaches 0.04°. Notably, even when the error reaches 13%, the maximum attitude error remains as low as 1.33, which is still considered a very small value in the context of attitude determination. The difference between the attitude errors under 1 and 13% disturbance is only 0.14°, further demonstrating the method's resilience to input uncertainties. These results collectively highlight the robustness of the proposed approach in effectively mitigating disturbances across a wide range of error levels.

4.7 Summary

This chapter presents the application of the IDWA method integrated with an MOOA to address an attitude determination task considering uncertainties. By introducing the interval uncertainty to the classical attitude determination problem, an interval attitude determination problem is addressed through the IDWA method. This narrow-bound theoretical approach effectively avoids overestimation commonly associated with classical interval perturbation techniques. A sophisticated evolutionary algorithm, SPEA/R, is utilized to tackle the bi-objective optimization task entailed in attitude determination. To validate the proposed methodology, two sets of vectors are employed as measurement and reference vectors. All feasible solutions can be identified along the Pareto front, and three representative solutions are selected for detailed analysis. The demonstrated capability and precision of the introduced approach are validated through comparison with MCS, particle filter, and FMUKF. The results confirm that the suggested approach is capable of achieving precise attitude determination under interval uncertainties. Notably, this method is particularly suitable for space applications where only limited data samples are available, making it a promising tool for robust attitude determination in real-world spacecraft operations.

References

Ahmed S, Kerrigan EC, Jaimoukha IM (2012) A semidefinite relaxation-based algorithm for robust attitude estimation. IEEE Trans Signal Process 60:3942–3952. https://doi.org/10.1109/TSP.2012.2198820

Ben-Haim Y (1994) A non-probabilistic concept of reliability. Struct Saf 14:227–245. https://doi.org/10.1016/0167-4730(94)90013-2

Cheng W, Hu S, Zhang M, Song H, Zhang G, Chen Y (2021) Near-vertical attitude determination based on balance correction. IEEE Trans Instrum Meas 70:1–9. https://doi.org/10.1109/TIM.2021.3070888

Djuric PM, Kotecha JH, Zhang J, Huang Y, Ghirmai T, Bugallo MF, Miguez J (2003) Particle filtering. IEEE Signal Process Mag 20:19–38

Fully multiplicative unscented Kalman filter for attitude estimation. J Guid Control Dyn

Gan Y, Sui L, Xiao G, Zhang Q, Wang L (2021) Real-time GNSS attitude determination by a direct approach with efficiency and robustness. Meas Sci Technol 32:115904. https://doi.org/10.1088/1361-6501/ac18a4

Ghadiri H, Esmaelzadeh R, Zardashti R (2022) Robust optimal attitude determination using interval analysis. Adv Space Res 69:2611–2617. https://doi.org/10.1016/j.asr.2021.12.045

Hasan MN, Haris M, Qin S (2023) Finite-time active fault-tolerant attitude control for flexible spacecraft with vibration suppression and anti-unwinding. Adv Space Res 71:3644–3660. https://doi.org/10.1016/j.asr.2022.04.002

He J, Sun C, Zhang B, Wang P (2021) Adaptive error-state Kalman Filter for attitude determination on a moving platform. IEEE Trans Instrum Meas 70:1–10. https://doi.org/10.1109/TIM.2021.3111971

Horn BKP (1987) Closed-form solution of absolute orientation using unit quaternions. JOSA A. 4:629–642. https://doi.org/10.1364/JOSAA.4.000629

Hsu T-H, Chen T-L (2022) A New Instantaneous method for attitude determination using GPS phase measurement. IEEE Trans Instrum Meas 71:1–13. https://doi.org/10.1109/TIM.2021.3137158

Jiang S, Yang S (2017) A strength pareto evolutionary algorithm based on reference direction for multiobjective and many-objective optimization. IEEE Trans Evol Comput 21:329–346. https://doi.org/10.1109/TEVC.2016.2592479

Jiang C, Han X, Lu GY, Liu J, Zhang Z, Bai YC (2011) Correlation analysis of non-probabilistic convex model and corresponding structural reliability technique. Comput Methods Appl Mech Eng 200:2528–2546. https://doi.org/10.1016/j.cma.2011.04.007

Keat JE (1977) Analysis of least-squares attitude determination routine DOAO. Zenodo

Lee D, Gallucci S (2023) Spacecraft attitude estimation under attitude tracking maneuver during close-proximity operations. Adv Space Res 71:4315–4331. https://doi.org/10.1016/j.asr.2023.01.004

Qiu Z, Elishakoff I (1998) Antioptimization of structures with large uncertain-but-non-random parameters via interval analysis. Comput Methods Appl Mech Eng 152:361–372. https://doi.org/10.1016/S0045-7825(96)01211-X

Sun X, Mao X, Chen P (2022) High-precision attitude determination using spaceborne gravity gradiometer and gyroscope. Acta Astronaut 200:213–225. https://doi.org/10.1016/j.actaastro.2022.08.008

Sun Z, Simo J, Gong S (2023) Satellite attitude identification and prediction based on neural network compensation. Space Sci Technol 3:0009. https://doi.org/10.34133/space.0009

Wahba G (1965) A least squares estimate of satellite attitude. SIAM Rev 7:409–409. https://doi.org/10.1137/1007077

Wang L, Xiong C, Wang X, Xu M, Li Y (2018) A dimension-wise method and its improvement for multidisciplinary interval uncertainty analysis. Appl Math Model 59:680–695. https://doi.org/10.1016/j.apm.2018.02.022

Wu S, Chu W, Ma X, Radice G, Wu Z (2018) Multi-objective integrated robust H∞ control for attitude tracking of a flexible spacecraft. Acta Astronaut 151:80–87. https://doi.org/10.1016/j.actaastro.2018.05.062

Xu M, Du J, Chen J, Wang C, Li Y (2018) An iterative dimension-wise approach to the structural analysis with interval uncertainties. Int J Comput Methods 15:1850044. https://doi.org/10.1142/S0219876218500445

Zhang K, Wu S, Wu Z (2021) Multibody dynamics and robust attitude control of a MW-level solar power satellite. Aerosp Sci Technol 111:106575. https://doi.org/10.1016/j.ast.2021.106575

Chapter 5
Interval Uncertainty-Based Optimal Attitude Control for Space Power Satellite

Abstract This chapter presents an optimal control framework tailored for space power satellites (SPS) subject to multi-source uncertainties and periodic disturbances. Addressing the limitations of probabilistic methods in data-sparse space environments, the dynamic modeling is grounded in non-probabilistic interval analysis, treating initial states and uncertain inertia as unknown-but-bounded (UBB) parameters. A hybrid control architecture is developed, integrating feedforward compensation for periodic disturbances with an interval Linear-Quadratic Regulator (LQR). A key theoretical contribution discussed is the formulation and solution of the interval algebraic Riccati equation (IARE), which allows for the direct computation of uncertain feedback gain bounds. Furthermore, to assess the system's safety over time, an interval-based time-dependent reliability (ITDR) model is constructed utilizing interval process theory and first passage concepts. The control parameters are refined through a constrained multi-objective optimization strategy. Numerical simulations on a classical Abacus architecture SPS demonstrate that this interval-based methodology achieves control precision comparable to Monte Carlo simulations (MCSs) but with significantly higher computational efficiency.

5.1 Introduction

The precipitous decline in conventional fossil fuel reserves has led to a serious global energy emergency. As a boundless and pollution-free resource, solar energy has the potential to fundamentally transform how humans obtain and utilize energy. Properly harnessing this energy can significantly reduce dependence on fossil fuels while improving environmental conditions. However, collecting solar energy on Earth is highly inefficient due to limitations such as geographical location, climate variability, and weather conditions. Furthermore, the cost of terrestrial solar collection systems must not exceed the economic benefits derived from power generation. To address these issues, Dr. Glaser introduced the idea of the space power satellite (SPS) in 1968 (Glaser 1968), offering a promising alternative for the continuous acquisition of Superior, enduring, and cost-effective solar energy (Mankins 1997; Mori et al. 2006).

C. Yang and Y. Xia, *Set Theory-Based Spacecraft Dynamics and Control*, Intelligent Control and Learning Systems 26, https://doi.org/10.1007/978-981-95-9179-4_5

Based on cost-effectiveness analyses, for SPS to be feasible in power generation, it must feature a solar array with an area of 10 km^2 and an antenna with a 1 km diameter. For such a large system, attitude control plays a crucial role in enabling the SPS to fulfill its intended functions. However, in the information-sparse space environment characterized by multi-source uncertainties, achieving accurate and reliable control of the SPS necessitates the quantification of parameter uncertainties. Consequently, uncertainty-based attitude control methods are essential for realizing reliable attitude control of the SPS. Existing control strategies, particularly those addressing uncertainty, are inadequate for managing systems with limited-sample uncertainty in the information-sparse space environment. Thus, this chapter is dedicated to addressing the limitations of existing approaches in SPS attitude control while proposing an interval uncertainty-based LQR attitude controller for SPS.

In this chapter, a dynamic model for the sun-pointing SPS is formulated as an interval time-varying system controlled by an interval LQR method; this construction is established using interval analysis (Wang et al. 2022; Qiu and Elishakoff 1998; Jiang et al. 2008) and incorporates multi-source uncertainties and external disturbance torques. These uncertain factors are treated as UBB variables (Liu et al. 2021). This formulation avoids the limitations of probabilistic methods that rely on sparse sampling (Chen and Li 2007; Castet and Saleh 2009; Lemaire 2013). As long as the boundaries of the UBB uncertain parameters are defined, the state-space equations describing the SPS attitude dynamics with interval uncertainties can be derived and solved through the time-dependent state-space equation with interval uncertainties.

Based on the SPS attitude dynamics with interval uncertainties, two sorts of control schemes, namely control mechanisms employing both feedforward action and uncertain feedback LQR principles, are designed to counteract disturbance torques and regulate system behavior (Wu et al. 2016; Zhang et al. 2021). To achieve uncertain attitude control for the SPS, an LQR controller is implemented, with its gains determined by utilizing the algebraic Riccati equation (ARE) with interval uncertainties, following the conceptual foundation of the nominal Riccati solution. The resulting methodology allows for obtaining the uncertain interval ranges pertinent to the LQR controller, attitude state variables, and the associated cost function. Moreover, a novel interval-based time-dependent reliability (ITDR) approach is formulated to evaluate the time-dependent integrity of the complete attitude control architecture. This model, grounded in interval process theory and first-passage analysis (Ditlevsen and Madsen 1996), addresses the inability of uncertain attitude states to directly represent system safety. Additionally, a coevolutionary constrained multi-objective optimization (CCMO) method (Tian et al. 2021) is adopted to construct the controller considering uncertainties for SPS sun-pointing control. The purpose is to reduce both the control cost and the variability of the state responses while satisfying ITDR constraints. Finally, the demonstrated capability of the introduced approach is validated through simulations of a large-scale SPS system based on the Abacus architecture, demonstrating its feasibility and superior performance in achieving optimal attitude control.

5.2 Nominal Attitude Dynamic Model of SPS

5.2.1 Architecture and Coordinate Systems

The Abacus architecture SPS, designed with a power capacity of 1.2 GW, consists of a 3.2 km × 3.2 km Solar panel array, a transmit antenna with a 500 m diameter, and a rotational reflector measuring 500 m × 700 m, as illustrated in Fig. 5.1. Since the reflector mass (0.8×10^6 kg) accounts for no more than 4% of the total mass of the arrays (21×10^6 kg), the mass as well as moment of inertia can be reasonably ignored. Consequently, instead of a multi-body system, the SPS is treated as a single rigid body, with its principal inertia axes aligned with the three rotational axes. In addition, due to the antenna and reflector being mounted at the extremity of the structure, the mass centroid and the mean pressure point are not coincident, resulting in a disturbing torque induced by solar radiation pressure.

5.2.2 Sun-Pointing Deterministic Attitude Dynamics

By modeling the SPS as a large rigid body, its attitude dynamics are governed by the following equations:

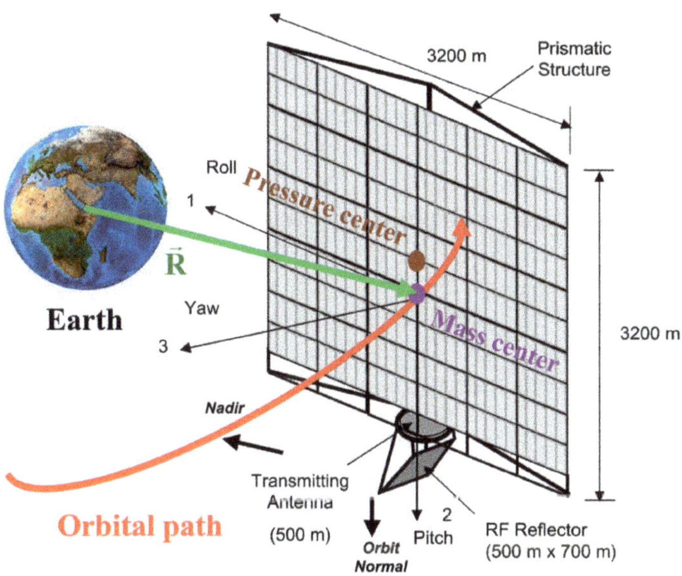

Fig. 5.1 The 1.2-GW Abacus SPS baseline structure, illustrated with its associated coordinate system and orbital path (Wie and Roithmayr 2005)

$$\begin{cases} I_1\dot{\omega}_1 - (I_2 - I_3)\omega_2\omega_3 = M_1 + u_1 + d_1 \\ I_2\dot{\omega}_2 - (I_3 - I_1)\omega_3\omega_1 = M_2 + u_2 + d_2 \\ I_3\dot{\omega}_3 - (I_1 - I_2)\omega_1\omega_2 = M_3 + u_3 + d_3 \end{cases} \tag{5.1}$$

The diagonal inertia entries and the body-rate components are, respectively, represented by I_i and ω_i $(i = 1, 2, 3)$. The variables M_i, u_i, and d_i $(i = 1, 2, 3)$ correspond to all acting torques (including control, gravity-gradient, and perturbations).

Assuming the Earth is modeled as a homogeneous, perfectly spherical body, the gravity-gradient torque is expressed as

$$\begin{cases} M_1 = 3\mu\big/R^3(I_3 - I_2)\alpha_2\alpha_3 \\ M_2 = 3\mu\big/R^3(I_1 - I_3)\alpha_1\alpha_3 \\ M_3 = 3\mu\big/R^3(I_2 - I_1)\alpha_1\alpha_2 \end{cases} \tag{5.2}$$

where μ denotes Earth's gravitational constant, while R represents the orbit radius. The parameters α_1, α_2, and α_3 describe the attitude parameters transformation along each principal axis. The Euler angles θ_1, θ_2, and θ_3 correspond to roll, pitch, and yaw, respectively, which define the SPS's attitude within the Earth-centered inertial (ECI) coordinate frame. Under small-angle and near-equilibrium assumptions, the relationships $\dot{\theta}_i = \omega_i$, $i = 1, 2, 3$ hold. Given the rotational order θ_2, θ_3, and θ_1, the transformation matrix from the inertial frame $(\mathbf{n}_1, \mathbf{n}_2, \mathbf{n}_3)$ to the body-fixed frame $(\mathbf{b}_1, \mathbf{b}_2, \mathbf{b}_3)$ can be written as

$$\begin{bmatrix} \mathbf{b}_1 \\ \mathbf{b}_2 \\ \mathbf{b}_3 \end{bmatrix} = \begin{bmatrix} C_2C_3 & S_3 & -S_2C_3 \\ -C_2S_3C_1 + S_1S_2 & C_3C_1 & S_2S_3C_1 + S_1C_2 \\ C_2S_3S_1 + C_1S_2 & -C_3S_1 & -S_2S_3C_1 + C_1C_2 \end{bmatrix} \begin{bmatrix} \mathbf{n}_1 \\ \mathbf{n}_2 \\ \mathbf{n}_3 \end{bmatrix} \tag{5.3}$$

where $C_i = \cos\theta_i$, and $S_i = \sin\theta_i$. $(\mathbf{b}_1, \mathbf{b}_2, \mathbf{b}_3)$ represent the unit vectors of the SPS body frame, configured with its origin at the mass centroid and oriented along the principal inertia axes. $(\mathbf{n}_1, \mathbf{n}_2, \mathbf{n}_3)$ form the ECI coordinate system, where $\mathbf{n}_1, \mathbf{n}_3$ lie in the plane of the equator and $\mathbf{n}_2$ is orthogonal to it. $\overrightarrow{\mathbf{R}}$ indicates the distance from Earth's mass centroid to the SPS mass centroid, as depicted in Fig. 5.1. These descriptions follow references (Wie and Roithmayr 2005).

Therefore, the expressions for α_i $(i = 1, 2, 3)$ in Eq. (5.2) are derived as

$$\begin{cases} \alpha_1 = -S_2C_3\cos(nt) - C_2C_3\sin(nt) \\ \alpha_2 = (S_2S_3C_1 + S_1C_2)\cos(nt) - (-C_2S_3C_1 + S_1S_2)\sin(nt) \\ \alpha_3 = (-S_2S_3C_1 + C_1C_2)\cos(nt) - (C_2S_3S_1 + C_1S_2)\sin(nt) \end{cases} \tag{5.4}$$

where $n = \sqrt{\mu\big/R^3}$ represents the mean orbital motion.

By substituting Eqs. (5.2) and (5.4) with the dependence $\dot{\theta}_i = \omega_i$ into Eq. (5.1) while omitting high-order contributions, the linear deterministic attitude dynamics of SPS for sun-pointing can be obtained as

$$\begin{cases} I_1\ddot{\theta}_1 - 3n^2(I_3 - I_2)\left[\cos^2(nt)\theta_1 + \sin(nt)\cos(nt)\theta_3\right] = u_1 + d_1 \\ I_2\ddot{\theta}_2 - 3n^2(I_3 - I_1)\left[\left(\cos^2(nt) - \sin^2(nt)\right)\theta_2 + \sin(nt)\cos(nt)\right] = u_2 + d_2 \\ I_3\ddot{\theta}_3 - 3n^2(I_1 - I_2)\left[\sin^2(nt)\theta_3 + \sin(nt)\cos(nt)\theta_1\right] = u_3 + d_3 \end{cases} \quad (5.5)$$

Equation (5.5) indicates that the pitch movement is independent of the roll and yaw movement, while these two rotation motions are mutually coupled due to the time-dependent gravity-gradient torques acting in both directions. In addition to these effects, other disturbances caused by radiation from the sun and microwave influence the sun-pointing stability of the SPS. These perturbations are represented as (Wie and Roithmayr 2005)

$$\begin{cases} d_1(t) = 12000 - 11900\cos(nt) \\ d_2(t) = 1200 \\ d_3(t) = -11900\sin(nt) \end{cases} \quad (5.6)$$

5.2.3 Deterministic State-Space Equation

The nominal attitude dynamic description of the SPS, considering the roll and yaw in Eq. (5.5), can be reformulated into two time-dependent state-space equations as follows:

$$\begin{cases} \dot{\mathbf{x}}_{13}(t) = \mathbf{A}_{13}(t)\mathbf{x}_{13}(t) + \mathbf{B}_{13}(t)\mathbf{f}_{13}(t) + \mathbf{B}_{13}(t)\mathbf{u}_{13}(t) \\ \mathbf{y}_{13}(t) = \mathbf{C}_{13}(t)\mathbf{x}_{13}(t) \end{cases} \quad (5.7)$$

where $\mathbf{x}_{13}(t) = \left[\theta_1\ \theta_3\ \dot{\theta}_1\ \dot{\theta}_3\right]^{\mathrm{T}}$, $\mathbf{f}_{13}(t) = \left[d_1(t)\ d_3(t)\right]^{\mathrm{T}}$, $\mathbf{u}_{13}(t) = \left[u_1\ u_3\right]^{\mathrm{T}}$, $\mathbf{B}_{13}(t) = \begin{bmatrix} 0 & 0 & I_1^{-1} & 0 \\ 0 & 0 & 0 & I_3^{-1} \end{bmatrix}^{\mathrm{T}}$, $\mathbf{C}_{13}(t) = \begin{bmatrix} 1 & 0 & 0 & 0 \\ 0 & 1 & 0 & 0 \end{bmatrix}$ and $\mathbf{A}_{13}(t) = \begin{bmatrix} 0 & 0 & 1 & 0 \\ 0 & 0 & 0 & 1 \\ I_1^{-1}3n^2(I_3 - I_2)\cos^2(nt) & I_1^{-1}3n^2(I_3 - I_2)\sin(nt)\cos(nt) & 0 & 0 \\ I_3^{-1}3n^2(I_1 - I_2)\sin(nt)\cos(nt) & I_3^{-1}3n^2(I_1 - I_2)\sin^2(nt) & 0 & 0 \end{bmatrix}$.

Similarly, for the pitch movement, the corresponding state-space representation can be expressed as

$$\begin{cases} \dot{\mathbf{x}}_2(t) = \mathbf{A}_2(t)\mathbf{x}_2(t) + \mathbf{B}_2(t)f_2(t) + \mathbf{B}_2(t)u_2(t) \\ \mathbf{y}_2(t) = \mathbf{C}_2(t)\mathbf{x}_2(t) \end{cases} \tag{5.8}$$

where $\mathbf{x}_2(t) = \left[\theta_2\ \dot{\theta}_2\right]^{\mathrm{T}}$, $f_2(t) = d_2 + 3n^2(I_3 - I_1)\sin(nt)\cos(nt)$, $\mathbf{u}_2(t) = u_2$, $\mathbf{B}_2(t) = \left[0\ I_2^{-1}\right]^{\mathrm{T}}$, $\mathbf{C}_2(t) = \left[1\ 0\right]$ and $\mathbf{A}_2(t) = \begin{bmatrix} 0 & 1 \\ I_2^{-1}3n^2(I_3 - I_1)\left(\cos^2(nt) - \sin^2(nt)\right) & 0 \end{bmatrix}$.

5.3 Interval Uncertainty-Based State-Space Equation of SPS Attitude System

5.3.1 Interval Uncertainty-Based State-Space Equation

In Eqs. (5.7) and (5.8), the complete dynamics for the attitude of the SPS can be expressed by 2 sets of state-space equations. Considering a linearized time-varying system, the general state-space form is given by

$$\dot{\mathbf{x}}(t) = \mathbf{A}(t)\mathbf{x}(t) + \mathbf{B}(t)\mathbf{u}(t) \qquad \mathbf{x}(t) \in \mathbb{R}^h, \mathbf{u}(t) \in \mathbb{R}^p \tag{5.9}$$

where $\mathbf{x}(t)=[x_1(t), \ldots, x_h(t)]$ denotes the h-dimensional state vector (with $h = 4$ for roll, yaw, and $h = 2$ for pitch movement), $\mathbf{u}(t)$ represents the p-dimensional control input (with $p = 2$ for roll, yaw, and $p = 1$ for pitch movement), $\mathbf{A}(t)$ is the $h \times h$ system matrix, and $\mathbf{B}(t)$ is the $h \times p$ input matrix.

The LQR approach yields an optimal control input of the following form:

$$\mathbf{u}(t) = -\mathbf{K}(t)\mathbf{x}(t) \tag{5.10}$$

where $\mathbf{K}$ represents the feedback gain matrix.

Substituting Eq. (5.10) into Eq. (5.9), the system is derived

$$\dot{\mathbf{x}}(t) = [\mathbf{A}(t) - \mathbf{B}(t)\mathbf{K}(t)]\mathbf{x}(t) = \mathbb{A}(t)\mathbf{x}(t) \tag{5.11}$$

where $\mathbb{A}(t) = \mathbf{A}(t) - \mathbf{B}(t)\mathbf{K}(t)$.

The nominal state-space representation in Eq. (5.9) assumes ideal conditions without uncertainty. However, in practical SPS attitude control under sun-pointing conditions, multiple sources of uncertainty exist. Therefore, the deterministic model in Eq. (5.9) can be extended into the formulation with interval parameters utilizing the interval uncertainty analysis approach, leading to

$$\dot{\mathbf{x}}^{\mathrm{I}}(t) = \left[\mathbf{A}\left(\mathbf{b}^{\mathrm{I}}, t\right) - \mathbf{B}\left(\mathbf{b}^{\mathrm{I}}, t\right)\mathbf{K}\left(\mathbf{b}^{\mathrm{I}}, t\right)\right]\mathbf{x}^{\mathrm{I}}(t) = \mathbb{A}\left(\mathbf{b}^{\mathrm{I}}, t\right)\mathbf{x}^{\mathrm{I}}(t) \tag{5.12}$$

The goal of solving this interval-based problem is to obtain the bound of $\mathbf{x}\left(\mathbf{b}^{\mathrm{I}},t\right)$.

When the uncertainty ratio $\Delta b_k / b_k^{\mathrm{c}}$ is small, the interval matrices in Eq. (5.12) can be estimated via a 1st- order Taylor expansion as

$$\begin{cases} \mathbf{A}\left(\mathbf{b}^{\mathrm{I}},t\right) = \mathbf{A}(\mathbf{b}^{\mathrm{c}},t) + \sum_{k=1}^{m} \frac{\partial \mathbf{A}(\mathbf{b}^{\mathrm{c}},t)}{\partial b_k}\left(b_k^{\mathrm{I}} - b_k^{\mathrm{c}}\right) = \mathbf{A}(\mathbf{b}^{\mathrm{c}},t) + \sum_{k=1}^{m} \frac{\partial \mathbf{A}(\mathbf{b}^{\mathrm{c}},t)}{\partial b_k}\Delta b_k \xi \\ \mathbf{B}\left(\mathbf{b}^{\mathrm{I}},t\right) = \mathbf{B}(\mathbf{b}^{\mathrm{c}},t) + \sum_{k=1}^{m} \frac{\partial \mathbf{B}(\mathbf{b}^{\mathrm{c}},t)}{\partial b_k}\left(b_k^{\mathrm{I}} - b_k^{\mathrm{c}}\right) = \mathbf{B}(\mathbf{b}^{\mathrm{c}},t) + \sum_{k=1}^{m} \frac{\partial \mathbf{B}(\mathbf{b}^{\mathrm{c}},t)}{\partial b_k}\Delta b_k \xi \\ \mathbf{K}\left(\mathbf{b}^{\mathrm{I}},t\right) = \mathbf{K}(\mathbf{b}^{\mathrm{c}},t) + \sum_{k=1}^{m} \frac{\partial \mathbf{K}(\mathbf{b}^{\mathrm{c}},t)}{\partial b_k}\left(b_k^{\mathrm{I}} - b_k^{\mathrm{c}}\right) = \mathbf{K}(\mathbf{b}^{\mathrm{c}},t) + \sum_{k=1}^{m} \frac{\partial \mathbf{K}(\mathbf{b}^{\mathrm{c}},t)}{\partial b_k}\Delta b_k \xi \end{cases} \tag{5.13}$$

The uncertain state is expressed as $\mathbf{x}^{\mathrm{I}}(t) = \mathbf{x}^{\mathrm{c}}(t) + \Delta\mathbf{x}^{\mathrm{I}}(t)$. Computing the time derivative and substituting Eq. (5.13) into Eq. (5.11), one obtains

$$\dot{\mathbf{x}}^{\mathrm{c}} + \Delta\dot{\mathbf{x}}^{\mathrm{I}} = \left[\begin{array}{l} \left(\mathbf{A}(\mathbf{b}^{\mathrm{c}},t) + \sum_{k=1}^{m} \frac{\partial \mathbf{A}(\mathbf{b}^{\mathrm{c}},t)}{\partial b_k}\Delta b_k \xi\right) - \\ \left(\mathbf{B}(\mathbf{b}^{\mathrm{c}},t) + \sum_{k=1}^{m} \frac{\partial \mathbf{B}(\mathbf{b}^{\mathrm{c}},t)}{\partial b_k}\Delta b_k \xi\right)\left(\mathbf{K}(\mathbf{b}^{\mathrm{c}},t) + \sum_{k=1}^{m} \frac{\partial \mathbf{K}(\mathbf{b}^{\mathrm{c}},t)}{\partial b_k}\Delta b_k \xi\right) \end{array}\right]\left(\mathbf{x}^{\mathrm{c}} + \Delta\mathbf{x}^{\mathrm{I}}\right) \tag{5.14}$$

By linearizing Eq. (5.14) through the 1st-order Taylor expansion, the interval state-space model with feedback uncertainty is derived as

$$\dot{\mathbf{x}}^{\mathrm{c}}(t) = \left[\mathbf{A}(\mathbf{b}^{\mathrm{c}},t) - \mathbf{B}(\mathbf{b}^{\mathrm{c}},t)\mathbf{K}(\mathbf{b}^{\mathrm{c}},t)\right]\mathbf{x}^{\mathrm{c}}(t) = \mathbb{A}(\mathbf{b}^{\mathrm{c}},t)\mathbf{x}^{\mathrm{c}}(t) \tag{5.15}$$

and

$$\begin{aligned} \Delta\dot{\mathbf{x}}^{\mathrm{I}}(t) = \sum_{k=1}^{m} & \left(\frac{\partial \mathbf{A}(\mathbf{b}^{\mathrm{c}},t)}{\partial b_k} - \mathbf{B}(\mathbf{b}^{\mathrm{c}},t)\frac{\partial \mathbf{K}(\mathbf{b}^{\mathrm{c}},t)}{\partial b_k} - \frac{\partial \mathbf{B}(\mathbf{b}^{\mathrm{c}},t)}{\partial b_k}\mathbf{K}(\mathbf{b}^{\mathrm{c}},t)\right) \\ & \Delta b_k \xi \mathbf{x}^{\mathrm{c}}(t) + \mathbb{A}(\mathbf{b}^{\mathrm{c}},t)\Delta\mathbf{x}^{\mathrm{I}}(t) \end{aligned} \tag{5.16}$$

Remark 5.1 Equations (5.15) and (5.16) can be combined into an order-extended matrix that describes both the nominal and uncertain dynamics of the system, as.

$$\begin{bmatrix} \dot{\mathbf{x}}^{\mathrm{c}}(t) \\ \Delta\dot{\mathbf{x}}^{\mathrm{I}}(t) \end{bmatrix} = \begin{bmatrix} \mathbb{A}(\mathbf{b}^{\mathrm{c}},t) & \mathbf{0} \\ \sum_{k=1}^{m}\left(\frac{\partial \mathbf{A}(\mathbf{b}^{\mathrm{c}},t)}{\partial b_k} - \mathbf{B}(\mathbf{b}^{\mathrm{c}},t)\frac{\partial \mathbf{K}(\mathbf{b}^{\mathrm{c}},t)}{\partial b_k} - \frac{\partial \mathbf{B}(\mathbf{b}^{\mathrm{c}},t)}{\partial b_k}\mathbf{K}(\mathbf{b}^{\mathrm{c}},t)\right)\Delta b_k \xi & \mathbb{A}(\mathbf{b}^{\mathrm{c}},t) \end{bmatrix}\begin{bmatrix} \mathbf{x}^{\mathrm{c}}(t) \\ \Delta\mathbf{x}^{\mathrm{I}}(t) \end{bmatrix} \tag{5.17}$$

which is equal to

$$\dot{\hat{\mathbf{x}}}^{\mathrm{I}}(t) = \hat{\mathbb{A}}\left(\mathbf{b}^{\mathrm{I}}, t\right)\hat{\mathbf{x}}^{\mathrm{I}}(t), \qquad \hat{\mathbf{x}} \in \mathbb{R}^{2h} \tag{5.18}$$

where

$$\hat{\mathbb{A}}\left(\mathbf{b}^{\mathrm{I}}, t\right) = \begin{bmatrix} \hat{\mathbb{A}}(\mathbf{b}^{\mathrm{c}}, t) & \mathbf{0} \\ \sum_{k=1}^{m}\left(\frac{\partial \mathbf{A}(\mathbf{b}^{\mathrm{c}},t)}{\partial b_k} - \mathbf{B}(\mathbf{b}^{\mathrm{c}}, t)\frac{\partial \mathbf{K}(\mathbf{b}^{\mathrm{c}},t)}{\partial b_k} - \frac{\partial \mathbf{B}(\mathbf{b}^{\mathrm{c}},t)}{\partial b_k}\mathbf{K}(\mathbf{b}^{\mathrm{c}}, t)\right)\Delta b_k \xi & \hat{\mathbb{A}}(\mathbf{b}^{\mathrm{c}}, t) \end{bmatrix}$$

and $\hat{\mathbf{x}}^{\mathrm{I}}(t) = \left[\mathbf{x}^{\mathrm{c}}(t)^{\mathrm{T}} \ \Delta\mathbf{x}^{\mathrm{I}}(t)^{\mathrm{T}}\right]^{\mathrm{T}}$.

5.3.2 Solution Method for Interval State-Space Equation

The preceding section established the continuous state-space model for the control system. To facilitate computational implementation and numerical simulation, this section presents the discrete state-space solution approach. The solution to the nominal linear time-varying (LTV) homogeneous system can be formulated as

$$\mathbf{x}\left[(j+1)\Delta t\right] = \mathbf{\Phi}\left[(j+1)\Delta t, j\Delta t\right]\mathbf{x}(j\Delta t) \tag{5.19}$$

where $\mathbf{\Phi}\left[(j+1)\Delta t, j\Delta t\right]$ represents the state transition matrix between consecutive time steps that can be written as

$$\mathbf{\Phi}\left[(j+1)\Delta t, j\Delta t\right] = \begin{cases} \exp\left[\int_{j\Delta t}^{(j+1)\Delta t} \mathrm{A}(\tau)\mathrm{d}\tau\right], & \text{if } \mathrm{A}\left[(j+1)\Delta t\right]\int_{j\Delta t}^{(j+1)\Delta t} \mathrm{A}(\tau)\mathrm{d}\tau = \int_{j\Delta t}^{(j+1)\Delta t} \mathrm{A}(\tau)\mathrm{d}\tau \mathrm{A}\left[(j+1)\Delta t\right] \\ \mathrm{I} + \int_{j\Delta t}^{(j+1)\Delta t} \mathrm{A}(\tau_0)\mathrm{d}\tau_0 + \int_{j\Delta t}^{(j+1)\Delta t} \mathrm{A}(\tau_0)\int_{j\Delta t}^{\tau_0} \mathrm{A}(\tau_1)\mathrm{d}\tau_1\mathrm{d}\tau_0 + \cdots, & \text{else} \end{cases} \tag{5.20}$$

Following the same principle, the uncertain LTV homogeneous system can be derived and formulated by extending the nominal approach. The nominal and uncertain components can be written using the Taylor series expansion as

$$\mathbf{x}^{\mathrm{c}}\left[(j+1)\Delta t\right] = \mathbf{\Phi}^{\mathrm{c}}\left[\mathbf{b}^{\mathrm{c}}, (j+1)\Delta t, j\Delta t\right]\mathbf{x}^{\mathrm{c}}(j\Delta t) \tag{5.21}$$

and

$$\begin{aligned}\Delta\mathbf{x}^{\mathrm{I}}\left[(j+1)\Delta t\right] &= \mathbf{\Phi}^{\mathrm{c}}\left[\mathbf{b}^{\mathrm{c}}, (j+1)\Delta t, j\Delta t\right]\Delta\mathbf{x}^{\mathrm{I}}(j\Delta t) \\ &+ \sum_{k=1}^{m}\frac{\partial\mathbf{\Phi}\left[\mathbf{b}^{\mathrm{c}}, (j+1)\Delta t, j\Delta t\right]}{\partial b_k}\mathbf{x}^{\mathrm{c}}(j\Delta t)\Delta b_k\xi\end{aligned} \tag{5.22}$$

Equations (5.21) and (5.22) can be combined as

$$\begin{bmatrix}\mathbf{x}^{\mathrm{c}}[(j+1)\Delta t] \\ \Delta\mathbf{x}^{\mathrm{I}}[(j+1)\Delta t]\end{bmatrix} = \begin{bmatrix}\boldsymbol{\Phi}^{\mathrm{c}}[\mathbf{b}^{\mathrm{c}},(j+1)\Delta t, j\Delta t] & \mathbf{0} \\ \sum_{k=1}^{m}\frac{\partial\boldsymbol{\Phi}[\mathbf{b}^{\mathrm{c}},(j+1)\Delta t, j\Delta t]}{\partial b_k}\Delta b_k\xi & \boldsymbol{\Phi}^{\mathrm{c}}[\mathbf{b}^{\mathrm{c}},(j+1)\Delta t, j\Delta t]\end{bmatrix}\begin{bmatrix}\mathbf{x}^{\mathrm{c}}[j\Delta t] \\ \Delta\mathbf{x}^{\mathrm{I}}[j\Delta t]\end{bmatrix} \tag{5.23}$$

With the definition $\hat{\mathbf{x}}^{\mathrm{I}} = \left[(\mathbf{x}^{\mathrm{c}})^{\mathrm{T}}\ (\Delta\mathbf{x}^{\mathrm{I}})^{\mathrm{T}}\right]^{\mathrm{T}}$, Eq. (5.23) is simplified as

$$\hat{\mathbf{x}}^{\mathrm{I}}[(j+1)\Delta t] = \hat{\Phi}^{\mathrm{I}}[\mathbf{b}^{\mathrm{c}},(j+1)\Delta t, j\Delta t]\hat{\mathbf{x}}^{\mathrm{I}}(j\Delta t), \qquad \hat{\mathbf{x}} \in \mathbb{R}^{2h} \tag{5.24}$$

where $\hat{\Phi}^{\mathrm{I}}[\mathbf{b}^{\mathrm{c}},(j+1)\Delta t, j\Delta t] = \begin{bmatrix}\boldsymbol{\Phi}^{\mathrm{c}}[\mathbf{b}^{\mathrm{c}},(j+1)\Delta t, j\Delta t] & \mathbf{0} \\ \sum_{k=1}^{m}\frac{\partial\boldsymbol{\Phi}[\mathbf{b}^{\mathrm{c}},(j+1)\Delta t, j\Delta t]}{\partial b_k}\Delta b_k\xi & \boldsymbol{\Phi}^{\mathrm{c}}[\mathbf{b}^{\mathrm{c}},(j+1)\Delta t, j\Delta t]\end{bmatrix}$.

The term $\frac{\partial\boldsymbol{\Phi}[\mathbf{b}^{\mathrm{c}},(j+1)\Delta t, j\Delta t]}{\partial b_k}$ can be obtained through calculating the partial derivatives of Eq. (5.20) in relation to b_k:

$$\frac{\partial\boldsymbol{\Phi}[\mathbf{b}^{\mathrm{c}},(j+1)\Delta t, j\Delta t]}{\partial b_k} = \begin{cases}\exp\left[\int_{j\Delta t}^{(j+1)\Delta t}\frac{\partial \mathrm{A}(\mathbf{b}^{\mathrm{c}},\tau)}{\partial b_k}\mathrm{d}\tau\right], \text{ if } \mathrm{A}[(j+1)\Delta t]\int_{j\Delta t}^{(j+1)\Delta t}\mathrm{A}(\tau)\mathrm{d}\tau = \int_{j\Delta t}^{(j+1)\Delta t}\mathrm{A}(\tau)\mathrm{d}\tau\mathrm{A}[(j+1)\Delta t] \\ \int_{j\Delta t}^{(j+1)\Delta t}\frac{\partial \mathrm{A}(\mathbf{b}^{\mathrm{c}},\tau_0)}{\partial b_k}\mathrm{d}\tau_0 + \int_{j\Delta t}^{(j+1)\Delta t}\frac{\partial \mathrm{A}(\mathbf{b}^{\mathrm{c}},\tau_0)}{\partial b_k}\int_{j\Delta t}^{\tau_0}\mathrm{A}(\tau_1)\mathrm{d}\tau_1\mathrm{d}\tau_0 + \int_{j\Delta t}^{(j+1)\Delta t}\mathrm{A}(\tau_0)\int_{j\Delta t}^{\tau_0}\frac{\partial \mathrm{A}(\mathbf{b}^{\mathrm{c}},\tau_1)}{\partial b_k}\mathrm{d}\tau_1\mathrm{d}\tau_0 + \cdots, \text{ else}\end{cases} \tag{5.25}$$

Consequently, the interval state can be determined as

$$\begin{cases}\left\langle\hat{\mathbf{x}}^{\mathrm{I}}[(j+1)\Delta t]\right\rangle = \left\langle\hat{\Phi}^{\mathrm{I}}[\mathbf{b}^{\mathrm{c}},(j+1)\Delta t, j\Delta t]\right\rangle\left\langle\hat{\mathbf{x}}^{\mathrm{I}}(j\Delta t)\right\rangle, \quad t \geq 0, \qquad \hat{\mathbf{x}} \in \mathbb{R}^{2h} \\ \left\langle\hat{\mathbf{x}}^{\mathrm{I}}(j\Delta t)\right\rangle\Big|_{j=0} = \left[[\mathbf{x}^{\mathrm{c}}(j\Delta t)]^{\mathrm{T}}, \left\langle\left[\Delta\mathbf{x}^{\mathrm{I}}(j\Delta t)\right]^{\mathrm{T}}\right\rangle\right]^{\mathrm{T}}\Big|_{j=0} = \left[[\mathbf{x}^{\mathrm{c}}(j\Delta t)]^{\mathrm{T}}, [\Delta\mathbf{x}(j\Delta t)]^{\mathrm{T}}\right]^{\mathrm{T}}\Big|_{j=0} \\ \qquad = \left[[\mathbf{x}^{\mathrm{c}}(0)]^{\mathrm{T}}, [\Delta\mathbf{x}(0)]^{\mathrm{T}}\right]^{\mathrm{T}}\end{cases} \tag{5.26}$$

Here $\langle\rangle$ indicates the absolute value applied only to the uncertainty elements of a matrix:

$$\left\langle\hat{\Phi}^{\mathrm{I}}[\mathbf{b}^{\mathrm{c}},(j+1)\Delta t, j\Delta t]\right\rangle = \begin{bmatrix}\boldsymbol{\Phi}^{\mathrm{c}}[\mathbf{b}^{\mathrm{c}},(j+1)\Delta t, j\Delta t] & \mathbf{0} \\ \sum_{k=1}^{m}\left|\frac{\partial\boldsymbol{\Phi}[\mathbf{b}^{\mathrm{c}},(j+1)\Delta t, j\Delta t]}{\partial b_k}\right|\Delta b_k\xi & \boldsymbol{\Phi}^{\mathrm{c}}[\mathbf{b}^{\mathrm{c}},(j+1)\Delta t, j\Delta t]\end{bmatrix} \tag{5.27}$$

$$\begin{aligned}\langle\hat{\mathbf{x}}[(j+1)\Delta t]\rangle &= \left[\mathbf{x}^{c}[(j+1)\Delta t]^{T}\ |\Delta\mathbf{x}^{I}[(j+1)\Delta t]|^{T}\right]^{T} \\ &= \left[\mathbf{x}^{c}[(j+1)\Delta t]^{T}\ \Delta\mathbf{x}[(j+1)\Delta t]^{T}\right]^{T}\end{aligned} \tag{5.28}$$

Thus, the UB and LB of the interval state vector can be written as

$$\begin{cases}\underline{\mathbf{x}}[(j+1)\Delta t] = \mathbf{x}^{c}[(j+1)\Delta t] - \Delta\mathbf{x}[(j+1)\Delta t] \\ \overline{\mathbf{x}}[(j+1)\Delta t] = \mathbf{x}^{c}[(j+1)\Delta t] + \Delta\mathbf{x}[(j+1)\Delta t]\end{cases} \tag{5.29}$$

Finally, the state vector considering interval uncertainties of SPS optimal attitude control $\mathbf{x}^{I}[(j+1)\Delta t] = [\underline{\mathbf{x}}[(j+1)\Delta t], \overline{\mathbf{x}}[(j+1)\Delta t]]$ can be derived.

5.4 Interval Uncertainty-Based Optimal Attitude Controller

5.4.1 *Two Types of Controllers*

Concerning the optimal attitude control strategy of the SPS, the three rotational axes control inputs $\mathbf{u}_{13}^{*}(t)$ and $u_{2}^{*}(t)$ are written as

$$\begin{cases}\mathbf{u}_{13}^{*}(t) = \mathbf{u}_{13}^{0}(t) + \mathbf{u}_{13}(t) \\ u_{2}^{*}(t) = u_{2}^{0} + u_{2}(t)\end{cases} \tag{5.30}$$

where $\mathbf{u}_{13}^{0}(t) = -\mathbf{f}_{13}(t)$ and $u_{2}^{0} = -f_{2}$ denote the feedforward components of rotation controllers, respectively. The feedback controllers can be written as $\mathbf{u}_{13}(t) = -\mathbf{K}_{13}(t)\mathbf{x}_{13}(t)$ and $u_{2}^{0}(t) = -K_{2}(t)x_{2}(t)$, both of which are designed using the LQR method.

As discussed above, each control channel contains both feedforward and feedback terms. The feedforward controllers are obtained directly from the disturbance torques described in Eq. (5.6). Consequently, the primary task is to design the feedback controllers based on the LQR method, after which both components are combined to compute the combined control torques.

5.4.2 *Control Law of LQR Controller*

The infinite-horizon LQR solution can be obtained through the ARE, which serves as the foundational control theory framework discussed in this chapter and will be further extended to uncertain interval systems.

Lemma 5.1 For specified weighting matrices $\mathbf{Q}$ and $\mathbf{R}$, the J representing control costs is written as.

$$J = \int_0^{\infty} \left(\mathbf{x}^{\mathrm{T}}\mathbf{Q}\mathbf{x} + \mathbf{u}^{\mathrm{T}}\mathbf{R}\mathbf{u}\right)\mathrm{d}t \tag{5.31}$$

The corresponding optimal feedback gain in Eq. (5.10) is expressed as

$$\mathbf{K} = \mathbf{R}^{-1}\mathbf{B}^{\mathrm{T}}\mathbf{P}(t) \tag{5.32}$$

where $\mathbf{P}(t)$ is a symmetric matrix satisfying the ARE:

$$\mathbf{Q} - \mathbf{P}\mathbf{B}\mathbf{R}^{-1}\mathbf{B}^{\mathrm{T}}\mathbf{P} + \mathbf{A}^{\mathrm{T}}\mathbf{P} + \mathbf{P}\mathbf{A} = \mathbf{0} \tag{5.33}$$

Since the complete attitude dynamics control are expressed by two sets of state-space equations in Eqs. (5.7) and (5.8), one of them can be represented as follows. The control costs J^* is defined by

$$J^* = \int_0^{\infty} \left(\mathbf{x}(t)^{\mathrm{T}}\mathbf{Q}\mathbf{x}(t) + \left(\mathbf{u}^*(t)\right)^{\mathrm{T}}\mathbf{R}\left(\mathbf{u}^*(t)\right)\right)\mathrm{d}t \tag{5.34}$$

Based on Eq. (5.30), J^* can be separated into two components as

$$J^* = \int_0^{\infty} \left(\mathbf{u}^0(t)\right)^{\mathrm{T}}\mathbf{R}\left(\mathbf{u}^0(t)\right)\mathrm{d}t + \int_0^{\infty} \left(\mathbf{x}^{\mathrm{T}}(t)\mathbf{Q}\mathbf{x}(t) + \mathbf{u}(t)^{\mathrm{T}}\mathbf{R}\mathbf{u}(t)\right)\mathrm{d}t = J^0 + J \tag{5.35}$$

where $J^0 = \int_0^{\infty} \left(\mathbf{u}^0(t)\right)^{\mathrm{T}}\mathbf{R}\left(\mathbf{u}^0(t)\right)\mathrm{d}t$ and $J = \int_0^{\infty} \left(\mathbf{x}^{\mathrm{T}}(t)\mathbf{Q}\mathbf{x}(t) + \mathbf{u}(t)^{\mathrm{T}}\mathbf{R}\mathbf{u}(t)\right)\mathrm{d}t$.

Since J^0 can be directly obtained from the known disturbances described in Eq. (5.6), the main objective of this chapter is to design the cost function J. However, it should be noted that Lemma 5.1 only applies to deterministic LQR systems and cannot be directly used to handle uncertain cases.

5.4.3 Interval Riccati Equation-Based LQR Controller

Accounting for the uncertainties $\mathbf{A}\left(\mathbf{b}^{\mathrm{I}}, t\right)$ and $\mathbf{B}\left(\mathbf{b}^{\mathrm{I}}, t\right)$ that is critical the optimal attitude control of the SPS, Eq. (5.33) can be reformulated within an interval form as

$$\mathbf{Q} - \mathbf{P}\left(\mathbf{b}^{\mathrm{I}}, t\right)\mathbf{B}\left(\mathbf{b}^{\mathrm{I}}, t\right)\mathbf{R}^{-1}\left(\mathbf{B}\left(\mathbf{b}^{\mathrm{I}}, t\right)\right)^{\mathrm{T}}\mathbf{P}\left(\mathbf{b}^{\mathrm{I}}, t\right)$$

$$+\left(\mathbf{A}\left(\mathbf{b}^{\mathrm{I}},t\right)\right)^{\mathrm{T}}\mathbf{P}\left(\mathbf{b}^{\mathrm{I}},t\right)+\mathbf{P}\left(\mathbf{b}^{\mathrm{I}},t\right)\mathbf{A}\left(\mathbf{b}^{\mathrm{I}},t\right)=\mathbf{0} \tag{5.36}$$

where

$$\mathbf{P}^{\mathrm{I}}(t)=\mathbf{P}^{\mathrm{c}}(t)+\Delta\mathbf{P}^{\mathrm{I}}(t) \tag{5.37}$$

Therefore, Eq. (5.36) can be decomposed into nominal and uncertain components as

$$\mathbf{Q}-\mathbf{P}^{\mathrm{c}}(t)\mathbf{B}(\mathbf{b}^{\mathrm{c}},t)\mathbf{R}^{-1}\mathbf{B}(\mathbf{b}^{\mathrm{c}},t)\mathbf{P}^{\mathrm{c}}(t)+\mathbf{A}(\mathbf{b}^{\mathrm{c}},t)^{\mathrm{T}}\mathbf{P}^{\mathrm{c}}(t)+\mathbf{P}^{\mathrm{c}}(t)\mathbf{A}(\mathbf{b}^{\mathrm{c}},t)=\mathbf{0} \tag{5.38}$$

and

$$\begin{aligned}
&-\mathbf{P}^{\mathrm{c}}(t)\mathbf{B}(\mathbf{b}^{\mathrm{c}},t)\mathbf{R}^{-1}(\mathbf{B}(\mathbf{b}^{\mathrm{c}},t))^{\mathrm{T}}\Delta\mathbf{P}^{\mathrm{I}}\\
&-\mathbf{P}^{\mathrm{c}}(t)\mathbf{B}(\mathbf{b}^{\mathrm{c}},t)\mathbf{R}^{-1}\left(\sum_{k=1}^{m}\frac{\partial\mathbf{B}(\mathbf{b}^{\mathrm{c}},t)}{\partial b_k}\Delta b_k\xi\right)^{\mathrm{T}}\mathbf{P}^{\mathrm{c}}(t)\\
&-\mathbf{P}^{\mathrm{c}}(t)\left(\sum_{k=1}^{m}\frac{\partial\mathbf{B}(\mathbf{b}^{\mathrm{c}},t)}{\partial b_k}\Delta b_k\xi\right)\mathbf{R}^{-1}(\mathbf{B}(\mathbf{b}^{\mathrm{c}},t))^{\mathrm{T}}\mathbf{P}^{\mathrm{c}}(t)\\
&-\Delta\mathbf{P}^{\mathrm{I}}\mathbf{B}(\mathbf{b}^{\mathrm{c}},t)\mathbf{R}^{-1}(\mathbf{B}(\mathbf{b}^{\mathrm{c}},t))^{\mathrm{T}}\mathbf{P}^{\mathrm{c}}(t)\\
&+\left(\sum_{k=1}^{m}\frac{\partial\mathbf{A}(\mathbf{b}^{\mathrm{c}},t)}{\partial b_k}\Delta b_k\xi\right)^{\mathrm{T}}\mathbf{P}^{\mathrm{c}}(t)+(\mathbf{A}(\mathbf{b}^{\mathrm{c}},t))^{\mathrm{T}}\Delta\mathbf{P}^{\mathrm{I}}+\Delta\mathbf{P}^{\mathrm{I}}\mathbf{A}(\mathbf{b}^{\mathrm{c}},t)\\
&+\mathbf{P}^{\mathrm{c}}(t)\left(\sum_{k=1}^{m}\frac{\partial\mathbf{A}(\mathbf{b}^{\mathrm{c}},t)}{\partial b_k}\Delta b_k\xi\right)=\mathbf{0}
\end{aligned} \tag{5.39}$$

As $\mathbf{Q}$, $\mathbf{R}$, and $\mathbf{P}^{\mathrm{I}}(t)$ are all real or interval symmetric matrices, which follows

$$\begin{aligned}
&\left[\mathbf{A}(\mathbf{b}^{\mathrm{c}},t)-\mathbf{B}(\mathbf{b}^{\mathrm{c}},t)\mathbf{R}^{-1}(\mathbf{B}(\mathbf{b}^{\mathrm{c}},t))^{\mathrm{T}}\mathbf{P}^{\mathrm{c}}(t)\right]^{\mathrm{T}}\\
&=(\mathbf{A}(\mathbf{b}^{\mathrm{c}},t))^{\mathrm{T}}-\mathbf{P}^{\mathrm{c}}(t)\mathbf{B}(\mathbf{b}^{\mathrm{c}},t)\mathbf{R}^{-1}(\mathbf{B}(\mathbf{b}^{\mathrm{c}},t))^{\mathrm{T}}
\end{aligned} \tag{5.40}$$

Thus, Eq. (5.39) can be expressed as the Lyapunov equation for the first-order perturbation $\Delta\mathbf{S}^{\mathrm{I}}$:

$$\begin{aligned}
&\left[(\mathbf{A}(\mathbf{b}^{\mathrm{c}},t))^{\mathrm{T}}-\mathbf{P}^{\mathrm{c}}(t)\mathbf{B}(\mathbf{b}^{\mathrm{c}},t)\mathbf{R}^{-1}(\mathbf{B}(\mathbf{b}^{\mathrm{c}},t))^{\mathrm{T}}\right]\Delta\mathbf{P}^{\mathrm{I}}\\
&+\Delta\mathbf{P}^{\mathrm{I}}\left[\mathbf{A}(\mathbf{b}^{\mathrm{c}},t)-\mathbf{B}(\mathbf{b}^{\mathrm{c}},t)\mathbf{R}^{-1}(\mathbf{B}(\mathbf{b}^{\mathrm{c}},t))^{\mathrm{T}}\mathbf{P}^{\mathrm{c}}(t)\right]\\
&+\left(\sum_{k=1}^{m}\frac{\partial\mathbf{A}(\mathbf{b}^{\mathrm{c}},t)}{\partial b_k}\Delta b_k\xi\right)^{\mathrm{T}}\mathbf{P}^{\mathrm{c}}(t)+\mathbf{P}^{\mathrm{c}}(t)\left(\sum_{k=1}^{m}\frac{\partial\mathbf{A}(\mathbf{b}^{\mathrm{c}},t)}{\partial b_k}\Delta b_k\xi\right)
\end{aligned}$$

$$- \mathbf{P}^{\mathrm{c}}(t)\mathbf{B}(\mathbf{b}^{\mathrm{c}}, t)\mathbf{R}^{-1}\left(\sum_{k=1}^{m} \frac{\partial \mathbf{B}(\mathbf{b}^{\mathrm{c}}, t)}{\partial b_k} \Delta b_k \xi\right)^{\mathrm{T}} \mathbf{P}^{\mathrm{c}}(t)$$

$$- \mathbf{P}^{\mathrm{c}}(t)\left(\sum_{k=1}^{m} \frac{\partial \mathbf{B}(\mathbf{b}^{\mathrm{c}}, t)}{\partial b_k} \Delta b_k \xi\right)\mathbf{R}^{-1}(\mathbf{B}(\mathbf{b}^{\mathrm{c}}, t))^{\mathrm{T}}\mathbf{P}^{\mathrm{c}}(t) = \mathbf{0} \tag{5.41}$$

Therefore, the nominal term $\mathbf{P}^{\mathrm{c}}(t)$ in $\mathbf{P}^{\mathrm{I}}(t)$ is already obtained from Eq. (5.33) and serves as a known quantity in Eq. (5.41). The remaining unknown term $\Delta\mathbf{S}(t)$ can thus be evaluated from Eq. (5.41) via $\Delta\mathbf{P}(t) = \left|\Delta\mathbf{P}^{\mathrm{I}}(t)\right|$ using the Lyapunov equation.

Similarly, the nominal optimal feedback gain $\mathbf{K}$ in Eq. (5.32) can be expanded to an uncertain form $\mathbf{K}^{\mathrm{I}}$:

$$\begin{aligned}\mathbf{K}^{\mathrm{I}}(t) &= \mathbf{K}^{\mathrm{c}}(t) + \Delta\mathbf{K}^{\mathrm{I}}(t)\\ &= -\mathbf{R}^{-1}\left(\mathbf{B}(\mathbf{b}^{\mathrm{c}}, t) + \sum_{k=1}^{m} \frac{\partial \mathbf{B}(\mathbf{b}^{\mathrm{c}}, t)}{\partial b_k} \Delta b_k \xi\right)^{\mathrm{T}} \left(\mathbf{P}^{\mathrm{c}}(t) + \Delta\mathbf{P}^{\mathrm{I}}(t)\right)\end{aligned} \tag{5.42}$$

The uncertain feedback gain can be separated into nominal and uncertain components as

$$\mathbf{K}^{\mathrm{c}}(t) = -\mathbf{R}^{-1}(\mathbf{B}(\mathbf{b}^{\mathrm{c}}, t))^{\mathrm{T}}\mathbf{P}^{\mathrm{c}}(t) \tag{5.43}$$

and

$$\Delta\mathbf{K}^{\mathrm{I}}(t) = -\mathbf{R}^{-1}\left[(\mathbf{B}(\mathbf{b}^{\mathrm{c}}, t))^{\mathrm{T}}\Delta\mathbf{P}^{\mathrm{I}}(t) + \left(\sum_{k=1}^{m} \frac{\partial \mathbf{B}(\mathbf{b}^{\mathrm{c}}, t)}{\partial b_k} \Delta b_k \xi\right)^{\mathrm{T}} \mathbf{P}^{\mathrm{c}}(t)\right] \tag{5.44}$$

The uncertainty radius of the feedback gain can be determined utilizing the natural interval extension as

$$\Delta\mathbf{K}(t) = \left|\mathbf{R}^{-1}\right|\left[\left|(\mathbf{B}(\mathbf{b}^{\mathrm{c}}, t))^{\mathrm{T}}\right|\Delta\mathbf{P}(t) + \sum_{k=1}^{m}\left|\left(\frac{\partial \mathbf{B}(\mathbf{b}^{\mathrm{c}}, t)}{\partial b_k}\right)^{\mathrm{T}} \mathbf{P}^{\mathrm{c}}\right|\Delta b\right] \tag{5.45}$$

Consequently, the bounds of the uncertain feedback gain are given by

$$\begin{cases}\underline{\mathbf{K}}(t) = \mathbf{K}^{\mathrm{c}}(t) - \Delta\mathbf{K}(t)\\ \overline{\mathbf{K}}(t) = \mathbf{K}^{\mathrm{c}}(t) + \Delta\mathbf{K}(t)\end{cases} \tag{5.46}$$

The precise interval-valued estimator targeting the robust optimal feedback coefficient, denoted $\mathbf{K}^{\mathrm{I}} = \left[\underline{\mathbf{K}}, \overline{\mathbf{K}}\right]$, is thus successfully determined. Subsequently, the interval control costs can be computed, yielding a foundational equation by combining Eqs. (5.10) and (5.32).

Thus, the complete interval estimator for the uncertain optimal feedback gain $\mathbf{K}^{\mathrm{I}} = \left[\underline{\mathbf{K}}, \overline{\mathbf{K}}\right]$ can be determined. By combining Eqs. (5.10) and (5.32) can obtain

$$\mathbf{Ru} = -\mathbf{B}^{\mathrm{T}}\mathbf{Px} \tag{5.47}$$

Substituting Eqs. (5.10), (5.43), (5.33), and (5.47) into Eq. (5.31) yields the deterministic cost function

$$J = \int_0^\infty \mathbf{x}^{\mathrm{T}}\left(2\mathbf{Q} + \mathbf{A}^{\mathrm{T}}\mathbf{P} + \mathbf{PA}\right)\mathbf{x}\mathrm{d}t \tag{5.48}$$

which can be extended to the interval form as

$$\begin{aligned} J^{\mathrm{I}} &= J^{\mathrm{c}} + \Delta J^{\mathrm{I}} \\ &= \int_0^\infty \left(\mathbf{x}^{\mathrm{c}}(t) + \Delta\mathbf{x}^{\mathrm{I}}(t)\right)^{\mathrm{T}} \left[\begin{array}{l} 2\mathbf{Q} + \left(\mathbf{A}(\mathbf{b}^{\mathrm{c}}, t) + \sum\limits_{k=1}^{m} \frac{\partial \mathbf{A}(\mathbf{b}^{\mathrm{c}}, t)}{\partial b_k}\Delta b_k \xi\right)^{\mathrm{T}} \left(\mathbf{P}^{\mathrm{c}}(t) + \Delta\mathbf{P}^{\mathrm{I}}(t)\right) \\ +\left(\mathbf{P}^{\mathrm{c}}(t) + \Delta\mathbf{P}^{\mathrm{I}}(t)\right)\left(\mathbf{A}(\mathbf{b}^{\mathrm{c}}, t) + \sum\limits_{k=1}^{m} \frac{\partial \mathbf{A}(\mathbf{b}^{\mathrm{c}}, t)}{\partial b_k}\Delta b_k \xi\right) \end{array}\right] \left(\begin{array}{c} \mathbf{x}^{\mathrm{c}}(t) \\ +\Delta\mathbf{x}^{\mathrm{I}}(t) \end{array}\right) \mathrm{d}t \end{aligned} \tag{5.49}$$

Therefore, the nominal and uncertain parts can be written as

$$J^{\mathrm{c}} = \int_0^\infty (\mathbf{x}^{\mathrm{c}}(t))^{\mathrm{T}}\left[2\mathbf{Q} + (\mathbf{A}(\mathbf{b}^{\mathrm{c}}, t))^{\mathrm{T}}\mathbf{P}^{\mathrm{c}}(t) + \mathbf{P}^{\mathrm{c}}(t)\mathbf{A}(\mathbf{b}^{\mathrm{c}}, t)\right]\mathbf{x}^{\mathrm{c}}(t)\mathrm{d}t \tag{5.50}$$

and

$$\Delta J^{\mathrm{I}} = \int_0^\infty \left(\begin{array}{l} (\mathbf{x}^{\mathrm{c}}(t))^{\mathrm{T}}\left[(\mathbf{A}(\mathbf{b}^{\mathrm{c}}, t))^{\mathrm{T}}\Delta\mathbf{P}^{\mathrm{I}}(t) + \left(\sum\limits_{k=1}^{m} \frac{\partial \mathbf{A}(\mathbf{b}^{\mathrm{c}}, t)}{\partial b_k}\Delta b_k \xi\right)^{\mathrm{T}} \mathbf{P}^{\mathrm{c}}(t)\right]\mathbf{x}^{\mathrm{c}}(t) \\ +(\mathbf{x}^{\mathrm{c}}(t))^{\mathrm{T}}\left[\mathbf{P}^{\mathrm{c}}(t)\sum\limits_{k=1}^{m} \frac{\partial \mathbf{A}(\mathbf{b}^{\mathrm{c}}, t)}{\partial b_k}\Delta b_k \xi + \Delta\mathbf{P}^{\mathrm{I}}(t)\mathbf{A}(\mathbf{b}^{\mathrm{c}}, t)\right]\mathbf{x}^{\mathrm{c}}(t) \\ +(\mathbf{x}^{\mathrm{c}}(t))^{\mathrm{T}}\left[(\mathbf{A}(\mathbf{b}^{\mathrm{c}}, t))^{\mathrm{T}}\mathbf{P}^{\mathrm{c}}(t) + \mathbf{P}^{\mathrm{c}}(t)\mathbf{A}(\mathbf{b}^{\mathrm{c}}, t)\right]\Delta\mathbf{x}^{\mathrm{I}}(t) \\ +\left(\Delta\mathbf{x}^{\mathrm{I}}(t)\right)^{\mathrm{T}}\left[(\mathbf{A}(\mathbf{b}^{\mathrm{c}}, t))^{\mathrm{T}}\mathbf{P}^{\mathrm{c}}(t) + \mathbf{P}^{\mathrm{c}}(t)\mathbf{A}(\mathbf{b}^{\mathrm{c}}, t)\right]\mathbf{x}^{\mathrm{c}}(t) \end{array}\right)\mathrm{d}t \tag{5.51}$$

Thus, ΔJ can be obtained from Eq. (5.51) using interval expansion.

However, since the estimation of ΔJ via this method may be excessive, a refined interval estimation approach is employed to reduce overestimation and improve accuracy.

Remark 5.2 To simplify representation and calculation, Let the matrices $\mathbf{M}(t)$ and $\mathbf{N}(t)$ be defined as

$$\mathbf{M}(t) = \mathbf{A}(t)^{\mathrm{T}}\mathbf{P}(t) + \mathbf{P}(t)\mathbf{A}(t) \tag{5.52}$$

and

$$\mathbf{N}(t) = \mathbf{B}(t)^{\mathrm{T}}\mathbf{P}(t) \tag{5.53}$$

Thus, Eq. (5.33) becomes

$$\mathbf{Q}(t) - \mathbf{N}(t)^{\mathrm{T}}\mathbf{R}^{-1}\mathbf{N}(t) + \mathbf{M}(t) = \mathbf{0} \tag{5.54}$$

Equation (5.54) can be generalized to the interval form as

$$\mathbf{Q} - \left(\mathbf{N}^{\mathrm{c}}(t) + \Delta\mathbf{N}^{\mathrm{I}}(t)\right)^{\mathrm{T}}\mathbf{R}^{-1}\left(\mathbf{N}^{\mathrm{c}}(t) + \Delta\mathbf{N}^{\mathrm{I}}(t)\right) + \mathbf{M}^{\mathrm{c}}(t) + \Delta\mathbf{M}^{\mathrm{I}}(t) = \mathbf{0} \tag{5.55}$$

The nominal and uncertain components can be separated as

$$\mathbf{Q} - (\mathbf{N}^{\mathrm{c}}(t))^{\mathrm{T}}\mathbf{R}^{-1}\mathbf{N}^{\mathrm{c}}(t) + \mathbf{M}^{\mathrm{c}}(t) = \mathbf{0} \tag{5.56}$$

and

$$\Delta\mathbf{M}^{\mathrm{I}}(t) = (\mathbf{N}^{\mathrm{c}}(t))^{\mathrm{T}}\mathbf{R}^{-1}\Delta\mathbf{N}^{\mathrm{I}}(t) + \left(\Delta\mathbf{N}^{\mathrm{I}}(t)\right)^{\mathrm{T}}\mathbf{R}^{-1}\mathbf{N}^{\mathrm{c}}(t) \tag{5.57}$$

Considering Eq. (5.52), the expression for the uncertain part becomes

$$\begin{aligned}\Delta\mathbf{M}^{\mathrm{I}}(t) = {} & (\mathbf{A}^{\mathrm{c}}(\mathbf{b}^{\mathrm{c}}, t))^{\mathrm{T}}\Delta\mathbf{P}^{\mathrm{I}}(t) + \left(\sum_{k=1}^{m}\frac{\partial\mathbf{A}(\mathbf{b}^{\mathrm{c}}, t)}{\partial b_k}\Delta b_k\xi\right)^{\mathrm{T}}\mathbf{P}^{\mathrm{c}}(t) \\ & + \mathbf{P}^{\mathrm{c}}(t)\sum_{k=1}^{m}\frac{\partial\mathbf{A}(\mathbf{b}^{\mathrm{c}}, t)}{\partial b_k}\Delta b_k\xi + \Delta\mathbf{P}^{\mathrm{I}}(t)\mathbf{A}^{\mathrm{c}}(\mathbf{b}^{\mathrm{c}}, t)\end{aligned} \tag{5.58}$$

Accordingly, Eq. (5.51) can be reformulated as

$$\Delta J^{\mathrm{I}} = \int_0^{\infty}\left((\mathbf{x}^{\mathrm{c}}(t))^{\mathrm{T}}\Delta\mathbf{M}^{\mathrm{I}}(t)\mathbf{x}^{\mathrm{c}}(t) + (\mathbf{x}^{\mathrm{c}}(t))^{\mathrm{T}}\mathbf{M}^{\mathrm{c}}(t)\Delta\mathbf{x}^{\mathrm{I}}(t) + \left(\Delta\mathbf{x}^{\mathrm{I}}(t)\right)^{\mathrm{T}}\mathbf{M}^{\mathrm{c}}(t)\mathbf{x}^{\mathrm{c}}(t)\right)\mathrm{d}t \tag{5.59}$$

It is evident that applying the interval formulation of Eq. (5.59) effectively minimizes the overestimation compared to Eq. (5.51). Consequently, the uncertain control cost radius can be computed through the interval expansion approach as

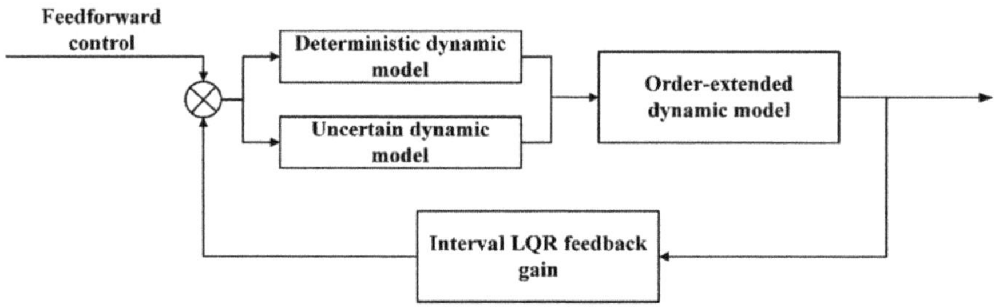

Fig. 5.2 The architectural sketch of the introduced control strategy

$$\Delta J = \left|\Delta J^{\mathrm{I}}\right| \tag{5.60}$$

The corresponding bounds of the control costs are then expressed as

$$\begin{cases} \underline{J} = J^{\mathrm{c}} - \Delta J \\ \overline{J} = J^{\mathrm{c}} + \Delta J \end{cases} \tag{5.61}$$

Therefore, the precise and compact interval estimator for the control costs $J^{\mathrm{I}} = \left[\underline{J}, \overline{J}\right]$ can be derived. This completes the interval analysis for the controller, from which the control system overall design schematic can be constructed, as illustrated in Fig. 5.2.

5.5 Set Theory-Based Time-Dependent Reliability

5.5.1 Interval Process Model

Two state response intervals, representing angle/angular velocity, are proposed to construct the ITDR model. The state responses $x(\mathbf{b}, t_1)$ and $x(\mathbf{b}, t_2)$ at two consecutive time instants ($t_2 = t_1 + \Delta t$, where Δt denotes the time step) can be transformed into typical intervals ξ_1 and ξ_2 expressed as

$$\xi_1 = \frac{x(\mathbf{b}, t_1) - x^{\mathrm{c}}(\mathbf{b}, t_1)}{\Delta x(\mathbf{b}, t_1)}, \ \xi_2 = \frac{x(\mathbf{b}, t_2) - x^{\mathrm{c}}(\mathbf{b}, t_2)}{\Delta x(\mathbf{b}, t_2)} \tag{5.62}$$

where $x^{\mathrm{c}}(\mathbf{b}, t_1)$ and $\Delta x(\mathbf{b}, t_1)$ represent the center and radius of the state interval at time t_1, while $x^{\mathrm{c}}(\mathbf{b}, t_2)$ and $\Delta x(\mathbf{b}, t_2)$ denote the corresponding quantities at time t_2.

Accordingly, the state variables $x(\mathbf{b}, t_1)$ and $x(\mathbf{b}, t_2)$ can be described as

$$\begin{cases} x(\mathbf{b}, t_1) \in \left[\underline{x}(\mathbf{b}, t_1), \overline{x}(\mathbf{b}, t_1)\right] = x^{\mathrm{c}}(\mathbf{b}, t_1) + \Delta x(\mathbf{b}, t_1)\xi_1 \\ x(\mathbf{b}, t_2) \in \left[\underline{x}(\mathbf{b}, t_2), \overline{x}(\mathbf{b}, t_2)\right] = x^{\mathrm{c}}(\mathbf{b}, t_2) + \Delta x(\mathbf{b}, t_2)\xi_2 \end{cases} \tag{5.63}$$

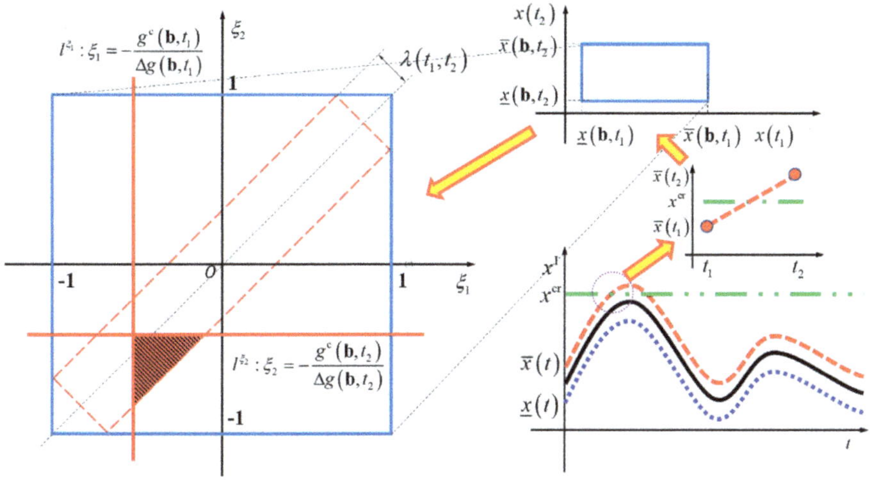

Fig. 5.3 Sketch highlighting the $\lambda(t_1, t_2)$ length and spatial extent of $E(j)$

The correlation coefficient $\rho(t_1, t_2)$ between $x(\mathbf{b}, t_1)$ and $x(\mathbf{b}, t_2)$ within the non-probabilistic interval framework is calculated by the following equation:

$$\begin{aligned}\rho(t_1, t_2) &= \rho(\xi_1, \xi_2)\\ &= \frac{\sum_{k=1}^{m} \frac{\partial x(\mathbf{b}^c, t_1)}{\partial b_k} \frac{\partial x(\mathbf{b}^c, t_2)}{\partial b_k} (\Delta b_k)^2}{\sqrt{\sum_{k=1}^{m} \left(\frac{\partial x(\mathbf{b}^c, t_1)}{\partial b_k} \Delta b_k\right)^2} \sqrt{\sum_{k=1}^{m} \left(\frac{\partial x(\mathbf{b}^c, t_2)}{\partial b_k} \Delta b_k\right)^2}} = 1 - \sqrt{2}\lambda(t_1, t_2), \qquad 0 \le \lambda(t_1, t_2) \le \sqrt{2}g(\mathbf{b}, t)\end{aligned} \tag{5.64}$$

where $\lambda(t_1, t_2)$ represents the half-length of the diagonal side of the oblique rectangle that characterizes the relationship between the two-dimensional intervals of $x(\mathbf{b}, t_1)$ and $x(\mathbf{b}, t_2)$, as illustrated in Fig. 5.3.

Thus, the value of $\lambda(t_1, t_2)$ can be determined from Eq. (5.64), which also serves as a basis for evaluating the system reliability.

5.5.2 *First Passage Theory*

The attitude control system failure in the SPS system is characterized by the controlled state violating its safe limits. Accordingly, the state function $g(\mathbf{b}, t)$ is formulated to represent such failure conditions as follows:

$$g(\mathbf{b}, t) = x^{\text{cr}} - x(\mathbf{b}, t) \tag{5.65}$$

where x^{cr} denotes the permissible boundary of the controlled state. When $g(\mathbf{b}, t) \geq 0$, the attitude control is considered to operate safely, otherwise, failure occurs. Thus, the ITDR of the sun-pointing attitude control, denoted by $R(T)$, within the time span $[0, T]$, is formulated as

$$R(T) = \mathrm{Pos}\{\forall 0 \leq t \leq T : g(\mathbf{b}, t) = x^{\mathrm{cr}} - x(\mathbf{b}, t)\} \tag{5.66}$$

where $\mathrm{Pos}\{\cdot\}$ represents the possibility measure.

Accounting for system uncertainties, the key parameters of $g(\mathbf{b}, t)$ can be expressed as

$$\begin{cases} g^{\mathrm{c}}(\mathbf{b}, t) = x^{\mathrm{cr}} - x^{\mathrm{c}}(\mathbf{b}, t) \\ \Delta g(\mathbf{b}, t) = \Delta x(\mathbf{b}, t) \\ g(\mathbf{b}, t) \in g^{\mathrm{I}}(\mathbf{b}, t) = g^{\mathrm{c}}(\mathbf{b}, t) + \Delta g^{\mathrm{I}}(\mathbf{b}, t) = g^{\mathrm{c}}(\mathbf{b}, t) + \Delta g(\mathbf{b}, t)\xi \end{cases} \tag{5.67}$$

where $g^{\mathrm{c}}(\mathbf{b}, t)$ and $\Delta g(\mathbf{b}, t)$ respectively correspond to the nominal and radius of the process $g(\mathbf{b}, t)$.

For the time-dependent process $g(\mathbf{b}, t)$ defined over the period $[0, T]$, a discretization yields a series using a suitable time step $\Delta t = {}^{T}/_{N}$, yielding:

$$[g(\mathbf{b}, 0), g(\mathbf{b}, \Delta t), \ldots, g(\mathbf{b}, j\Delta t), \ldots, g(\mathbf{b}, N\Delta t)] \tag{5.68}$$

5.5.3 Non-probabilistic Time-Dependent Reliability Assessment

The out-crossing rate of the attitude angle, denoted as $\nu(j\Delta t)$, can be formulated as

$$\nu(j\Delta t) = \frac{\mathrm{Pos}\{E(j)\}}{\Delta t} = \frac{\mathrm{Pos}\{g(\mathbf{b}, t_1) > 0 \cap g(\mathbf{b}, t_2) < 0\}}{\Delta t} \tag{5.69}$$

where $\mathrm{Pos}\{E(j)\}$ represents the probability of failure in the attitude control for $g(\mathbf{b}, t)$ within the interval $[t_1, t_2]$. The failure possibility of the SPS attitude control can be expressed as the geometric proportion between the interfering shaded region and the entire oblique rectangular space, as illustrated in Fig. 5.3, and is expressed as

$$\mathrm{Pos}\{E_{p_x}(j)\} = \frac{S^j_{\mathrm{interferece}}}{S^j_{\mathrm{total}}} l^{\xi_1} l^{\xi_2} \tag{5.70}$$

In Fig. 5.3, the variables l^{ξ_1} and l^{ξ_2} are given by

$$\begin{cases} l^{\xi_1} : \xi_1 = -\dfrac{g^{\mathrm{c}}(\mathbf{b}, t_1)}{\Delta g(\mathbf{b}, t_1)} \\ l^{\xi_2} : \xi_2 = -\dfrac{g^{\mathrm{c}}(\mathbf{b}, t_2)}{\Delta g(\mathbf{b}, t_2)} \end{cases} \tag{5.71}$$

Hence, the overall ITDR of the SPS sun-pointing attitude control $Rel(T)$ is determined by

$$Rel(T) = 1 - \left\{ \mathrm{Pos}(0) + \sum_{j=1}^{N} \left(E_{p_x}(j)\right) \right\} = 1 - \left\{ \mathrm{Pos}(0) + \sum_{j=1}^{N} \frac{S^j_{\text{interferece}}}{S^j_{\text{total}}} \right\} \tag{5.72}$$

where $\mathrm{Pos}(0)$ represents the rate of the initial possibility of failure, which is usually equal to 0.

Thus, the final expression for the ITDR used to assess the attitude state of the SPS is acquired.

5.6 Uncertain Multi-objective Optimization with Reliability Constraint

5.6.1 Uncertain Multi-objective Optimization Problem

The introduced multi-objective optimization problem considering interval uncertainties for attitude control under reliability constraints is designed to minimize both the control costs and the variation of the attitude angle of the SPS. This optimization problem can be formulated as

$$\begin{aligned} &\text{find } \mathbf{Q}, \mathbf{R} \\ &\min \left(J^*\right)^{\mathrm{c}} = J^0 + J^{\mathrm{c}} \\ &\min \|\Delta\theta\| \\ &\text{s.t. } Rel(T) \geq \text{poss} \\ &\bar{\mathbf{b}} \leq \mathbf{b} \leq \bar{\mathbf{b}} \end{aligned} \tag{5.73}$$

where $(J^*)^{\mathrm{c}}$ represents the nominal value of control costs that combines the nominal feedforward component J^0 and the feedback component J^{c}. The term poss denotes the predefined reliability threshold for the SPS sun-pointing attitude controller.

5.6.2 Coevolution Constrained Multi-objective Optimization Algorithm

To address this problem efficiently, a novel coevolutionary CCMO algorithm (Tian et al. 2021) is employed. This algorithm offers several advantages for solving uncertain optimization problems of this nature. By adopting a weak cooperative mechanism between two interacting populations, the CCMO algorithm provides improved performance over traditional genetic algorithms in handling multi-objective optimization problems considering constraints such as the one described in Eq. (5.73). In the CCMO framework, two distinct populations are utilized: one focuses on solving the original optimization problem, while the other addresses a helper problem derived from it. Through limited cooperation and information exchange between these populations, the CCMO accelerates convergence and enhances search efficiency. In this chapter, by discarding the dominated solutions and preserving the non-dominated ones, the resulting solutions are organized into Pareto fronts based on dominance relationships. Depending on the imposed constraints, each solution is classified as feasible or infeasible. The CCMO algorithm is adopted in this chapter to address the multi-objective optimization problem based on LQR considering interval uncertainties for SPS attitude control, due to its strong global search capability.

5.7 Numerical Example

According to the structural configuration and coordinate framework of the SPS, the inertia moments along the three coordinate axes, I_1, I_2, and I_3 are 2.8×10^{13}, 1.8×10^{13}, and 4.6×10^{13} kgm^2, respectively. The corresponding interval uncertainties ΔI_1, ΔI_2, and ΔI_3 are respectively 5.6×10^{11}, 3.6×10^{11}, and 9.2×10^{11} $\text{kg}\cdot\text{m}^2$. Accounting for the original state of the SPS, the nominal and uncertain initial values of the attitude angles θ_1, θ_2, θ_3 and angular velocities ω_1, ω_2, ω_3 are defined as: $\hat{\mathbf{x}}_{13}^{\mathrm{I}}(0) = [\theta_1, \theta_3, \omega_1, \omega_3, \Delta\theta_1, \Delta\theta_3, \Delta\omega_1, \Delta\omega_3]^{\mathrm{T}}$and $\hat{\mathbf{x}}_2^{\mathrm{I}}(0) = [\theta_2, \omega_2, \Delta\theta_2, \Delta\omega_2]^{\mathrm{T}}$ with initial values of $[10, -10, -1.1\times10^{-3}, 2.3\times10^{-3}, 0.6, 0.2, 6.6\times10^{-5}, 4.6\times10^{-5}]$ and $\left[10, -2.9\times10^{-3}, 0.2, 5.8\times10^{-5}\right]$. The elements of θ and ω are measured in degrees and degrees per second. To achieve precise attitude control of large-scale SPS platforms, multiple actuation components comprising electric-propulsion thrusters, momentum wheels, and control moment gyros are required. The development of high-momentum-density flywheels is also essential for future applications.

The interval LQR control strategy is employed to enhance system performance and ensure stable attitude response. For roll/yaw control, weighting matrices $\mathbf{Q}_{13}$ and $\mathbf{R}_{13}$ are used, while $\mathbf{Q}_2$ and $\mathbf{R}_2$ are applied to pitch control. The feedback gains are derived through the interval algebraic Riccati equation (IARE), forming the primary component of the control input, while the auxiliary component compensates for external time-dependent disturbances. The interval control input $\mathbf{u}$ is yielded by the solution to the IARE. The total nominal cost function $(J^*)^c$ combines the contributions from both feedforward and feedback controls.

Within this framework, the optimization procedure seeks to minimize both the nominal control costs $(J^*)^c$ and the variation of the attitude angle $\|\Delta\boldsymbol{\theta}\| = \|[\Delta\theta_1, \Delta\theta_2, \Delta\theta_3]\|$, while satisfying reliability constraints. The system reliability for the response θ_3 exceeding 5.84° is set at 0.9 to ensure operational safety. The interval dynamics for rotational motions are constructed, and the interval response of θ_3 is obtained. Additionally, the bounds of θ_3 from 1,000 MCS are compared with the interval outcome to verify the precision of the introduced approach.

The Pareto front, generated with an initial population size of 50 and at most 50,000 function evaluations, is shown in Fig. 5.4. Two representative optimal solutions are presented in Table 5.1. Using the introduced ITDR approach, the calculated reliabilities of the response θ_3, as shown in Table 5.1, are 0.9637, and 1, all exceeding the target value of 0.9. The MCS results yield corresponding reliabilities of 0.9640, and 1, respectively, confirming the accuracy of the interval-based reliability estimation. Two main sources of error are observed. The first arises from the conservative nature of the interval control approach, which, however, benefits from low requirements on uncertainty quantification and rapid computation. Using a system with a CPU of 3.90 GHz (16 cores) and 16 GB RAM, the computing times for MCS are 46.12 and 47.07 s, whereas the introduced interval approach requires only 0.164, and 0.311 s, corresponding to merely 0.3, and 0.6‰ of the MCS time. This significant reduction in computation demonstrates the superior expediency of the introduced interval approach. The second source of discrepancy originates from the tight roll-yaw coupling, leading to conservative optimization results when using the same $\mathbf{Q}$ and $\mathbf{R}$ parameters. The close agreement between the LB and UB response bounds of θ_3 obtained via both the interval method and MCS, as shown in Fig. 5.4, confirms the method's precision and reliability (Fig. 5.5).

Except for the constrained response θ_3, the interval envelopes for other responses $\theta_1, \theta_2, \omega_1, \omega_2$, and ω_3 under the two optimized solution sets are illustrated in Figs. 5.6, 5.7, 5.8, 5.9 and 5.10. These results indicate that the interval control method can accurately estimate the bounds of state responses when system uncertainties are characterized. All state responses converge to zero. Although minor variations exist among the results from different optimized parameter sets, their trends are similar. As shown in Fig. 5.6, the interval width of θ_1 narrows as the control input increases, leading to faster convergence. A comparable trend is evident in Fig. 5.7, where the lower bound of θ_2 gradually increases as the control torque rises.

The interval control method also provides precise and rapid estimates of the angular velocity response range. As shown in Figs. 5.8, 5.9 and 5.10, the response behavior of ω exhibits slight differences across the two representative Pareto solution sets for control, with ω_1 in Fig. 5.8 showing the most distinct pattern. When the applied control torque is relatively low in numerical terms, the amplitude of ω_1 gradually drops down before any excessive response beyond the setpoint occurs. As the control input grows, the amplitude of ω_1 surpasses its initial value, leading to a faster control response and enabling the system to reach attitude stability more rapidly.

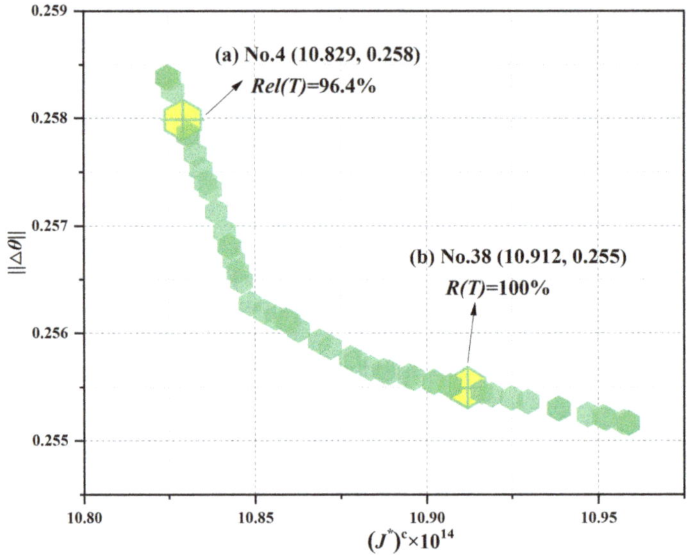

Fig. 5.4 Pareto-optimal set for attitude control

Table 5.1 System parameters for the optimized Q and R sets

	$J^*(\times 10^{14})$	$\|\Delta\boldsymbol{\theta}\|$	Reliability	Relative error of the proposed method against MCS (‰)
(a)	10.829	0.258	0.9637	0.3
(b)	10.912	0.255	1	/

Overall, derived from the multi-objective optimization process, the optimized **Q** and **R** enable the attitude angle θ_3 to meet the reliability requirement while converging rapidly to zero. Due to the periodic nature of the roll/yaw disturbance torque described by Eq. (5.6), the steady-state control inputs along these axes are not constant but oscillatory, synchronized with the disturbance period, as illustrated in Fig. 5.11. Moreover, the two optimal solutions exhibit nearly identical control inputs for pitch and yaw, while variations are mainly observed in roll control torques. Figure 5.12 presents the interval bounds of the control costs calculated utilizing the introduced interval optimal control approach. The variation of the control cost remains within the predicted range under varying parameters, demonstrating that the interval method effectively provides accurate, rapid output feedback and state reconstruction while reducing the complexity of uncertainty quantification.

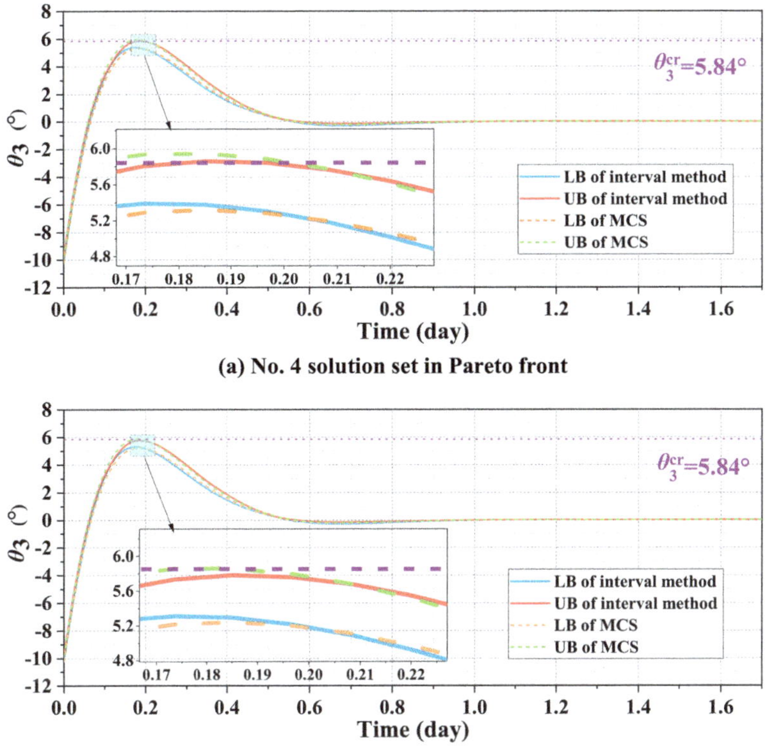

(a) No. 4 solution set in Pareto front

(b) No. 38 solution set in Pareto front

Fig. 5.5 θ_3 response envelope of two typical solutions using interval analysis and MCS method

5.8 Summary

This chapter presents a set theory-based approach to optimal attitude control for SPS under multi-source uncertainties. By modeling these uncertainties as UBB quantities, an interval time-varying attitude dynamics model is formulated. A composite control strategy combining feedforward compensation for disturbances and an interval-based LQR feedback controller derived from an IARE is developed to manage system responses in the presence of uncertainty. The resulting framework yields bounds on both the state response and control cost, while a novel ITDR metric, grounded in interval process theory and first-passage criteria, quantifies the safety of the controlled attitude over time. To optimize both control cost and system response fluctuation under reliability constraints, a CCMO algorithm is employed. Numerical simulations demonstrate that the proposed method achieves comparable accuracy to MCS but with significantly reduced computational cost, which requires less than 1% of the time, making it well-suited for uncertainty-involved and reliability-constrained applications.

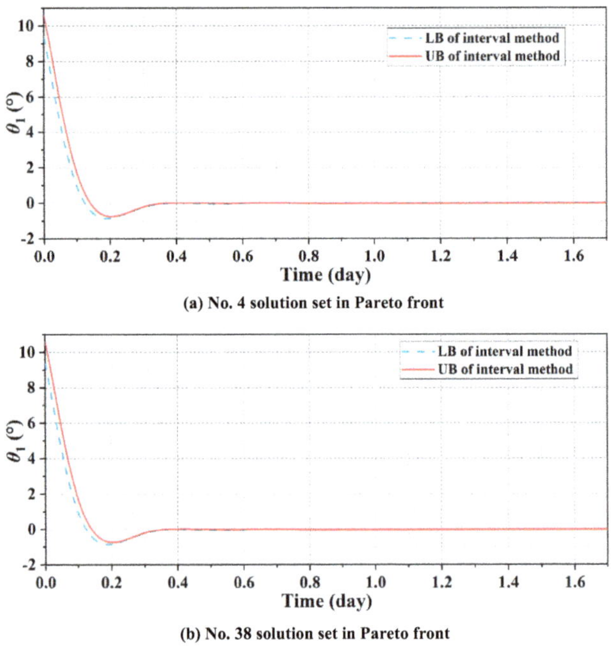

Fig. 5.6 θ_1 response envelope of two typical solutions using interval analysis

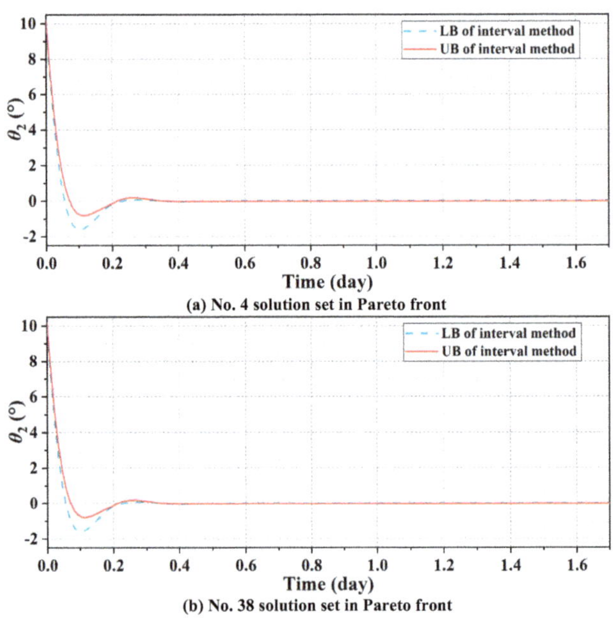

Fig. 5.7 θ_2 response envelope of two typical solutions using interval analysis

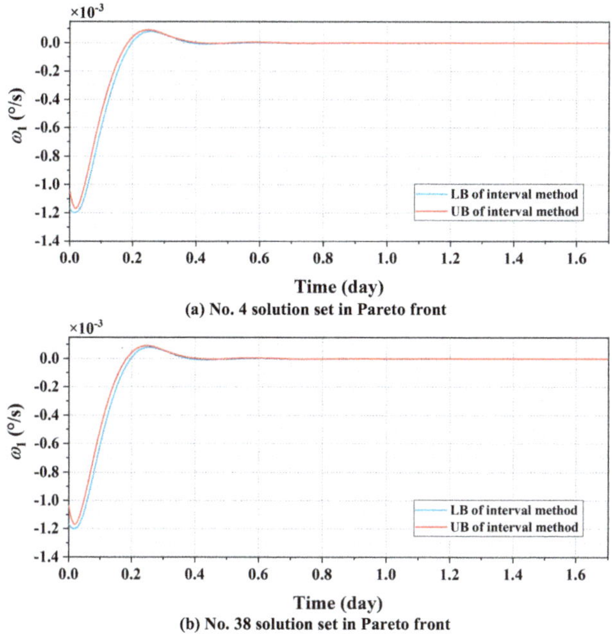

Fig. 5.8 ω_1 response envelope of two typical solutions using interval analysis

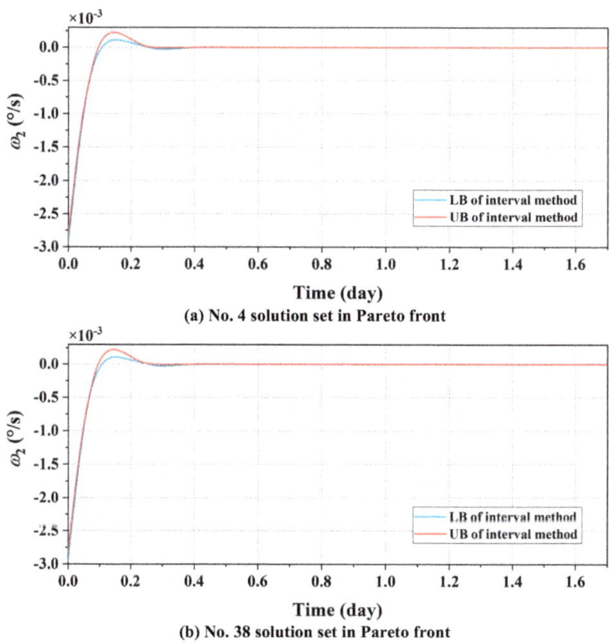

Fig. 5.9 ω_2 response envelope of two typical solutions using interval analysis

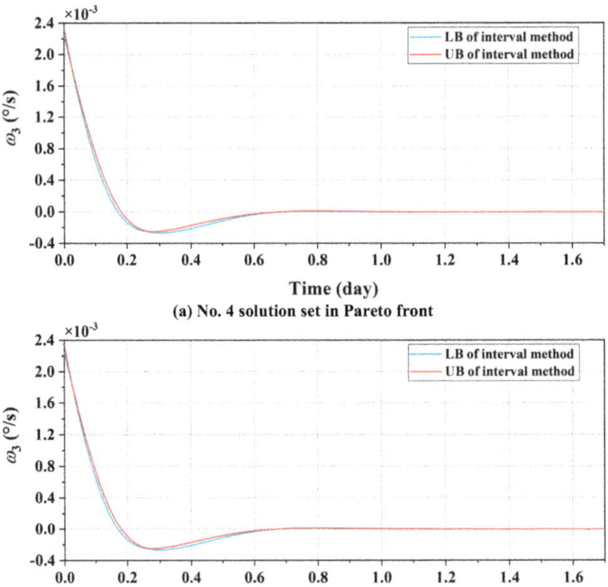

(a) No. 4 solution set in Pareto front

(b) No. 38 solution set in Pareto front

Fig. 5.10 ω_3 response envelope of two typical solutions using interval analysis

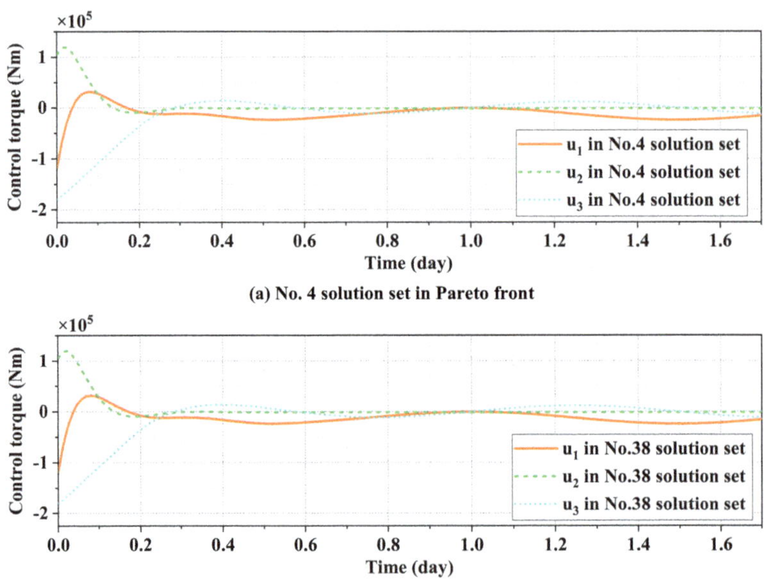

(a) No. 4 solution set in Pareto front

(b) No. 38 solution set in Pareto front

Fig. 5.11 Control input in two typical solutions

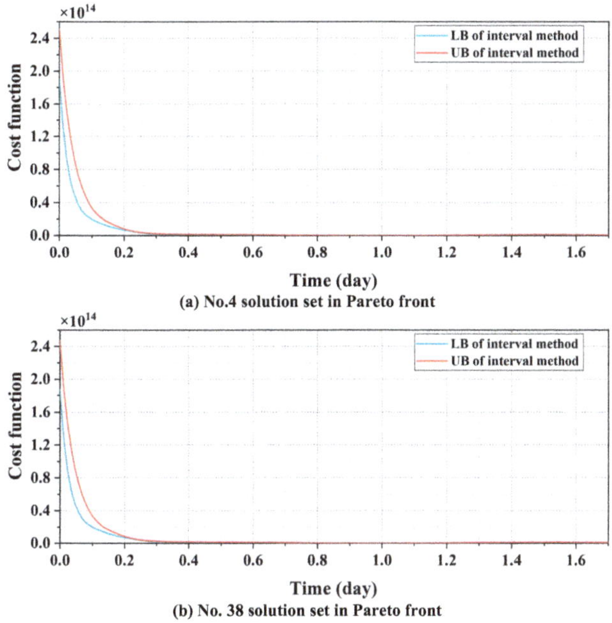

Fig. 5.12 Cost function envelope of two typical solutions using the interval analysis

References

Castet J-F, Saleh JH (2009) Satellite and satellite subsystems reliability: statistical data analysis and modeling. Reliab Eng Syst Saf 94:1718–1728. https://doi.org/10.1016/j.ress.2009.05.004

Chen J-B, Li J (2007) The extreme value distribution and dynamic reliability analysis of nonlinear structures with uncertain parameters. Struct Saf 29:77–93. https://doi.org/10.1016/j.strusafe.2006.02.002

Ditlevsen O, Madsen HO (1996) Structural reliability methods. Wiley, New York

Glaser PE (1968) Power from the sun: its future. Science 162:857–861. https://doi.org/10.1126/science.162.3856.857

Jiang C, Han X, Liu GR, Liu GP (2008) A nonlinear interval number programming method for uncertain optimization problems. Eur J Oper Res 188:1–13. https://doi.org/10.1016/j.ejor.2007.03.031

Liu J, Ouyang H, Han X, Liu G (2021) Optimal sensor placement for uncertain inverse problem of structural parameter estimation. Mech Syst SIGNAL Process 160. https://doi.org/10.1016/j.ymssp.2021.107914

Lemaire M (2013) Structural reliability. John Wiley & Sons

Mankins J (1997) A fresh look at space solar power: new architectures, concepts and technologies. ACTA Astronaut 41:347–359. https://doi.org/10.1016/S0094-5765(98)00075-7

Mori M, Kagawa H, Saito Y (2006) Summary of studies on space solar power systems of Japan Aerospace Exploration Agency (JAXA). ACTA Astronaut 59:132–138. https://doi.org/10.1016/j.actaastro.2006.02.033

Qiu Z, Elishakoff I (1998) Antioptimization of structures with large uncertain-but-non-random parameters via interval analysis. Comput Methods Appl Mech Eng 152:361–372. https://doi.org/10.1016/S0045-7825(96)01211-X

Tian Y, Zhang T, Xiao J, Zhang X, Jin Y (2021) A coevolutionary framework for constrained multiobjective optimization problems. IEEE Trans Evol Comput 25:102–116. https://doi.org/10.1109/TEVC.2020.3004012

Wang L, Zhou Z, Liu J (2022) Interval-based optimal trajectory tracking control method for manipulators with clearance considering time-dependent reliability constraints. Aerosp Sci Technol 128. https://doi.org/10.1016/j.ast.2022.107745

Wu S, Zhang K, Peng H, Wu Z, Radice G (2016) Robust optimal sun-pointing control of a large solar power satellite. ACTA Astronaut 127:226–234. https://doi.org/10.1016/j.actaastro.2016.05.019

Wie B, Roithmayr C (2005) Attitude and orbit control of a very large geostationary solar power satellite. J Guid CONTROL Dyn 28:439–451. https://doi.org/10.2514/1.6813

Zhang K, Wu S, Wu Z (2021) Multibody dynamics and robust attitude control of a MW-level solar power satellite. Aerosp Sci Technol 111. https://doi.org/10.1016/j.ast.2021.106575

Chapter 6
Convex-Set-Based LQR Controller for SPS Attitude with Safety Assessment

Abstract This chapter addresses the challenge of space power stations (SPS) attitude control under multi-source uncertainties that are both bounded and correlated. Instead of relying on traditional probabilistic methods, which require extensive sampling, a non-probabilistic framework based on convex set theory is established to quantify these uncertainties. The chapter first constructs the uncertain attitude dynamics using an extended-order state-space formulation, explicitly capturing the correlation effects among system parameters. A novel convex set-based linear quadratic regulator (CSLQR) is then developed. By deriving and solving a convex set-based algebraic Riccati equation (CSARE) alongside the Lyapunov equation, the proposed method analytically determines the bounds of feedback gains and control costs. To further ensure operational safety, a convex set-based time-dependent reliability (CSTDR) index is introduced as a dynamic assessment metric. This reliability index is subsequently integrated as a key constraint within an optimization framework to balance control performance against system safety. Numerical simulations verify that this approach effectively mitigates the impact of nonlinear dynamics and uncertainties, offering a computationally efficient alternative to Monte Carlo simulations (MCSs).

6.1 Introduction

As introduced in Chap. 5, the SPS demands exceptional precision in attitude control. Its massive configuration (Ni et al. 2022) and dual-pointing requirements subjected to a complex environment introduce nonlinear and uncertain dynamics (Kruse et al. 2024; Yang et al. 2024a; Wang et al. 2024; Liu et al. 2024). Even minor deviations in orientation can significantly degrade power transmission efficiency and threaten mission stability, underscoring the necessity for uncertainty-aware control methodologies.

Building upon the previous exploration of the interval-based LQR controller in SPS, this chapter extends the discussion to the convex set-based linear quadratic regulator (CSLQR) controller. While interval analysis provides an effective means of quantifying bounded uncertainties along independent variables, many practical

C. Yang and Y. Xia, *Set Theory-Based Spacecraft Dynamics and Control*, Intelligent Control and Learning Systems 26, https://doi.org/10.1007/978-981-95-9179-4_6

aerospace systems exhibit correlated and coupled uncertainties that cannot be fully captured by interval bounds alone (Jiang et al. 2011; Zhang et al. 2023). The CSB approach presented in this chapter is designed to overcome these limitations (Liao 2013; Fu et al. 2022). By characterizing uncertainties through convex sets defined by known bounds and correlations, the proposed method captures both the range and dependence of uncertain parameters without assuming specific probability distributions.

When obtaining the CSLQR controller, the reliability of the system should be taken into account. Proposed in recent years, NPTDR theory is a precise and efficient approach to quantify reliability considering uncertainty (Jiang et al. 2008; Wang et al. 2022; Yang et al. 2024b). A further advancement introduced in this chapter is the convex set-based time-dependent reliability (CSTDR), which assesses the safety of the control system over time. By embedding CSTDR into the optimization process, the controller design ensures not only minimal cost but also a guaranteed reliability level that evolves dynamically with mission duration (Hu et al. 2022). Based on CSTDR, a reliability-constrained optimization can be performed to balance control performance against system safety. Thus, an integrated framework of modeling, analysis, and optimal control of large-scale space systems under complex uncertainty structures is established.

Comprehensive numerical simulations validate the efficacy of the CSB framework in handling bounded and correlated uncertainties. Compared with traditional probabilistic methods such as MCS, the proposed method achieves a desirable trade-off between precision, robustness, and computational speed, making it particularly suitable for applications where uncertainty must be managed under limited data conditions.

6.2 Convex Set Perturbation-Based LQR Control

The deterministic dynamics model and LQR control system for the SPS satellite have been established in Chap. 5, with reference to Eqs. (5.1)–(5.11).

The conventional LQR is only applicable to nominal systems. When the system exhibits correlated uncertainties, the classical LQR becomes inadequate. In Chap. 5, an interval uncertainty-based LQR controller is established to address this limitation. Considering the aforementioned discussions on the strengths of interval and convex set uncertainty analysis, in this Chapter, a CSLQR controller is introduced.

Based on the propagation of CSB uncertainties, Eq. (5.33) can be separated into nominal and uncertain components, which has been discussed in detail in Chap. 3. The uncertainty terms arise from $\mathbf{A}^{\mathrm{CS}}(\mathbf{b}, t)$ and $\mathbf{B}^{\mathrm{CS}}(\mathbf{b}, t)$. And the CSB uncertain dynamic model can be obtained as an extended-order matrix form as

$$\dot{\mathbb{X}}(t) = \mathbb{A}(\mathbf{b})\mathbb{X}(t) + \mathbb{B}(\mathbf{b})\mathbb{U}(\mathbf{b}, t), \qquad \mathbb{X} \in \mathbb{R}^{2n} \tag{6.1}$$

where $\dot{\mathbb{X}}(t) = \left[(\dot{\mathbf{x}}^{c}(t))^{T}, \left(\partial\dot{\mathbf{x}}^{c}(t)/\partial b_k\right)^{T}\right]^{T}$, $\mathbb{X}(t) = \left[(\mathbf{x}^{c}(t))^{T}, \left(\partial\mathbf{x}^{c}(t)/\partial b_k\right)^{T}\right]^{T}$, $\mathbb{A}(\mathbf{b}) = \begin{bmatrix} \mathbf{A}^{c}(\mathbf{b}) & \mathbf{0} \\ \partial\mathbf{A}^{c}(\mathbf{b})/\partial b_k & \mathbf{A}^{c}(\mathbf{b}) \end{bmatrix}$, $\mathbb{B}(\mathbf{b}) = \begin{bmatrix} \mathbf{B}^{c}(\mathbf{b}) & \mathbf{0} \\ \partial\mathbf{B}^{c}(\mathbf{b})/\partial b_k & \mathbf{B}^{c}(\mathbf{b}) \end{bmatrix}$, and $\mathbb{U}(\mathbf{b}, t) = \left[(\mathbf{u}^{c}(\mathbf{b}, t))^{T}, \left(\partial\mathbf{u}^{c}(\mathbf{b}, t)/\partial b_k\right)^{T}\right]^{T}$.

Based on the uncertain dynamic model, the CSLQR controller can be obtained. The deterministic part of the Ricatti Equation of the LQR controller is expressed as

$$\mathbf{Q} - \mathbf{P}^{c}(t)\mathbf{B}^{c}(\mathbf{b}, t)\mathbf{R}^{-1}(\mathbf{B}^{c}(\mathbf{b}, t))^{T}\mathbf{P}^{c}(t) + (\mathbf{A}^{c}(\mathbf{b}, t))^{T}\mathbf{P}^{c}(t) + \mathbf{P}^{c}(t)\mathbf{A}^{c}(\mathbf{b}, t) = \mathbf{0} \tag{6.2}$$

And the uncertain partial derivative term is represented as

$$\begin{aligned} &\frac{\partial(\mathbf{A}^{c}(\mathbf{b}, t))^{T}}{\partial b_k}\mathbf{P}^{c}(t) + (\mathbf{A}^{c}(\mathbf{b}, t))^{T}\frac{\partial\mathbf{P}^{c}(t)}{\partial b_k} + \frac{\partial\mathbf{P}^{c}(t)}{\partial b_k}\mathbf{A}^{c}(\mathbf{b}, t) + \mathbf{P}^{c}(t)\frac{\partial\mathbf{A}^{c}(\mathbf{b}, t)}{\partial b_k} \\ &- \frac{\partial\mathbf{P}^{c}(t)}{\partial b_k}\mathbf{B}^{c}(\mathbf{b}, t)\mathbf{R}^{-1}(\mathbf{B}^{c}(\mathbf{b}, t))^{T}\mathbf{P}^{c}(t) - \mathbf{P}^{c}(t)\frac{\partial\mathbf{B}^{c}(\mathbf{b}, t)}{\partial b_k}\mathbf{R}^{-1}(\mathbf{B}^{c}(\mathbf{b}, t))^{T}\mathbf{P}^{c}(t) \\ &- \mathbf{P}^{c}(t)\mathbf{B}^{c}(\mathbf{b}, t)\mathbf{R}^{-1}\frac{\partial(\mathbf{B}^{c}(\mathbf{b}, t))^{T}}{\partial b_k}\mathbf{P}^{c}(t) - \mathbf{P}^{c}(t)\mathbf{B}^{c}(\mathbf{b}, t)\mathbf{R}^{-1}(\mathbf{B}^{c}(\mathbf{b}, t))^{T}\frac{\partial\mathbf{P}^{c}(t)}{\partial b_k} = 0 \end{aligned} \tag{6.3}$$

This expression can be compactly written as

$$\mathcal{A}_1 + \mathcal{A}_2 + \mathcal{A}_3 = 0 \tag{6.4}$$

where

$$\begin{cases} \mathcal{A}_1 = \frac{\partial(\mathbf{A}^{c}(\mathbf{b},t))^{T}}{\partial b_k}\mathbf{P}^{c}(t) + \left(\mathbf{A}^{c}(\mathbf{b}, t)\right)^{T}\frac{\partial\mathbf{P}^{c}(t)}{\partial b_k} \\ \mathcal{A}_2 = \frac{\partial\mathbf{P}^{c}(t)}{\partial b_k}\mathbf{A}^{c}(\mathbf{b}, t) + \mathbf{P}^{c}(t)\frac{\partial\mathbf{A}^{c}(\mathbf{b},t)}{\partial b_k} \\ \mathcal{A}_3 = -\dfrac{\partial\mathbf{P}^{c}(t)}{\partial b_k}\mathbf{B}^{c}(\mathbf{b}, t)\mathbf{R}^{-1}\left(\mathbf{B}^{c}(\mathbf{b}, t)\right)^{T}\mathbf{P}^{c}(t) - \mathbf{P}^{c}(t)\mathbf{B}^{c}(\mathbf{b}, t)\mathbf{R}^{-1}\left(\mathbf{B}^{c}(\mathbf{b}, t)\right)^{T}\dfrac{\partial\mathbf{P}^{c}(t)}{\partial b_k} \\ \quad -\mathbf{P}^{c}(t)\dfrac{\partial\mathbf{B}^{c}(\mathbf{b}, t)}{\partial b_k}\mathbf{R}^{-1}\left(\mathbf{B}^{c}(\mathbf{b}, t)\right)^{T}\mathbf{P}^{c}(t) - \mathbf{P}^{c}(t)\mathbf{B}^{c}(\mathbf{b}, t)\mathbf{R}^{-1}\dfrac{\partial\left(\mathbf{B}^{c}(\mathbf{b}, t)\right)^{T}}{\partial b_k}\mathbf{P}^{c}(t) \end{cases} \tag{6.5}$$

Accordingly, Eq. (6.3) can be rewritten as

$$\begin{aligned} &\mathbf{P}^{c}(t)\partial\mathbf{A}^{c}(\mathbf{b}, t)/\partial b_k + \partial\mathbf{P}^{c}(t)/\partial b_k\left(\mathbf{A}^{c}(\mathbf{b}, t) - \mathbf{B}^{c}(\mathbf{b}, t)\mathbf{R}^{-1}\left(\mathbf{B}^{c}(\mathbf{b}, t)\right)^{T}\mathbf{P}^{c}(t)\right) \\ &- \mathbf{P}^{c}(t)\partial\mathbf{B}^{c}(\mathbf{b}, t)/\partial b_k\mathbf{R}^{-1}\left(\mathbf{B}^{c}(\mathbf{b}, t)\right)^{T}\mathbf{P}^{c}(t) - \mathbf{P}^{c}(t)\mathbf{B}^{c}(\mathbf{b}, t)\mathbf{R}^{-1}\partial\left(\mathbf{B}^{c}(\mathbf{b}, t)\right)^{T}/\partial b_k\mathbf{P}^{c}(t) \\ &+\partial\left(\mathbf{A}^{c}(\mathbf{b}, t)\right)^{T}/\partial b_k\mathbf{P}^{c}(t) + \left(\mathbf{A}^{c}(\mathbf{b}, t) - \mathbf{B}^{c}(\mathbf{b}, t)\mathbf{R}^{-1}\left(\mathbf{B}^{c}(\mathbf{b}, t)\right)^{T}\mathbf{P}^{c}(t)\right)^{T}\partial\mathbf{P}^{c}(t)/\partial b_k = \mathbf{0} \end{aligned} \tag{6.6}$$

After obtaining the solutions to both the deterministic ARE and the uncertain state-space equations, the quantities $\partial \mathbf{A}^{c}(\mathbf{b},t)^{T} / \partial b_k$, $\partial (\mathbf{B}^{c}(\mathbf{b},t))^{T} / \partial b_k$, and $\mathbf{P}^{c}(t)$ can be determined. Hence, $\partial \mathbf{P}^{c}(t) / \partial b_k$ is derived by solving Eq. (6.6) as a Lyapunov equation.

Similarly, based on Eq. (5.32), the feedback gain $\mathbf{K}$ is given by

$$\begin{cases} \mathbf{K}^{c}(t) = \mathbf{R}^{-1}(\mathbf{B}^{c}(\mathbf{b},t))^{T}\mathbf{P}^{c}(t) \\ \partial \mathbf{K}^{c}(t) / \partial b_k = \mathbf{R}^{-1}(\mathbf{B}^{c}(\mathbf{b},t))^{T}\partial \mathbf{P}^{c}(t) / \partial b_k + \mathbf{R}^{-1}\partial (\mathbf{B}^{c}(\mathbf{b},t))^{T} / \partial b_k \mathbf{P}^{c}(t) \end{cases} \tag{6.7}$$

The uncertainty envelopes are then precisely estimated utilizing the central perturbation–Lagrange approach, yielding

$$\begin{cases} K_{ij}^{r} = \sqrt{\mathbf{h}_{K_{ij}}^{T}\mathbf{M}^{-1}\left(\mathbf{b}^{r}, \boldsymbol{\rho}_{\mathbf{b}}\right)\mathbf{h}_{K_{ij}}} \\ K_{ij}^{L} = K_{ij}^{c} - K_{ij}^{r} \\ K_{ij}^{U} = K_{ij}^{c} + K_{ij}^{r} \\ K_{ij}^{CS} = K_{ij}^{c} + \varepsilon K_{ij}^{r} \in \left[K_{ij}^{L}, K_{ij}^{U}\right] \end{cases} \tag{6.8}$$

where $\mathbf{h}_{K_{ij}} = \left[\partial K_{ij}^{c} / \partial b_1, \cdots, \partial K_{ij}^{c} / \partial b_k, \cdots, \partial K_{ij}^{c} / \partial b_m\right]^{T}$, and factor $\varepsilon \in [-1, 1]$.

By substituting Eqs. (6.8) and (3.23) into (5.10) yields

$$\begin{cases} u_{i}^{r} = \sqrt{\mathbf{h}_{u_i}^{T}\mathbf{M}^{-1}\left(\mathbf{b}^{r}, \boldsymbol{\rho}_{\mathbf{b}}\right)\mathbf{h}_{u_i}} \\ u_{i}^{L} = u_{i}^{c} - u_{i}^{r} \\ u_{i}^{U} = u_{i}^{c} + u_{i}^{r} \\ u_{i}^{CS} = u_{i}^{c} + \varepsilon u_{i}^{r} \in \left[u_{i}^{L}, u_{i}^{U}\right] \end{cases} \tag{6.9}$$

where $\mathbf{h}_{u_i} = \left[\partial u_{i}^{c} / \partial b_1, \cdots, \partial u_{i}^{c} / \partial b_k, \cdots, \partial u_{i}^{c} / \partial b_m\right]^{T}$.

Combining Eq. (5.20) and Eq. (5.32) gives

$$\mathbf{R}\mathbf{u}(t) = -\mathbf{B}^{T}(\mathbf{b},t)\mathbf{P}(t)\mathbf{x}(t) \tag{6.10}$$

By substituting Eqs. (5.10), (5.32), (5.33), and (6.10) into Eq. (5.31) can obtain

$$J = \int_{0}^{\infty} \mathbf{x}^{T}\left(2\mathbf{Q} + \mathbf{A}^{T}\mathbf{P} + \mathbf{P}\mathbf{A}\right)\mathbf{x}\mathrm{d}t \tag{6.11}$$

Similarly, Eq. (6.11) can be transformed into its nominal and uncertain derivative parts, as

$$\begin{cases} J^{\mathrm{c}} = \int_0^\infty \left(\mathbf{x}^{\mathrm{c}}(t)\right)^{\mathrm{T}} (2\mathbf{Q} + \mathcal{L}_3)\mathbf{x}^{\mathrm{c}}(t)\mathrm{d}t \\ \partial J^{\mathrm{c}} / \partial b_k = \int_0^\infty \left(\left(\mathbf{x}^{\mathrm{c}}(t)\right)^{\mathrm{T}} (\mathcal{A}_1 + \mathcal{A}_2)\mathbf{x}^{\mathrm{c}}(t) + \left(\mathbf{x}^{\mathrm{c}}(t)\right)^{\mathrm{T}} \mathcal{L}_3 \partial \mathbf{x}^{\mathrm{c}}(t) / \partial b_k + \partial \left(\mathbf{x}^{\mathrm{c}}(t)\right)^{\mathrm{T}} \Big/ \partial b_k \mathcal{L}_3 \mathbf{x}^{\mathrm{c}}(t) \right) \mathrm{d}t \end{cases} \tag{6.12}$$

where $\mathcal{L}_3 = (\mathbf{A}^{\mathrm{c}}(\mathbf{b}, t))^{\mathrm{T}} \mathbf{P}^{\mathrm{c}}(t) + \mathbf{P}^{\mathrm{c}}(t)\mathbf{A}^{\mathrm{c}}(\mathbf{b}, t)$.

Thus, the CSB representation is presented as

$$\begin{cases} J^{\mathrm{r}} = \sqrt{\mathbf{h}_J^{\mathrm{T}} \mathbf{M}^{-1}(\mathbf{b}^{\mathrm{r}}, \boldsymbol{\rho}_{\mathrm{b}}) \mathbf{h}_J} \\ J^{\mathrm{L}} = J^{\mathrm{c}} - J^{\mathrm{r}} \\ J^{\mathrm{U}} = J^{\mathrm{c}} + J^{\mathrm{r}} \\ J^{\mathrm{CS}} = J^{\mathrm{c}} + \varepsilon J^{\mathrm{r}} \in \left[J^{\mathrm{L}}, J^{\mathrm{U}}\right] \end{cases} \tag{6.13}$$

where $\mathbf{h}_J = \left[\, \partial J^{\mathrm{c}} / \partial b_1, \cdots, \partial J^{\mathrm{c}} / \partial b_k, \cdots, \partial J^{\mathrm{c}} / \partial b_m \,\right]^{\mathrm{T}}$.

6.3 CSB Time-Dependent Reliability

When the attitude of the spacecraft exceeds the allowable threshold, the control performance becomes unreliable. To assess the reliability of the attitude stabilization of SPS, a CSTDR analysis method is introduced.

6.3.1 Definition of CSTDR

If the system state surpasses its critical limit, the control process should be regarded as unsuccessful. Accordingly, system reliability is defined as the probability that the state remains below the critical boundary x_i^{cri}. The time-dependent security function is introduced as

$$g_i\left(\mathbf{b}, t, x_i^{\mathrm{cri}}\right) = x_i(t) - x_i^{\mathrm{cri}} \tag{6.14}$$

Here $g_i\left(\mathbf{b}, t, x_i^{\mathrm{cri}}\right)$ denotes the security function, and the condition $g_i\left(\mathbf{b}, t, x_i^{\mathrm{cri}}\right) > 0$ implies that the system is within a reliable range.

Considering the presence of uncertainties, the security function can be decomposed into deterministic and uncertain components. The nominal component of $g_i\left(\mathbf{b}, t, x_i^{\mathrm{cri}}\right)$ is expressed as

$$g_i^{\mathrm{c}}\left(\mathbf{b}, t, x_i^{\mathrm{cri}}\right) = x_i^{\mathrm{c}} - x_i^{\mathrm{cri}} \tag{6.15}$$

where $g_i^{\mathrm{c}}\left(\mathbf{b}, t, x_i^{\mathrm{cri}}\right)$ represents the nominal value of the security function.

The CSB uncertain part of $g_i(\mathbf{b}, t, x_i^{\text{cri}})$ can be characterized based on convex set theory as follows:

$$\begin{cases} g_i^{\text{r}}(\mathbf{b}, t, x_i^{\text{cri}}) = \sqrt{\mathbf{h}_{g_i}^{\text{T}} \mathbf{M}^{-1}(\mathbf{b}^{\text{r}}, \boldsymbol{\rho}_{\mathbf{b}}) \mathbf{h}_{g_i}} \\ g_i^{\text{L}}(\mathbf{b}, t, x_i^{\text{cri}}) = g_{\text{i}}^{\text{c}}(\mathbf{b}, t, x_i^{\text{cri}}) - g_i^{\text{r}}(\mathbf{b}, t, x_i^{\text{cri}}) \\ g_i^{\text{U}}(\mathbf{b}, t, x_i^{\text{cri}}) = g_{\text{i}}^{\text{c}}(\mathbf{b}, t, x_i^{\text{cri}}) + g_i^{\text{r}}(\mathbf{b}, t, x_i^{\text{cri}}) \\ g_i^{\text{CS}}(\mathbf{b}, t, x_i^{\text{cri}}) = g_{\text{i}}^{\text{c}}(\mathbf{b}, t, x_i^{\text{cri}}) + \varepsilon g_i^{\text{r}}(\mathbf{b}, t, x_i^{\text{cri}}) \in \left[g_i^{\text{L}}(\mathbf{b}, t, x_i^{\text{cri}}), g_i^{\text{U}}(\mathbf{b}, t, x_i^{\text{cri}})\right] \end{cases} \tag{6.16}$$

where $\mathbf{h}_{g_i} = \left[\partial g_i^{\text{c}} / \partial b_1, \cdots, \partial g_i^{\text{c}} / \partial b_k, \cdots, \partial g_i^{\text{c}} / \partial b_m\right]^{\text{T}}$.

The system reliability over the time interval $[0, T]$ can be expressed as

$$Rel_i(T) = \text{Pos}\left\{\forall t \in [0, T] : g_i(\mathbf{b}, t, x_i^{\text{cri}}) = x_i(t) - x_i^{\text{cri}} > 0\right\} \tag{6.17}$$

where $\text{Pos}\{\cdot\}$ denotes the possibility.

To facilitate computation, Eq. (6.17) can be discretized by dividing the time span $[0, T]$ into small intervals of length $\Delta t = T / N$. The corresponding discrete reliability expression becomes

$$Rel_i(N, T) = \text{Pos}\left\{\forall k \in [0, N] : g_i(\mathbf{b}, t_k, x_i^{\text{cri}}) = x_i(t_k) - x_i^{\text{cri}} > 0\right\} \tag{6.18}$$

6.3.2 CSTDR Assessment

Accounting for the propagation of uncertainties, the failure rate over a time interval can be expressed as

$$\nu(j\Delta t) = \frac{\text{Pos}\{E(k)\}}{\Delta t} = \frac{\text{Pos}\left\{g^{\text{CS}}(t_k) > 0 \cap g^{\text{CS}}(t_{k+1}) < 0\right\}}{\Delta t} \tag{6.19}$$

where $E(k)$ represents the event that a failure occurs within the interval $[t_k, t_{k+1}]$. This corresponds to the situation where the control remains effective at the current sampling moment t_k but fails at the next sampling moment t_{k+1}. The possibility of this event is given by

$$\text{Pos}\{E(k)\} = S_{i,k}^{\text{int}} / S_{i,k}^{\text{total}} \tag{6.20}$$

where the interference region $S_{i,k}^{\text{int}}$ characterizes the occurrence of $E(k)$, while the circular region $S_{i,k}^{\text{total}}$ represents the standard reference area constructed by the normalized variables ξ_1 and ξ_2, which can be calculated as

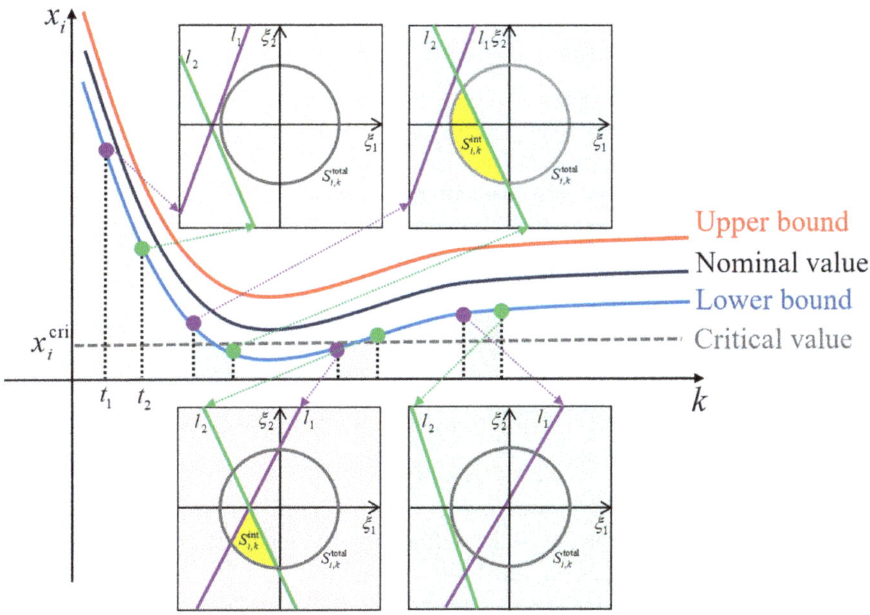

Fig. 6.1 Graphical representation of CSTDR

$$\begin{cases} \xi_1 = \left(x_i^{CS}(t_k) - x_i^{c}(t_k)\right) / x_i^{r}(t_k) \\ \xi_2 = \left(x_i^{CS}(t_{k+1}) - x_i^{c}(t_{k+1})\right) / x_i^{r}(t_{k+1}) \end{cases} \quad (6.21)$$

The spatial interpretation of this relationship is shown in Fig. 6.1.

In Fig. 6.1, l_1 and l_2 can be expressed as

$$\begin{cases} l_1 : r_1\zeta_1 - r_2\zeta_2 + \sqrt{2}g_i^{c}\left(\mathbf{b}, t_k, x_i^{cri}\right) / g_i^{r}\left(\mathbf{b}, t_k, x_i^{cri}\right) = 0 \\ l_2 : r_1\zeta_1 + r_2\zeta_2 + \sqrt{2}g_i^{c}\left(\mathbf{b}, t_{k+1}, x_i^{cri}\right) / g_i^{r}\left(\mathbf{b}, t_{k+1}, x_i^{cri}\right) = 0 \end{cases} \quad (6.22)$$

Consequently, Eq. (6.17) can be reformulated as

$$Rel_i(N, T) = \max\left\{1 - \text{Pos}_i(0) - \sum_{k=0}^{N-1} \left(S_{i,k}^{int} / S_{i,k}^{total}\right), 0\right\} \quad (6.23)$$

where $\text{Pos}_i(0)$ is a constant equal to zero, representing the probability of an initial failure condition.

6.4 CSB Reliability-Constrained Attitude Control Optimization

To optimize the SPS dynamics and control, the variables of the CSLQR controller are selected as the decision variables. Moreover, the CSTDR is employed as a constraint condition. The optimization objective is to minimize the overall control cost, which includes both the feedforward and feedback control efforts. Accordingly, a single-objective optimization problem that accounts for multiple uncertainties in the SPS can be formulated as

$$\begin{aligned} &\text{find } \mathbf{Q}, \ \mathbf{R} \\ &\min \ J^{\mathrm{c}} = J_{\mathrm{fw}} + J_{\mathrm{bk}}^{\mathrm{c}} \\ &\text{s.t.} \, Rel(T) \geq Rel_{\mathrm{th}} \\ &\mathbf{b}^{\mathrm{L}} \leq \mathbf{b} \leq \mathbf{b}^{\mathrm{U}} \end{aligned} \tag{6.24}$$

where J^{c}, J_{fw}, and $J_{\mathrm{bk}}^{\mathrm{c}}$ denote the total, feedforward, and feedback costs, respectively. The parameter R_{th} represents the reliability threshold.

Differential Evolution (DE) (Banerjee and Abu-Mahfouz 2014) is well-suited for single-objective problems. It adopts an efficient mutation and selection mechanism that allows rapid exploration of the search space, thereby providing an effective means for solving LQR optimization tasks. In addition, DE exhibits strong robustness and adaptability, enabling its application to a wide variety of optimization problems regardless of the specific characteristics of the objective function. Therefore, the reliability constraints of the system can be properly fulfilled using this method.

6.5 Numerical Example

This section demonstrates the numerical verification of the optimal attitude control framework for the 1.2 GW SPS. The parameters defining the nominal SPS control system are given in Tables 6.1, while 6.2 shows the quantified uncertain variables.

The LQR controller is adopted to achieve precise control. For the roll/yaw and pitch motions, $\mathbf{Q}_{13}$, $\mathbf{R}_{13}$, $\mathbf{Q}_{2}$, and $\mathbf{R}_{2}$ are designed separately. The controller is composed of two components: a feedforward control term that compensates for additional time-dependent disturbances, and a state feedback term obtained by CSARE.

Table 6.1 Nominal parameters of the Abacus SPS

Symbol	Random variables	Value
R_o	SPS orbital radius	4.2157×10^7 **m**
μ	Gravitational constant	$3.986004418 \times 10^{14} \mathrm{m}^3 / \mathrm{s}^2$
I_1	First principal inertia moment	2.8×10^{13} kg · m^2
I_2	Second principal inertia moment	1.8×10^{13} kg · m^2
I_3	Third principal inertia moment	4.5×10^{13} kg · m^2
θ_1	First initial angle	**10°**
θ_2	Second initial angle	**10°**
θ_3	Third initial angle	**10°**
ω_1	First initial angular velocity	-1.1×10^{-3} .° /s
ω_2	Second initial angular velocity	-2.9×10^{-3} .° /s
ω_3	Third initial angular velocity	2.3×10^{-3} .° /s

Table 6.2 Initial random variables with quantified uncertainty

Symbol	Random variables	Value
b_1^r	Uncertain radius I_1	1.4×10^{12} kg · m^2
b_2^r	Uncertain radius of I_2	1×10^{12} kg · m^2
b_3^r	Uncertain radius of I_3	1.4×10^{12} kg · m^2
ρ_{12}	The correlation coefficient between I_1 and I_2	**0**
ρ_{13}	The correlation coefficient between I_1 and I_3	**0**
ρ_{23}	The correlation coefficient between I_2 and I_3	**0**

To assess the validity of the CSB method, 10,000 MCS are conducted, from which the LB/UB of the response curves are extracted and compared with the results obtained using the CSB method. These bounded response curves are employed to assess the performance and robustness of SPS attitude control. Furthermore, quantitative reliability is calculated utilizing the CSTDR analysis method. The constraint value of reliability is set as 0.90.

The nominal total cost of control J^c, consisting of the feedforward part J_{fw} and the feedback part J_{bk}^c, is chosen as the optimization objective. The reliability bounds are imposed as constraints that determine the optimal control parameters under reliability requirements.

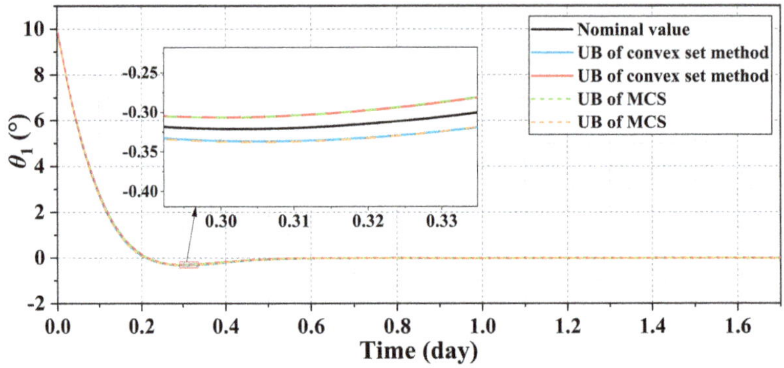

(a) Response curves of θ_1

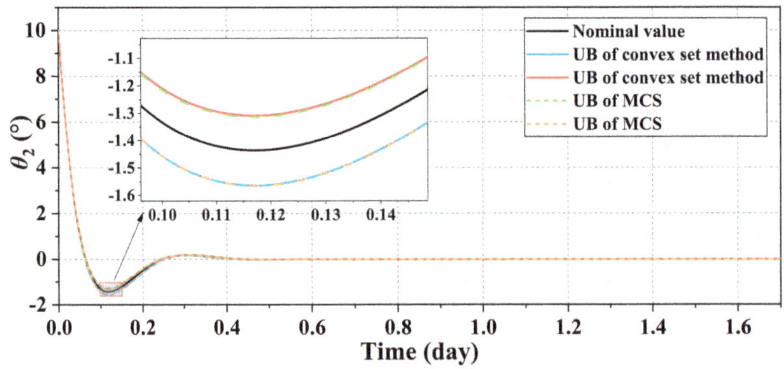

(b) Response curves of θ_2

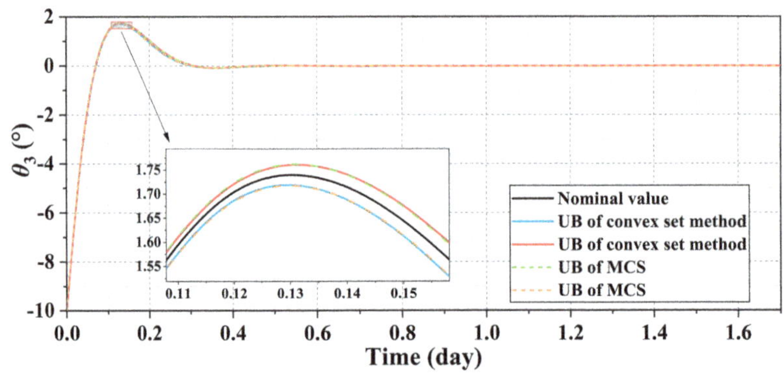

(c) Response curves of θ_3

Fig. 6.2 $\boldsymbol{\theta}$ response based on the convex set method and MCS

The initial population size is set to 10, and a maximum of 100 evaluations is performed to obtain the optimization results. By setting the critical value $\theta_3^{cri} = 1.769°$, the reliability can be validated, as illustrated in Fig. 6.2c. The computed CSTDR value of θ_3 is 0.9652, as shown in Table. 6.3, satisfying the constraint condition of 0.90. The corresponding reliability obtained via MCS is 0.9855.

The state responses, including θ_1, θ_2, θ_3, ω_1, ω_2, and ω_3, are shown in Figs. 6.2 and 6.3. It can be observed that the introduced approach provides accurate estimates of the state-response bounds.

According to Eq. (5.6), the disturbance exhibits periodic characteristics. Figure 6.4 displays the variation curves of the control input with the **Q**,**R** been optimized. As observed in Fig. 6.4, the control input also varies periodically with the same frequency as the disturbance. Consequently, the control input effectively compensates for the influence of the disturbance.

Figure 6.5 demonstrates that the variations in the cost function remain within the predicted range, even when the control system experiences fluctuations in specific parameters, as illustrated in Figs. 6.2 and 6.3. This suggests that the proposed method can substantially enhance control performance and provide accurate and rapid estimation and feedback while maintaining high precision.

Table 6.3 Results of SPS control system with optimization

Method	CSB	MCS
UB of J	3.5821×10^{14}	3.6522×10^{14}
LB of J	2.0200×10^{14}	2.1072×10^{14}
Reliability	**0.9652**	**0.9855**
Runtime(s)	**0.976**	**857.458**

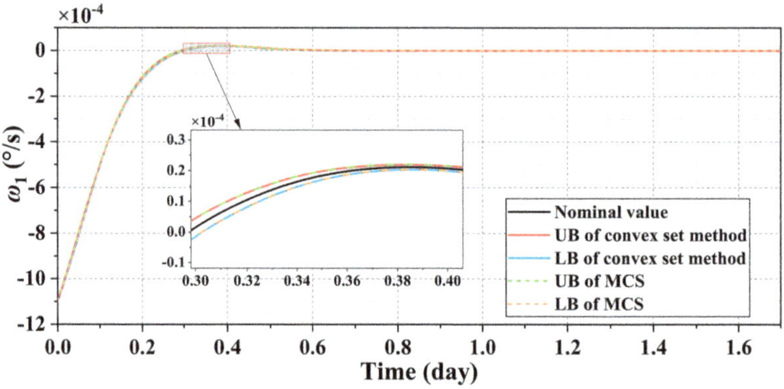

(a) Response curves of ω_1

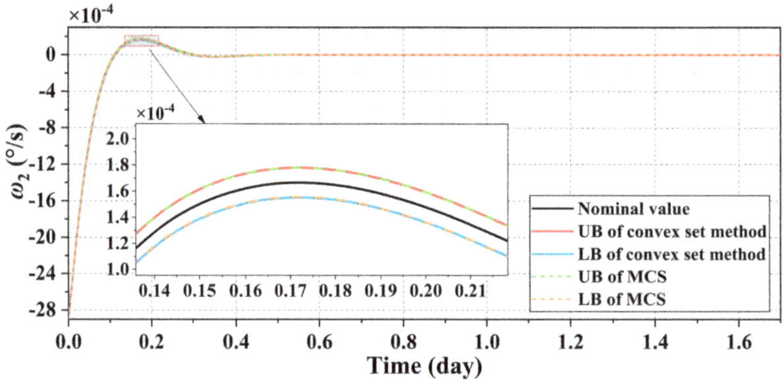

(b) Response curves of ω_2

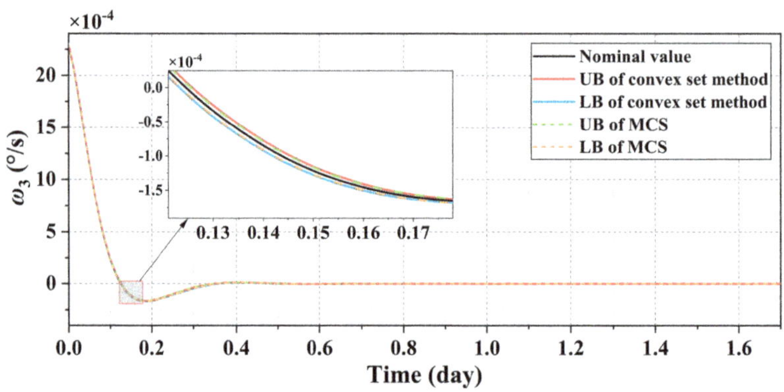

(c) Response curves of ω_3

Fig. 6.3 $\boldsymbol{\omega}$ response curves based on the convex set method and MCS

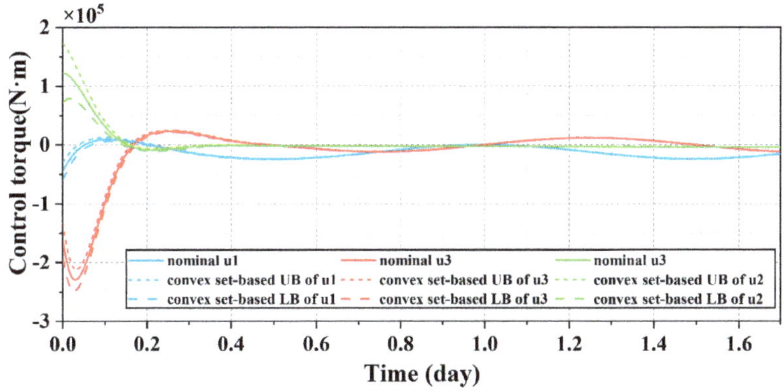

Fig. 6.4 Dynamic Curves of Input **u**

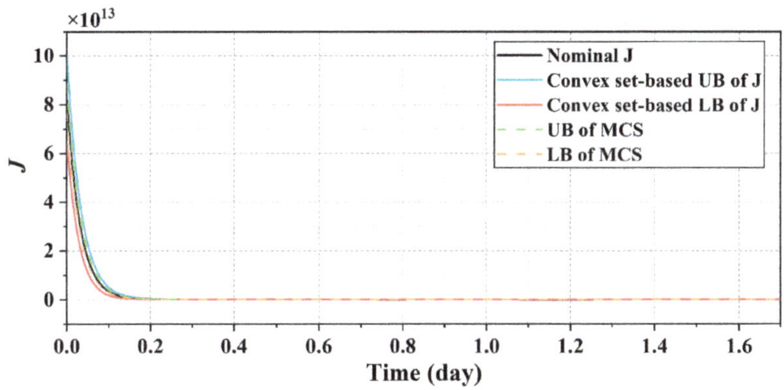

Fig. 6.5 Envelop curves of the control cost

6.6 Summary

In this chapter, a CSB optimal attitude control strategy for SPS was developed. The method employs convex set modeling to characterize bounded and correlated uncertainties, overcoming the limitations of traditional interval analysis, which cannot fully capture parameter dependencies. By integrating CSB Riccati optimization with the proposed CSTDR, the controller ensures both optimal performance and a guaranteed level of reliability that evolves throughout the mission duration. The approach is rooted in non-probabilistic convex set theory and optimal control principles. To optimize control cost under reliability constraints, a DE algorithm is employed. Numerical simulations verify that the proposed method attains accuracy comparable to MCS while substantially lowering computational expense.

References

Banerjee A, Abu-Mahfouz I (2014) A comparative analysis of particle swarm optimization and differential evolution algorithms for parameter estimation in nonlinear dynamic systems. Chaos Solitons Fractals 58:65–83. https://doi.org/10.1016/j.chaos.2013.11.004

Fu C, Zheng Z, Zhu W, Lu K, Yang Y (2022) Non-intrusive frequency response analysis of nonlinear systems with interval uncertainty: a comparative study. Chaos Solitons Fractals 165. https://doi.org/10.1016/j.chaos.2022.112815

Hu R, Zhang D, Gu X (2022) Reliability analysis of a class of stochastically excited nonlinear Markovian jump systems. Chaos Solitons Fractals 155. https://doi.org/10.1016/j.chaos.2021.111737

Jiang C, Han X, Liu GR, Liu GP (2008) A nonlinear interval number programming method for uncertain optimization problems. Eur J Oper Res 188:1–13. https://doi.org/10.1016/j.ejor.2007.03.031

Jiang C, Han X, Lu GY, Liu J, Zhang Z, Bai YC (2011) Correlation analysis of non-probabilistic convex model and corresponding structural reliability technique. Comput Methods Appl Mech Eng 200:2528–2546. https://doi.org/10.1016/j.cma.2011.04.007

Kruse N, Wallner H, Dittus A, Boettcher L, Barke I, Speller S, Starke J, Just W (2024) Large basins of attraction for control-based continuation of unstable periodic states. Nonlinear Dyn 112:19809–19823. https://doi.org/10.1007/s11071-024-10119-7

Liao S (2013) On the numerical simulation of propagation of micro-level inherent uncertainty for chaotic dynamic systems. Chaos Solitons Fractals 47:1–12. https://doi.org/10.1016/j.chaos.2012.11.009

Liu Q-Z, Zhang L, Sun B, Xiao Y, Fan G-W (2024) Fixed-time disturbance observer-based attitude prescribed performance predictive control for flexible spacecraft. IEEE Trans Aerosp Electron Syst 60:3209–3220. https://doi.org/10.1109/TAES.2024.3359591

Ni Z, Wu C, Wu S (2022) Recursive identification of inertia tensor parameters of space solar power satellite based on distributed placement of attitude sensors. Aerosp Sci Technol 130. https://doi.org/10.1016/j.ast.2022.107872

Wang L, Zhou Z, Liu J (2022) Interval-based optimal trajectory tracking control method for manipulators with clearance considering time-dependent reliability constraints. Aerosp Sci Technol 128. https://doi.org/10.1016/j.ast.2022.107745

Wang J, Zhao Y, Zhang X, Li Z, Yao W (2024) Model-free intelligent control for space soft robotic manipulators. SPACE-Sci Technol 4. https://doi.org/10.34133/space.0120

Yang C, Fang Z, Ren H, Lu W, Xia Y (2024a) Interval uncertainty-oriented impedance force control for space manipulator with time-dependent reliability. ACTA Astronaut 222:207–218. https://doi.org/10.1016/j.actaastro.2024.06.002

Yang C, Wang Q, Lu W, Li Y (2024) Integrated uncertain optimal design strategy for truss configuration and attitude-vibration control in rigid-flexible coupling structure with interval uncertainties. Nonlinear Dyn. https://doi.org/10.1007/s11071-024-10291-w

Zhang L, Li C, Su H, Wang X, Gong C (2023) A novel linear uncertainty propagation method for nonlinear dynamics with interval process. Nonlinear Dyn 111:4425–4450. https://doi.org/10.1007/s11071-022-08084-0

Chapter 7
Set-Theory-Based Iterative Optimal Attitude Control for Periodic Satellites

Abstract This chapter establishes an optimal control framework tailored for periodic satellites operating under bounded parameter uncertainties. Addressing the limitations of conventional design methods, the methodology adopts a set-theoretical perspective to extend the periodic linear quadratic regulator (PLQR) into the interval domain. By leveraging the Chebyshev interval method, the governing periodic Riccati differential equation is reconstructed to explicitly account for parameter variations, enabling the analytical computation of interval bounds for both the periodic feedback gain and the control cost function. The aforementioned interval-based time-dependent reliability (ITDR) index is used for reliability assessment. To solve this complex constrained problem, a novel set-based iterative optimization algorithm is developed. Numerical validations demonstrate that this approach effectively mitigates the impact of periodic disturbances and uncertainties, achieving a superior balance between control precision and system reliability compared to Monte Carlo simulations (MCSs).

7.1 Introduction

Building upon the work presented in the previous chapter on uncertain LQR control for a space power satellite (SPS), this chapter focuses on a different but crucial aspect of spacecraft attitude regulation: periodicity (Zhang et al. 2020; Bai et al. 2024). Many space missions, particularly those involving multi-functional satellites that must alternately point toward the Sun and the Earth, exhibit attitude changes that follow naturally recurring cycles. These periodic dynamics create challenges that differ fundamentally from those encountered in non-periodic control scenarios, underscoring the need for specialized control strategies tailored to cyclic behavior (Wu et al. 2019; Zhu et al. 2022).

In recent years, periodic attitude control has gained notable attention due to its potential to enhance both accuracy and efficiency compared to conventional non-periodic approaches (Wie and Roithmayr 2005; Wu et al. 2016; Jiang et al. 2022; Zuo et al. 2023; Wang et al. 2024). However, embedding periodicity within control

C. Yang and Y. Xia, *Set Theory-Based Spacecraft Dynamics and Control*, Intelligent Control and Learning Systems 26, https://doi.org/10.1007/978-981-95-9179-4_7

frameworks introduces additional complexities, especially when dealing with system uncertainties. Although much of the existing research has advanced the control of non-periodic systems, the combined issue of periodic dynamics under uncertain conditions remains insufficiently addressed (Pang et al. 2019; Liu et al. 2021; Han et al. 2024). Conventional techniques, such as convex optimization, sequential convex programming, and model predictive control, often overlook the inherent periodicity of satellite motion, despite its prevalence in real-world missions (Malyuta et al. 2021).

To bridge this gap, this chapter proposes an interval-based periodic linear quadratic regulator (I-PLQR) designed specifically for uncertain SPS attitude control systems exhibiting periodic dynamics. The method leverages interval analysis to describe uncertainties as bounded parameters, thereby avoiding the drawbacks of probabilistic methods that may become unreliable with limited data samples or significant measurement noise. The control system is further strengthened by integrating adaptive polynomial-based interval decoupling analysis (AP-IDA), which improves the estimation of uncertain parameter bounds. Together, these techniques support an iterative periodic optimal control algorithm that delivers both high performance and robust behavior under uncertainty.

Through the exploration of periodic attitude dynamics in the presence of uncertainties, this chapter extends the interval uncertainty-based optimal control principles introduced earlier, offering a systematic framework to meet the cyclical operational demands of SPS missions. The proposed methods provide a pathway toward more precise and dependable satellite attitude control, addressing both deterministic and uncertain factors inherent in periodic spaceflight operations.

7.2 Uncertain Iterative Algorithm for Periodic LQR

Although the dynamic model in the previous chapter utilized a periodic disturbance, it did not explicitly consider the inherent periodicity of the system dynamics. In this chapter, a periodic control strategy is developed to account for the periodic nature of SPS operations. As SPS continuously alternates between sun-pointing and Earth-pointing modes during its orbit, the control system must be designed to handle these recurring dynamic states effectively. By incorporating the periodicity of the system into the control model, this chapter aims to enhance the precision and reliability of attitude control under uncertain conditions.

7.2.1 Nominal Iterative Algorithm for Periodic LQR

The quadratic performance index for a general discrete-time control system is defined as

$$J = \sum_{k=0}^{\infty} \left(\mathbf{x}_k^{\mathrm{T}} \mathbf{Q} \mathbf{x}_k + \mathbf{u}_k^{\mathrm{T}} \mathbf{R} \mathbf{u}_k \right) \tag{7.1}$$

where **Q** and **R** denote symmetric positive definite weighting matrices associated with the state variables and control inputs. If the discrete-time periodic linear system is stabilizable, the unique optimal control law can be calculated as (Wu et al. 2019)

$$\mathbf{u}_k^* = -\left(\mathbf{R} + \mathbf{H}_k^{\mathrm{T}} \mathbf{P}_{k+1} \mathbf{H}_k\right)^{-1} \mathbf{H}_k^{\mathrm{T}} \mathbf{P}_{k+1} \mathbf{G}_k \mathbf{x}_k \tag{7.2}$$

where $\mathbf{P}_k$ satisfies the discrete-time periodic algebraic Riccati equation (DPARE) in the infinite horizon form:

$$\mathbf{G}_k^{\mathrm{T}} \mathbf{P}_{k+1} \mathbf{G}_k - \mathbf{P}_k - \mathbf{G}_k^{\mathrm{T}} \mathbf{P}_{k+1} \mathbf{H}_k \left(\mathbf{R} + \mathbf{H}_k^{\mathrm{T}} \mathbf{P}_{k+1} \mathbf{H}_k\right)^{-1} \mathbf{H}_k^{\mathrm{T}} \mathbf{P}_{k+1} \mathbf{G}_k = -\mathbf{Q} \tag{7.3}$$

For periodic discrete systems where $\mathbf{G}_{k+T_{\mathrm{d}}} = \mathbf{G}_k$ and $\mathbf{H}_{k+T_{\mathrm{d}}} = \mathbf{H}_k$, the matrix $\mathbf{P}_k$ also exhibits periodicity:

$$\mathbf{P}_{k+T_{\mathrm{d}}} = \mathbf{P}_k \tag{7.4}$$

According to Eq. (7.4), Wu et al. (2019) proposed a forward iterative algorithm expressed as

$$\begin{cases} \mathbf{P}_k(0) = \mathbf{0}, \\ \mathbf{P}_k(q+1) = \mathbf{G}_k^{\mathrm{T}} \mathbf{P}_{k+1}(q) \mathbf{G}_k - \mathbf{G}_k^{\mathrm{T}} \mathbf{P}_{k+1}(q) \mathbf{H}_k \left(\mathbf{R} + \mathbf{H}_k^{\mathrm{T}} \mathbf{P}_{k+1}(q) \mathbf{H}_k\right)^{-1} \mathbf{H}_k^{\mathrm{T}} \mathbf{P}_{k+1}(q) \mathbf{G}_k + \mathbf{Q}, k \in [1, T_{\mathrm{d}} - 1], \\ \mathbf{P}_{T_{\mathrm{d}}}(q+1) = \mathbf{G}_{T_{\mathrm{d}}}^{\mathrm{T}} \mathbf{P}_1(q+1) \mathbf{G}_{T_{\mathrm{d}}} - \mathbf{G}_{T_{\mathrm{d}}}^{\mathrm{T}} \mathbf{P}_1(q+1) \mathbf{H}_{T_{\mathrm{d}}} \left(\mathbf{R} + \mathbf{H}_{T_{\mathrm{d}}}^{\mathrm{T}} \mathbf{P}_1(q+1) \mathbf{H}_{T_{\mathrm{d}}}\right)^{-1} \mathbf{H}_{T_{\mathrm{d}}}^{\mathrm{T}} \mathbf{P}_1(q+1) \mathbf{G}_{T_{\mathrm{d}}} + \mathbf{Q}, \\ \mathbf{P}_{T_{\mathrm{d}}+1} = \mathbf{P}_1 \end{cases} \tag{7.5}$$

where q denotes the iteration index. The convergence criterion for this process is defined as

$$\|\mathbf{P}_k(q+1) - \mathbf{P}_k(q)\| / \|\mathbf{P}_k(q)\| \le \varepsilon, \ \forall k \in [1, T_{\mathrm{d}} - 1] \tag{7.6}$$

where $\| \ \|$ represents the norm operator, and ε is the predefined tolerance for convergence. When this condition is satisfied, the iteration converges and yields the solution $\mathbf{P}_k$ of Eq. (7.5); otherwise, the algorithm proceeds with further iterations by setting $q = q + 1$.

7.2.2 Adaptive Polynomial-Based Interval Decoupling Analysis

Due to the propagation of uncertainties, system parameters, including the state variables $\mathbf{x}_k$, the solution $\mathbf{P}_k$ of DPARE, and the control cost J, all exhibit uncertain characteristics. The classical interval perturbation method described in Sect. 2.3 provides good accuracy for problems involving small uncertainties. However, when uncertainties become significant, this traditional approach may fail to accurately capture the uncertainty bounds because it neglects the influence of high-order terms. To tackle this limitation, this chapter proposes an AP-IDA method designed to overcome the deficiencies of the classical approach. Without loss of generality, the procedure of AP-IDA for determining the interval bounds of uncertain variables is illustrated using the example of the kth sampling instant of a generic state variable x_k.

First, the initial uncertain parameter vector is defined as $\mathbf{b}^{\mathrm{I}} = \left[b_1^{\mathrm{I}}, \ldots, b_i^{\mathrm{I}}, \ldots, b_m^{\mathrm{I}}\right]$. There exists an implicit function between the initial uncertainty of the input parameters and the output parameter, which is described by Eq. (7.7):

$$x_k = \mathbf{\Gamma}(\mathbf{b}) \tag{7.7}$$

The implicit function in Eq. (7.7) defines the actual relationship between the input parameter $\mathbf{b}$ and the output parameter x_k. In most cases, this function is highly nonlinear or difficult to express in explicit form, making direct analysis impractical. To address this challenge, the AP-IDA method uses adaptive polynomials to approximate the response curves for each input–output dimension pair, thereby obtaining an approximate polynomial representation.

A total of s sample points are selected from the true response curves for each input dimension across the corresponding output dimension. The input sampling follows an equidistant strategy given by Eq. (7.8):

$$b_i^{(j)} = \underline{b_i} + 2\frac{j-1}{s-1}\Delta b_i,\ j = 1, 2, \ldots s \tag{7.8}$$

where the superscript $^{(j)}$ denotes the jth sampled point.

Each row in Eq. (7.9) represents a sampled input vector for the ith input parameter, where the sampled value $b_i^{(j)}$ replaces the central value of the corresponding component, while the remaining parameters retain their central values:

$$\mathbf{B}_i = \begin{bmatrix} \mathbf{b}_i^{(1)} \\ \vdots \\ \mathbf{b}_i^{(s)} \end{bmatrix} = \begin{bmatrix} b_1^{\mathrm{c}} & \cdots & b_{i-1}^{\mathrm{c}} & b_i^{(1)} & b_{i+1}^{\mathrm{c}} & \cdots & b_m^{\mathrm{c}} \\ \vdots & & \vdots & \vdots & \vdots & \vdots & \vdots \\ b_1^{\mathrm{c}} & \cdots & b_{i-1}^{\mathrm{c}} & b_i^{(s)} & b_{i+1}^{\mathrm{c}} & \cdots & b_m^{\mathrm{c}} \end{bmatrix} \tag{7.9}$$

Substituting the sampled input vectors into Eq. (7.7) produces the corresponding sampled output values:

$$x_k^{(j)} = \Gamma\left(\mathbf{b}_i^{(j)}\right), j = 1, 2, \dots s \tag{7.10}$$

Once the input–output sampling points are obtained, adaptive orthogonal polynomials are used to approximate the true functional relationship. The recurrence relations for the adaptive orthogonal polynomials are described as

$$\begin{cases} p_{-1}(x) = 0,\ p_1(x) = 1 \\ p_{n+1}(x) = (x - a_n)p_n(x) - b_n p_{n-1}(x),\ n = 0, 1, 2, \dots \end{cases} \tag{7.11}$$

where $p_n(x)$ is the nth order adaptive orthogonal polynomial with x as the independent variable. The coefficients a_n and b_n are calculated as

$$\begin{cases} a_n = \frac{(xp_n(x), p_n(x))}{(p_n(x), p_n(x))},\ n = 0, 1, 2, \dots \\ b_n = \frac{(p_n(x), p_n(x))}{(p_{n-1}(x), p_{n-1}(x))},\ n = 1, 2, 3 \dots \end{cases} \tag{7.12}$$

where (,) denotes the inner product operation.

Combining Eqs. (7.11) and (7.12) yields the nth order adaptive polynomial. Based on this, Eq. (7.13) defines the approximate polynomial model that relates the input parameter b_i to the output parameter x_k:

$$\hat{x}_k = \hat{\Gamma}(b_i) = c_0 p_0(b_i) + c_1 p_1(b_i) + \dots + c_n p_n(b_i) \tag{7.13}$$

where the superscript $\wedge$ indicates an approximation, and $c_0, \dots, c_n$ are the polynomial coefficients calculated as

$$c_k = \frac{(x_k, p_k(b_i))}{(p_k(b_i), p_k(b_i))},\ k = 0, 1, \dots, n \tag{7.14}$$

By substituting Eqs. (7.14) into (7.13), an approximate polynomial model is constructed. The local extremal points of the approximate polynomial can be found by solving $\frac{d\hat{x}_k}{db_i} = 0$, yielding $n - 1$ approximate local extremal points $\hat{\mathbf{b}}_{i,\text{root}} = \left[\hat{b}_{i,\text{root}}^{(1)}, \hat{b}_{i,\text{root}}^{(2)}, \dots \hat{b}_{i,\text{root}}^{(n-1)}\right]$ for the output parameter x_k relative to the input parameter b_i. The approximate global extrema of x_k with respect to b_i are then determined using the local extremal points and the input parameter boundaries as given by Eq. (7.15):

$$\begin{cases} \hat{\Gamma}\left(\hat{b}_{i,\max}\right) = \max\limits_{b_i \in \{\underline{b_i}, \hat{\mathbf{b}}_{i,root}, \overline{b_i}\}} \hat{\Gamma}(b_i) \\ \hat{\Gamma}\left(\hat{b}_{i,\min}\right) = \min\limits_{b_i \in \{\underline{b_i}, \hat{\mathbf{b}}_{i,root}, \overline{b_i}\}} \hat{\Gamma}(b_i) \end{cases} \tag{7.15}$$

Equation (7.15) thus provides the approximate global extremal relationship between the ith input and the kth output parameter. By applying this procedure to all input parameters, the approximate global extremal points for the output parameter are obtained as

$$\begin{cases} \hat{\mathbf{b}}_{\max} = \left[\hat{b}_{1,\max}, \hat{b}_{2,\max}, \ldots, \hat{b}_{m,\max}\right] \\ \hat{\mathbf{b}}_{\min} = \left[\hat{b}_{1,\min}, \hat{b}_{2,\min}, \ldots, \hat{b}_{m,\min}\right] \end{cases} \tag{7.16}$$

The UB/LB of the kth output parameter x_k can then be computed as

$$\overline{x_k} = \Gamma\left(\hat{\mathbf{b}}_{\max}\right), \; \underline{x_k} = \Gamma\left(\hat{\mathbf{b}}_{\min}\right) \tag{7.17}$$

By sequentially determining the UB/LB for all output dimensions, the interval boundaries of the response vector $\mathbf{x}$ can be defined as

$$\begin{aligned} \overline{\mathbf{x}} &= \left[\overline{x_1}, \overline{x_2}, \ldots, \overline{x_N}\right] \\ \underline{\mathbf{x}} &= \left[\underline{x_1}, \underline{x_2}, \ldots, \underline{x_N}\right] \end{aligned} \tag{7.18}$$

Using the same approach, the interval boundaries of other system quantities, including the solution $\mathbf{P}_k$ of DPARE and the control cost J, can be calculated.

7.2.3 Interval-Uncertainty-Based Periodic LQR

Considering both uncertainty and periodicity in the system, an I-PLQR is developed. According to interval theory, the DPARE defined in Eq. (7.5) is extended to incorporate interval uncertainty as follows:

$$\mathbf{P}_k^{\mathrm{I}} = \left(\mathbf{G}_k^{\mathrm{I}}\right)^{\mathrm{T}} \mathbf{P}_{k+1}^{\mathrm{I}} \mathbf{G}_k^{\mathrm{I}} - \left(\mathbf{G}_k^{\mathrm{I}}\right)^{\mathrm{T}} \mathbf{P}_{k+1}^{\mathrm{I}} \mathbf{H}_k^{\mathrm{I}} \left(\mathbf{R} + \left(\mathbf{H}_k^{\mathrm{I}}\right)^{\mathrm{T}} \mathbf{P}_{k+1}^{\mathrm{I}} \mathbf{H}_k^{\mathrm{I}}\right)^{-1} \left(\mathbf{H}_k^{\mathrm{I}}\right)^{\mathrm{T}} \mathbf{P}_{k+1}^{\mathrm{I}} \mathbf{G}_k^{\mathrm{I}} + \mathbf{Q} \tag{7.19}$$

where $\mathbf{G}_k^{\mathrm{I}}$ and $\mathbf{H}_k^{\mathrm{I}}$ represent the interval values of the uncertain matrices $\mathbf{G}_k$ and $\mathbf{H}_k$, respectively. The term $\mathbf{P}_k^{\mathrm{I}}$ denotes the infinite-horizon solution of the DPARE under interval uncertainty, which is computed using the iterative process:

$$\begin{cases} \mathbf{P}_k^{\mathrm{I}}(0) = \mathbf{0} \Leftrightarrow \begin{cases} \mathbf{P}_k^{\mathrm{c}}(0) = \mathbf{0} \\ \Delta\mathbf{P}_k(0) = \mathbf{0} \end{cases}, \\ \mathbf{P}_k^{\mathrm{I}}(q+1) = \left(\mathbf{G}_k^{\mathrm{I}}\right)^{\mathrm{T}}\mathbf{P}_{k+1}^{\mathrm{I}}(q)\mathbf{G}_k^{\mathrm{I}} \\ \quad -\left(\mathbf{G}_k^{\mathrm{I}}\right)^{\mathrm{T}}\mathbf{P}_{k+1}^{\mathrm{I}}(q)\mathbf{H}_k^{\mathrm{I}}\left(\mathbf{R}+\left(\mathbf{H}_k^{\mathrm{I}}\right)^{\mathrm{T}}\mathbf{P}_{k+1}^{\mathrm{I}}(q)\mathbf{H}_k^{\mathrm{I}}\right)^{-1}\left(\mathbf{H}_k^{\mathrm{I}}\right)^{\mathrm{T}}\mathbf{P}_{k+1}^{\mathrm{I}}(q)\mathbf{G}_k^{\mathrm{I}} + \mathbf{Q}, i \in [1, T_{\mathrm{d}}-1], \\ \mathbf{P}_{T_{\mathrm{d}}}^{\mathrm{I}}(q+1) = \left(\mathbf{G}_{T_{\mathrm{d}}}^{\mathrm{I}}\right)^{\mathrm{T}}\mathbf{P}_1^{\mathrm{I}}(q+1)\mathbf{G}_{T_{\mathrm{d}}}^{\mathrm{I}} \\ \quad -\left(\mathbf{G}_{T_{\mathrm{d}}}^{\mathrm{I}}\right)^{\mathrm{T}}\mathbf{P}_1^{\mathrm{I}}(q+1)\mathbf{H}_{T_{\mathrm{d}}}^{\mathrm{I}}\left(\mathbf{R}+\left(\mathbf{H}_{T_{\mathrm{d}}}^{\mathrm{I}}\right)^{\mathrm{T}}\mathbf{P}_1^{\mathrm{I}}(q+1)\mathbf{H}_{T_{\mathrm{d}}}^{\mathrm{I}}\right)^{-1}\left(\mathbf{H}_{T_{\mathrm{d}}}^{\mathrm{I}}\right)^{\mathrm{T}}\mathbf{P}_1^{\mathrm{I}}(q+1)\mathbf{G}_{T_{\mathrm{d}}}^{\mathrm{I}} + \mathbf{Q}, \\ \mathbf{P}_{T_{\mathrm{d}}+1}^{\mathrm{I}} = \mathbf{P}_1^{\mathrm{I}} \Leftrightarrow \begin{cases} \mathbf{P}_{T_{\mathrm{d}}+1}^{\mathrm{c}} = \mathbf{P}_1^{\mathrm{c}} \\ \Delta\mathbf{P}_{T_{\mathrm{d}}+1} = \Delta\mathbf{P}_1^{\mathrm{c}} \end{cases} \end{cases} \tag{7.20}$$

Inspired by the nominal convergence condition in Eq. (7.6), the convergence criterion for the proposed uncertain iterative algorithm in periodic optimal attitude control is defined as

$$\frac{\left\|\overline{\mathbf{P}_k}(q+1) - \overline{\mathbf{P}_k}(q)\right\| + \left\|\underline{\mathbf{P}_k}(q+1) - \underline{\mathbf{P}_k}(q)\right\|}{2\left\|\mathbf{P}_k^{\mathrm{c}}(q)\right\|} \le \varepsilon, \ \forall k \in [1, T_{\mathrm{d}}-1] \tag{7.21}$$

where $\underline{\mathbf{P}_k}(q)$ and $\overline{\mathbf{P}_k}(q)$ denote the LB and UB of $\mathbf{P}_k^{\mathrm{I}}(q)$, respectively, and $\mathbf{P}_k^{\mathrm{c}}(q) = \left(\overline{\mathbf{P}_k}(q) + \underline{\mathbf{P}_k}(q)\right)/2$ is the nominal value of $\mathbf{P}_k^{\mathrm{I}}(q)$. The tolerance ε is consistent with that defined in Eq. (7.6).

As the iterations progress, both $\underline{\mathbf{P}_k}(q)$ and $\overline{\mathbf{P}_k}(q)$ converge gradually. When Eq. (7.21) is first satisfied, the infinite-horizon interval solution of the DPARE described in Eq. (7.19) is achieved:

$$\mathbf{P}_k^{\mathrm{I}} = \left[\underline{\mathbf{P}_k}(q+1), \overline{\mathbf{P}_k}(q+1)\right], \ \forall k \in [1, N] \tag{7.22}$$

Here N denotes the total count of sampling steps.

Once $\mathbf{P}_k^{\mathrm{I}}$ is obtained, the interval state-feedback gain matrix $\mathbf{K}_k^{\mathrm{I}}$ can be calculated as

$$\mathbf{K}_k^{\mathrm{I}} = \left(\mathbf{R} + (\mathbf{H}_k^{\mathrm{I}})^{\mathrm{T}}\mathbf{P}_{k+1}^{\mathrm{I}}\mathbf{H}_k^{\mathrm{I}}\right)^{-1}(\mathbf{H}_k^{\mathrm{I}})^{\mathrm{T}}\mathbf{P}_{k+1}^{\mathrm{I}}\mathbf{G}_k^{\mathrm{I}} \tag{7.23}$$

The iterative equation for the interval state variable $\mathbf{x}_k^{\mathrm{I}}$ is expressed as

$$\mathbf{x}_{k+1}^{\mathrm{I}} = (\mathbf{G}_k^{\mathrm{I}} - \mathbf{H}_k^{\mathrm{I}}\mathbf{K}_k^{\mathrm{I}})\mathbf{x}_k^{\mathrm{I}} \tag{7.24}$$

The feedback control law with interval uncertainty is then obtained as

$$\mathbf{u}_k^{*,\mathrm{I}} = -\mathbf{K}_k^{\mathrm{I}}\mathbf{x}_k^{\mathrm{I}} \tag{7.25}$$

Accordingly, the interval of the control cost is expressed by

$$J^{\mathrm{I}} = \sum_{k=0}^{\infty}\left(\left(\mathbf{x}_k^{\mathrm{I}}\right)^{\mathrm{T}}\mathbf{Q}\mathbf{x}_k^{\mathrm{I}} + \left(\mathbf{u}_k^{\mathrm{I}}\right)^{\mathrm{T}}\mathbf{R}\mathbf{u}_k^{\mathrm{I}}\right) \tag{7.26}$$

where $\mathbf{u}_k^{\mathrm{I}} = \mathbf{u}_k^0 + \mathbf{u}_k^{*,\mathrm{I}}$ represents the total input of the system, which incorporates interval uncertainty.

7.3 Design of Reliability-Based Optimal Attitude Control

7.3.1 Optimization Problem

By employing the reliability obtained through the ITDR method as a constraint, treating the weighting matrices of the LQR approach as decision variables, and aiming to minimize the total control cost from both feedforward and feedback components, an optimization model for periodic SPS attitude control is formulated as

$$\begin{aligned} &\text{find} && \mathbf{Q}, \mathbf{R} \\ &\min && J^{\mathrm{c}} = J^0 + \left(J^*\right)^{\mathrm{c}} \\ &\text{s.t.} && Rel(N) \geq \text{poss} \end{aligned} \tag{7.27}$$

where J^0 and $(J^*)^{\mathrm{c}}$ denote the nominal control costs arising from the feedforward and feedback control actions, respectively, while J^{c} represents their combined nominal control cost. The term poss specifies the required reliability threshold.

7.3.2 Uncertain Single-Objective Optimization

In recent advances in single-objective optimization methods, the improved multi-operator differential evolution (IMODE) algorithm (Zuo et al. 2023), as an enhanced multi-operator differential evolution approach, has shown strong convergence and diverse performance across various benchmark test sets for unconstrained single-objective optimization problems. Compared with traditional differential evolution methods, IMODE introduces improved multi-operators and adjusts their application ratios dynamically according to their effectiveness at each iteration, enabling the algorithm to adapt well to different types of optimization tasks. Furthermore, IMODE gradually reduces the population size to accelerate convergence and employs the sequential quadratic programming local search strategy during the later stages of iterations to refine solutions, thereby improving accuracy. In this chapter, because constraints are present, each candidate solution is classified as either feasible or infeasible. Owing to its robust global search capability, the IMODE algorithm is suitable for addressing uncertainty-based periodic LQR single-objective optimization problems in spacecraft attitude control.

7.4 Numerical Example

Abacus SPS is introduced as a test case to show the effectiveness of the given periodic optimal attitude control strategy. The parameters of the 1.2 GW Abacus SPS are described as follows: The square-shaped solar array has a side length of 3.2 km and a mass of 21×10^6 kg. The antenna has a diameter of 500 m, while the rotating reflector measures 500×700 m and has a mass of 0.8×10^6 kg. The system's principal moments of inertia are $I_1 = 2.8 \times 10^{13}$ kg.m^2, $I_2 = 1.8 \times 10^{13}$ kg.m^2, and $I_3 = 4.5 \times 10^{13}$ kg.m^2, respectively. The associated uncertainty values are $\Delta I_1 = 5.6 \times 10^{11}$kg.m^2, $\Delta I_2 = 3.6 \times 10^{11}$ kg.m^2, and $\Delta I_3 = 9 \times 10^{11}$kg.m^2, respectively. The initial states are specified as $\hat{\mathbf{x}}_{13}^{c}(0) = [\theta_1, \theta_3, \omega_1, \omega_3]^{\mathrm{T}} = \left[10, -10, -1.2 \times 10^{-3}, 2.3 \times 10^{-3}\right]^{\mathrm{T}}$, and $\hat{\mathbf{x}}_{2}^{c}(0) = [\theta_2, \omega_2]^{\mathrm{T}} = \left[10, -2.8 \times 10^{-3}\right]^{\mathrm{T}}$, where θ and ω are expressed in degrees and degrees per second, respectively. The initial state variables are assumed to be free of uncertainties. The system operates with a sampling step size of $T/100$, and the total duration of the simulation is set to $3.5T$.

In the AP-IDA approach, five sampling points are selected, and a third-order adaptive polynomial is utilized for function fitting. The state variable θ_3 is chosen as the observable state, with an allowable limit of $\theta_3^{\mathrm{cr}} = 3.3°$. To ensure safety, the reliability threshold is defined as 0.9. The convergence tolerance ε in Eqs. (7.6) and (7.20) is specified as 1×10^{-6}. In this work, the I-PLQR method is adopted and compared against the interval-based non-periodic linear quadratic regulator (I-NPLQR) proposed in Chap. 5 and 10,000 MCS. In designing the optimal controller, I-NPLQR applies a non-periodic LQR strategy, whereas MCS employs the nominal periodic LQR approach. The reliability of both I-PLQR and I-NPLQR is evaluated using the ITDR method. The computational experiments are conducted on a system equipped with a 3.00 GHz Intel Core i9-13900KF 32-core processor. By applying the IMODE algorithm to optimize the model presented in Eq. (7.27), the final parameters of the LQR optimal control design are determined as follows: $\boldsymbol{Q}_{13} = \mathrm{diag}(4.50, 4.92, 5.09, 2.97) \times 10^{11}$, $\boldsymbol{R}_{13} = \mathrm{diag}(0.9, 0.9)$, $\boldsymbol{Q}_2 = diag(1.0, 1.0) \times 10^{10}$ and $R_2 = 0.01$.

According to Eq. (7.5), the I-PLQR method requires a coupled iterative procedure across all time steps when solving $\mathbf{P}_i(m)$. For each time step, the iteration continues until the solutions at all time steps reach convergence. Figure 7.1 depicts the convergence progress of the iterations at each time step. Since 100 sampling points are used for attitude control within each period, it can be observed from Fig. 7.1 that the number of convergence steps for each time step also exhibits periodic behavior with a period of 100. The entire system achieves convergence after 141 iterations. To further demonstrate the convergence of the matrix $\mathbf{P}_i(m)$, four representative

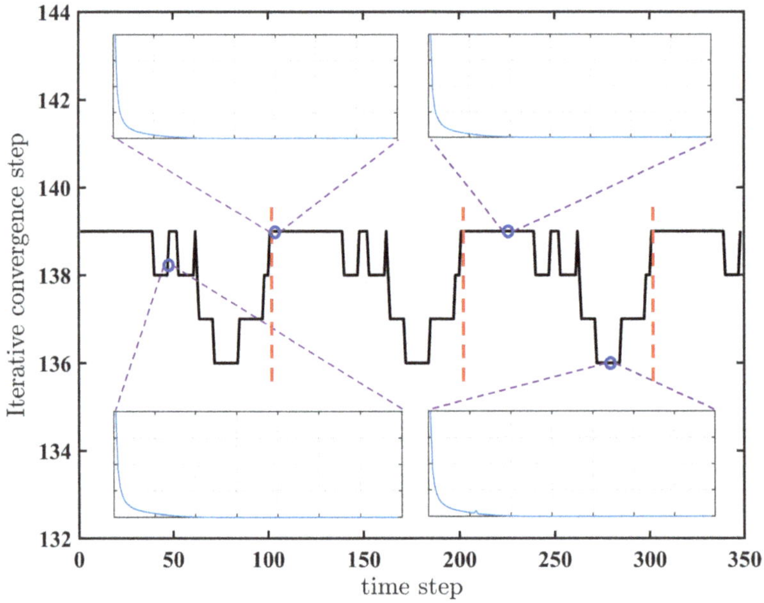

Fig. 7.1 Iterative convergence steps for each time step

sampling points are selected: the 1st, 26th, 51st, and 76th points within each period. Their corresponding convergence curves are presented in Fig. 7.1. The results show that all selected time steps under the I-PLQR method converge at a comparable rate and progressively meet the specified error constraints.

Using the final optimal matrices $\boldsymbol{Q}$ and $\boldsymbol{R}$, the results produced by the I-PLQR, I-NPLQR, and MCS methods are summarized in Table 7.1. The quantified reliabilities for both I-PLQR and I-NPLQR are evaluated using the I-TVR method. The corresponding state responses of θ_3 obtained from these three approaches are depicted in Fig. 7.2. From Table 7.1, it is observed that the uncertainty interval bounds of the control cost derived from the I-PLQR method are highly aligned with those obtained by the MCS method. The reliability calculated by the I-PLQR method is 0.9210, slightly lower than the value of 0.9968 from the MCS method. This difference is mainly due to the assumption in I-PLQR that the response curve of θ_3 is uniformly distributed across its uncertainty interval. In contrast, the MCS results indicate that the response of θ_3 is concentrated near its central value and relatively sparse near the boundaries. As a result, the reliability obtained by I-PLQR is marginally lower than that of the MCS method. Nevertheless, the reliability computed by I-PLQR still exhibits good accuracy and practical reference value, with both values exceeding the initial reliability threshold of 0.9. Because of the periodic characteristics of the system, the I-PLQR and MCS methods employing the periodic LQR strategy show better overall performance compared to the nominal results produced by the

I-NPLQR method, which uses a non-periodic LQR approach. Table 7.1 further indicates that the nominal control cost obtained through I-NPLQR is higher than that from I-PLQR. Moreover, Fig. 7.2 reveals that the peak value in the θ_3 response calculated using I-NPLQR proposed in Chap. 5 is greater than that obtained by I-PLQR and surpasses the allowable limit θ_3^{cr}. Therefore, the I-NPLQR results fail to meet the reliability requirement, confirming the superior control performance of the periodic iterative approach.

Regarding algorithm execution time, both I-PLQR and I-NPLQR adopt a serial computation strategy, whereas MCS is executed in parallel using 32 cores. As shown in Table 7.1, the computation time of the I-PLQR method is longer than that of the I-NPLQR method. This difference can be attributed to two primary factors. First, the I-PLQR method applies periodic LQR control, which requires iterative calculations of the matrix $\mathbf{P}_i(m)$ across all time steps. The time steps are interdependent,

Table 7.1 The main indicators obtained by I-PLQR, I-NPLQR, and MCS, respectively

		I-PLQR	I-NPLQR	MCS
$J^*(\times 10^{14})$	Upper bound	4.9903	5.1504	4.9851
	Nominal value	4.7900	4.8181	4.7900
	Lower bound	4.5967	4.4805	4.6055
	Reliability	0.9210	Failure	0.9968
	Runtime (s)	19.2547	0.6584	4125.3

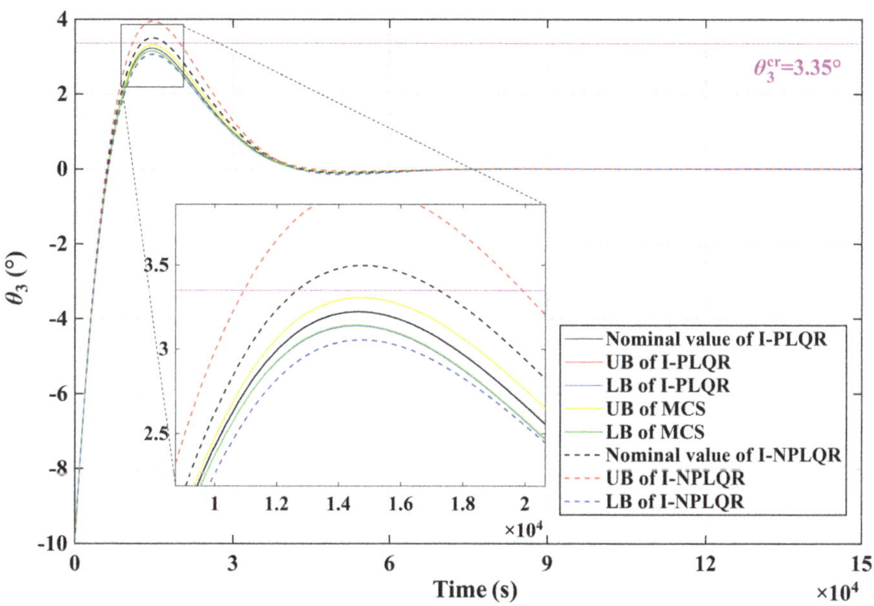

Fig. 7.2 Comparison of the system response curves θ_3 using I-PLQR, I-NPLQR, and MCS

and convergence must be achieved simultaneously for all of them. In contrast, the I-NPLQR method employs the conventional LQR approach, which does not require iterative calculations for each time step. While the periodic LQR method in I-PLQR leads to higher computational effort, it achieves lower control costs and superior control performance. Second, the I-PLQR method incorporates the AP-IDA procedure, where each output dimension requires adaptive polynomial fitting. This fitting process is computationally demanding and contributes to the increased computation time. By comparison, the I-NPLQR method uses the interval perturbation technique to estimate uncertainty intervals more rapidly. However, as illustrated in Fig. 7.2, the conservativeness of the interval perturbation method and the strong system coupling result in wider estimated intervals for I-NPLQR. The AP-IDA approach mitigates these issues and provides more accurate interval estimations. Despite the slightly longer computation time, the I-PLQR method demonstrates high efficiency. Under comparable control performance and uncertainty estimation accuracy, its execution time is only 0.5% of the 4125.3 s required by 10,000 MCS with parallel processing. In summary, while the I-PLQR method increases computation time to some extent, it delivers substantial improvements in both control performance and the precision of uncertainty estimation.

Apart from the constrained state variable θ_3, the responses of the other state variables, namely θ_1, θ_2, ω_1, ω_2, and ω_3, are presented in Figs. 7.3, 7.4, 7.5, 7.6 and 7.7. The results show that under uncertainties in the attitude dynamics model, the responses of all these state variables eventually converge to zero. The response intervals calculated using the I-PLQR method fully cover the state responses obtained by the MCS method, and the intervals demonstrate very high accuracy. The disturbance torques $d_1(t)$ and $d_3(t)$ are periodic. Consequently, the control inputs $\mathbf{u}_1(t)$ and $\mathbf{u}_3(t)$ derived using the I-PLQR method for these two state-space equations do not converge to zero; instead, they oscillate periodically, following the same pattern as the disturbances.

Figure 7.8 illustrates the input curves generated by the proposed method using the optimized $\boldsymbol{Q}$ and $\boldsymbol{R}$ parameters. Furthermore, corresponding to the control cost intervals summarized in Table 7.1, Fig. 7.9 presents the control cost curves for the

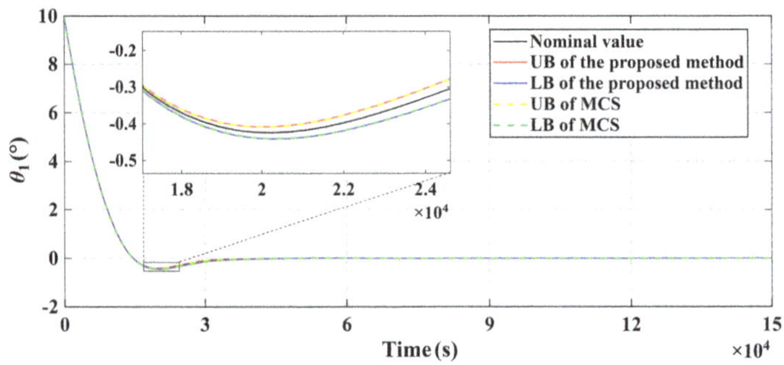

Fig. 7.3 Interval response curve of state variable θ_1

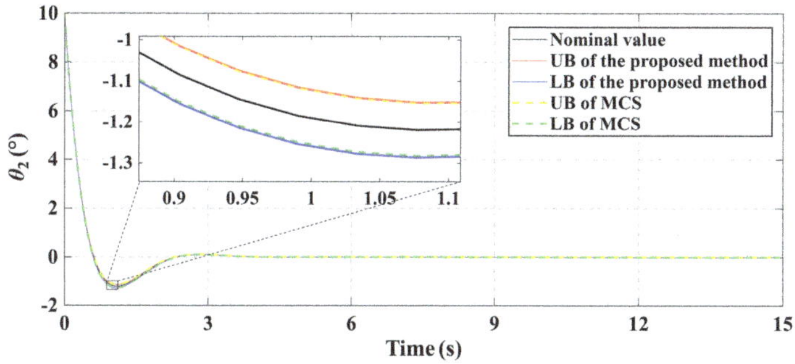

Fig. 7.4 Interval response curve of state variable θ_2

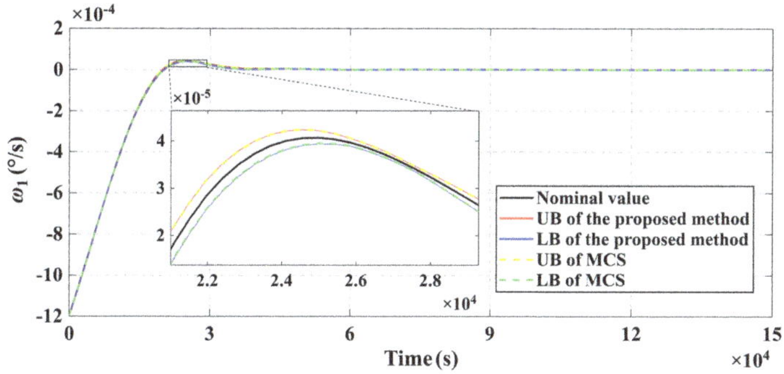

Fig. 7.5 Interval response curve of state variable ω_1

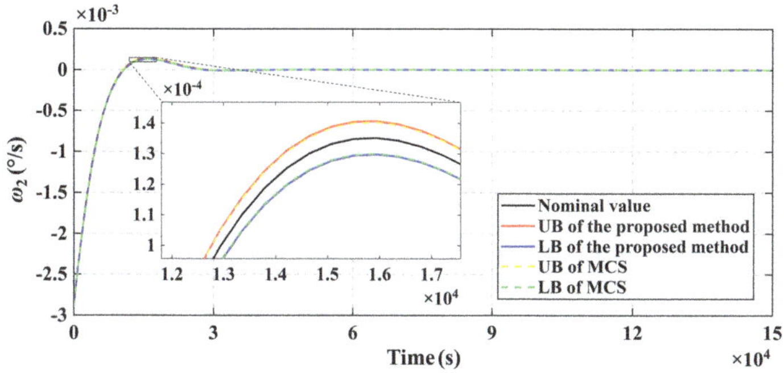

Fig. 7.6 Interval response curve of state variable ω_2

proposed I-PLQR method and the I-NPLQR method. The UB and LB of the proposed approach are computed using the AP-IDA technique. As observed in Fig. 7.9, the use of the periodic iterative LQR method leads to lower control costs. Additionally, when compared with the MCS results, the control cost boundaries estimated by the AP-IDA method display an exceptionally high degree of accuracy.

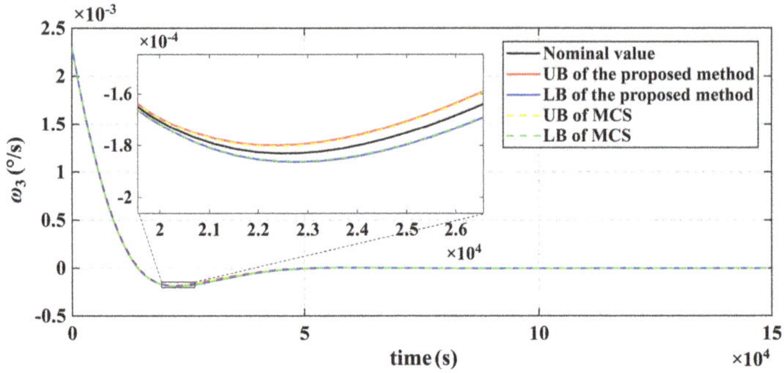

Fig. 7.7 Interval response curve of state variable ω_3

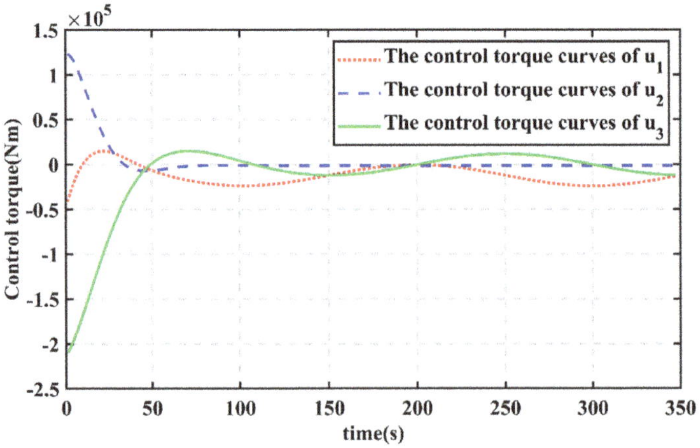

Fig. 7.8 The control torque curves of $\mathbf{u}_1$, $\mathbf{u}_2$, and $\mathbf{u}_3$

7.5 Summary

This chapter presents a novel approach to attitude control for SPS under bounded interval uncertainties and reliability constraints, focusing on the periodic nature of satellite operations. By utilizing an I-PLQR method, the chapter addresses both uncertainty and periodicity in satellite dynamics and control. An AP-IDA is introduced to accurately predict the uncertainty bounds of the period LQR controller. To quantify the safety of the attitude response, the ITDR method is introduced. The IMODE algorithm is utilized for reliability-constrained optimization of the control system. Compared to conventional probabilistic methods and non-periodic interval uncertainty-based strategies, the proposed approach offers improved control precision, reduced costs, and enhanced reliability, verified through numerical simulations and practical applications. This framework provides an efficient solution to the challenges of optimal control in periodic SPS systems with limited sample data.

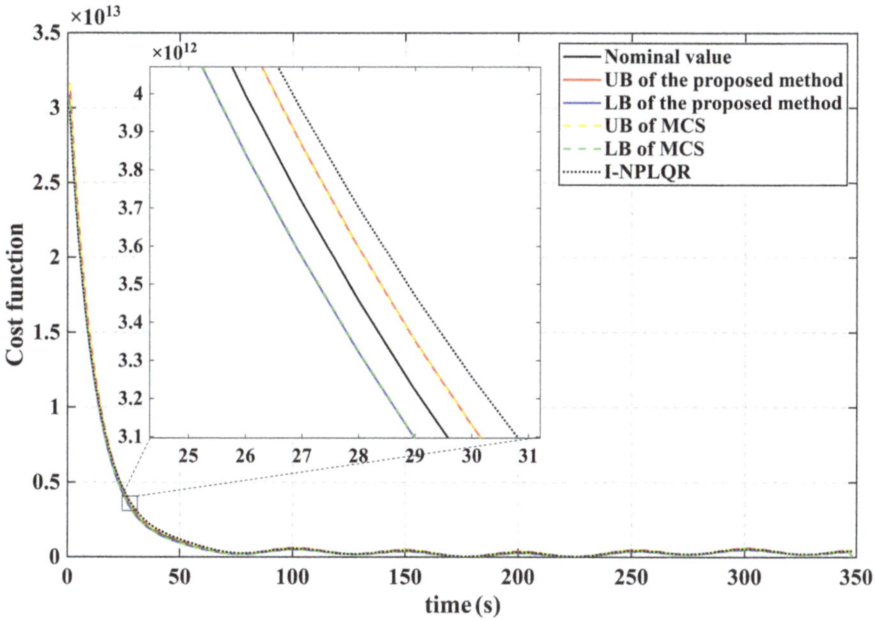

Fig. 7.9 Bound curves of the cost function J^*

References

Bai S, Chen Y, Zhang Y (2024) Prediction method of agile satellite's in-view periods based on analytical bounding curves. IEEE Trans Aerosp Electron Syst 60:1660–1678. https://doi.org/10.1109/TAES.2023.3344397

Han H, Zhang J, Hou Y, Qiao J (2024) Double-closed-loop robust optimal control for uncertain nonlinear systems. IEEE Trans. Cybern. 54:2332–2344. https://doi.org/10.1109/TCYB.2023.3266391

Jiang L, Jia H, Xu X, Rong W, Jiang W, Wang Q, Chen G, Xue X (2022) Numerical study on aerodynamic performance of mars parachute models with geometric porosities. SPACE-Sci Technol 2022. https://doi.org/10.34133/2022/9851982

Liu D, Liu H, Valavanis KP (2021) Fully distributed time-varying formation control for multiple uncertain missiles. IEEE Trans Aerosp Electron Syst 57:1646–1656. https://doi.org/10.1109/TAES.2020.3046329

Malyuta D, Yu Y, Elango P, Acikmese B (2021) Advances in trajectory optimization for space vehicle control. Annu Rev Control 52:282–315. https://doi.org/10.1016/j.arcontrol.2021.04.013

Pang ZJ, Zhao Z, Wang QT, Du ZH (2019) On the mechanics of periodic motion and control of tethered satellite system in elliptical orbit. Int J Appl Mech 11. https://doi.org/10.1142/S1758825119500571

Sallam KM, Elsayed SM, Chakrabortty RK, Ryan MJ (2020) Improved multi-operator differential evolution algorithm for solving unconstrained problems. In: 2020 ieee congress on evolutionary computation (cec). IEEE; IEEE Computat Intelligence Soc

Wang Z, Zhou X, Sun B, Zhang W, Su X, Shi J, Huang Q (2024) Model construction and dominant mechanism analysis of li-ion batteries under periodic excitation. SPACE-Sci Technol 4. https://doi.org/10.34133/space.0129

Wie B, Roithmayr C (2005) Attitude and orbit control of a very large geostationary solar power satellite. J Guid Control Dyn 28:439–451. https://doi.org/10.2514/1.6813

Wu S, Zhang K, Peng H, Wu Z, Radice G (2016) Robust optimal sun-pointing control of a large solar power satellite. ACTA Astronaut 127:226–234. https://doi.org/10.1016/j.actaastro.2016.05.019

Wu A-G, Zhang Y, Jiang T-Y (2019) Iterative algorithms for discrete periodic Riccati matrix equations. Int J Syst Sci 50:2102–2112. https://doi.org/10.1080/00207721.2019.1646345

Zhang K, Wu S, Wu Z (2020) Robust enhanced control strategy of a solar power satellite using multiple sensors. J Guid Control Dyn 43:338–346. https://doi.org/10.2514/1.G004733

Zhu W, Mu J, Shao C, Hu J, Wang B, Wen Z, Han F, Li S (2022) System design for pose determination of spacecraft using time-of-flight sensors. SPACE-Sci Technol 2022. https://doi.org/10.34133/2022/9763198

Zuo X, Xu M, Huang M, Li M, Peng N (2023) Maintenance strategy for elliptical orbit satellite with air-breathing electric propulsion. IEEE Trans Aerosp Electron Syst 59:6863–6877. https://doi.org/10.1109/TAES.2023.3282611

Chapter 8
Reliability-Constrained Uncertain Spacecraft Sliding Mode Attitude Tracking Control

Abstract This chapter presents an uncertainty-based sliding mode control (SMC) framework designed for spacecraft attitude tracking subject to incomplete information and interval parameter uncertainties. To address the challenge of data scarcity in aerospace missions, the methodology employs the set theory-based interval analysis strategy. The aforementioned interval dimension-wise analysis (IDWA) method is utilized for uncertainty bound estimation. The control architecture is established by transforming a nominal SMC into an interval-valued controller capable of managing trajectory boundaries. To rigorously evaluate spacecraft operational safety, an interval-based time-dependent reliability model is constructed, employing time-varying critical thresholds to adapt to dynamic mission phases. Furthermore, a multi-objective optimization scheme is implemented to tune the control parameters under these strict reliability constraints. Numerical validations confirm that this interval-based approach achieves tracking precision and safety assessment capabilities comparable to Monte Carlo simulations (MCSs), while offering significantly higher computational efficiency.

8.1 Introduction

With the advancement of modern aerospace technology, spacecraft attitude control has emerged as a critical capability for ensuring the success of space missions (Zhou 2015). Especially for missions involving fast maneuvers and target locking, the demand has shifted from simple stabilization to high-precision attitude tracking. Several kinds of spacecraft design (Zhou 2019; Zhou and Zhang 2024) and control methods have been used to solve this problem (Sari et al. 2019; Bello et al. 2023; Hu et al. 2022). However, the control problem remains highly challenging (Yin et al. 2016) due to the strong coupling and nonlinear nature of the spacecraft attitude control system, which are more pronounced when tracking rapidly time-varying reference trajectories. In their on-orbit operations, spacecraft are subjected to various uncertainties that can all adversely affect the attitude control system. Therefore, developing

C. Yang and Y. Xia, *Set Theory-Based Spacecraft Dynamics and Control*, Intelligent Control and Learning Systems 26, https://doi.org/10.1007/978-981-95-9179-4_8

attitude tracking control strategies capable of maintaining high precision and reliability in complex environments is essential for improving spacecraft performance and mission success rates. While advanced control methods can deliver satisfactory results in certain scenarios, their effectiveness may be compromised by variations in system parameters and external disturbances.

Owing to its strong robustness, fast response, and straightforward design, SMC has drawn considerable attention in the presence of disturbances (Nixon and Shtessel 2022; Nguyen et al. 2022; Yadegari et al. 2022). However, the practical implementation of these methods often requires precise knowledge of numerous uncertain parameters—an assumption that is not only expensive but also challenging to satisfy in real-world scenarios—potentially limiting their applicability to spacecraft attitude control.

In real-world conditions, spacecraft are subjected to a range of disturbances, such as control system uncertainties, environmental perturbations, and measurement errors. Consequently, it is crucial to account for these multi-source uncertainties in spacecraft attitude control design. However, conventional probabilistic approaches require extensive sampling data to accurately estimate the probability distributions of these uncertainties, which is an impractical demand in many engineering applications due to the high cost and difficulty of acquiring large volumes of precise data. This highlights the need for non-probabilistic uncertainty analysis methods, particularly when only limited sample information is available (Wang et al. 2022, 2023; Liu et al. 2023; Fravolini et al. 2020).

To enhance the performance of spacecraft attitude tracking control in the presence of various uncertainties, this chapter proposes a time-dependent reliability-constrained uncertain SMC strategy, leveraging interval analysis to characterize and manage these uncertainties. To more effectively constrain the system response—ensuring that performance requirements are met throughout different time stages while simultaneously improving response speed—this chapter introduces a time-dependent-threshold-based safety function as a reliability constraint. Furthermore, to identify optimal control parameters and further boost overall system performance, an interval SMC optimization framework is developed, built upon the multi-objective optimization algorithm CCMO. The control system is then subjected to a constrained multi-objective optimization process that explicitly incorporates these reliability constraints.

8.2 General Spacecraft Attitude Equations

The spacecraft attitude control system exhibits nonlinear behavior, strong coupling, and a multi-input multi-output structure, making it a highly complex nonlinear system. To tackle the spacecraft's tracking control problem, developing an accurate spacecraft model is a necessary first step.

Within the body-fixed coordinate system, the rotational motion of the spacecraft is governed by the following dynamic equation:

$$\mathbf{J}\dot{\Omega} = -\mathbf{\Omega}^{\times}\mathbf{J}\Omega + \mathbf{u} + \mathbf{z} \tag{8.1}$$

where $\mathbf{J} \in \Re^{3\times 3}$ represents the inertia matrix of the spacecraft rotating about its mass center, as

$$\mathbf{J} = \begin{bmatrix} J_{xx} & -J_{xy} & -J_{xz} \\ -J_{yx} & J_{yy} & -J_{yz} \\ -J_{zx} & -J_{zy} & J_{zz} \end{bmatrix} \tag{8.2}$$

in which the diagonal elements of the matrix denote the moments of inertia, while the other elements denote the products of inertia. $\mathbf{\Omega} = \begin{bmatrix} \omega x \ \omega y \ \omega z \end{bmatrix}^{T}$ denotes the angular velocity of the spacecraft expressed in its body-fixed frame, and $\mathbf{\Omega}^{\times}$ represents the skew-symmetric matrix constructed from $\mathbf{\Omega}$, which is defined as

$$\mathbf{\Omega}^{\times} = \begin{bmatrix} 0 & -\omega_{z} & \omega_{y} \\ \omega_{z} & 0 & -\omega_{x} \\ -\omega_{y} & \omega_{x} & 0 \end{bmatrix} \tag{8.3}$$

and $\mathbf{u} = \begin{bmatrix} u_1 \ u_2 \ u_3 \end{bmatrix}^{T}$ represents the control torque, and $\mathbf{z} = \begin{bmatrix} z_1 \ z_2 \ z_3 \end{bmatrix}^{T}$ represents the external disturbances.

When a spacecraft rotates about its mass center in the sequence of pitch, roll, and yaw, its kinematic equations can be written as

$$\dot{\theta} = \mathbf{R}(\boldsymbol{\theta})\mathbf{\Omega} \tag{8.4}$$

where $\boldsymbol{\theta} = \begin{bmatrix} \theta_1 \ \theta_2 \ \theta_3 \end{bmatrix}^{T}$ represents the spacecraft attitude angle, where θ_1, θ_2 and θ_3 represents the roll, yaw, and pitch angles, respectively. The rotation matrix $\mathbf{R}(\boldsymbol{\theta})$ can be derived based on the attitude angle as follows:

$$\mathbf{R}(\boldsymbol{\theta}) = \begin{bmatrix} 1 & \tan\theta_2 \sin\theta_1 & \tan\theta_2 \cos\theta_1 \\ 0 & \cos\theta_1 & -\sin\theta_1 \\ 0 & \frac{\sin\theta_1}{\cos\theta_2} & \frac{\cos\theta_1}{\cos\theta_2} \end{bmatrix} \tag{8.5}$$

Thus, the attitude dynamics of the spacecraft can be fully described as

$$\begin{cases} \mathbf{J}\dot{\Omega} = -\mathbf{\Omega}^{\times}\mathbf{J}\Omega + \mathbf{u} + \mathbf{z} \\ \dot{\theta} = \mathbf{R}(\boldsymbol{\theta})\mathbf{\Omega} \\ \mathbf{y} = \boldsymbol{\theta} \end{cases} \tag{8.6}$$

Based on Eqs. (8.4) and (8.6), one can obtain that

$$\ddot{\theta} = \dot{\mathbf{R}}(\boldsymbol{\theta})\mathbf{\Omega} + \mathbf{R}(\boldsymbol{\theta})\dot{\Omega} \tag{8.7}$$

which can be transformed to

$$\dot{\Omega} = \mathbf{R}^{-1}(\boldsymbol{\theta})\big(\ddot{\theta} - \dot{\mathbf{R}}(\boldsymbol{\theta})\boldsymbol{\Omega}\big) \tag{8.8}$$

Thus, considering Eq. (8.6) and Eq. (8.8), the expression for $\ddot{\theta}$ can be obtained as follows:

$$\ddot{\theta} = \dot{\mathbf{R}}(\boldsymbol{\theta})\boldsymbol{\Omega} + \mathbf{R}(\boldsymbol{\theta})\mathbf{J}^{-1}\big(-\boldsymbol{\Omega}^{\times}\mathbf{J}\Omega + \mathbf{u} + \mathbf{z}\big) \tag{8.9}$$

From Eq. (8.9), it can be observed that if the desired trajectory is denoted by $\boldsymbol{\theta}_{\mathrm{i}} = \left[\,\theta_{1\mathrm{i}}\ \theta_{2\mathrm{i}}\ \theta_{3\mathrm{i}}\,\right]^{\mathrm{T}}$, it is feasible to design an appropriate control input torque $\mathbf{u}$ such that the system is able to track $\boldsymbol{\theta}_{\mathrm{i}}$, implying that $\boldsymbol{\theta} \to \boldsymbol{\theta}_{\mathrm{i}}$.

8.3 Spacecraft Sliding Mode Attitude Tracking Control with Interval Uncertainties

8.3.1 The Nominal Sliding Mode Attitude Controller Design

In spacecraft attitude control, the main goal is to enable the spacecraft to accurately track a desired reference attitude trajectory. To lower the system's relative degree and effectively suppress disturbances, a double-loop SMC scheme can be adopted. In this architecture, the inner loop handles disturbance rejection, while the outer loop ensures accurate attitude tracking. The block diagram of the proposed control method is featured as in Fig. 8.1.

As illustrated in Fig. 8.1, the outer-loop SMC law is employed to track the desired attitude angle $\boldsymbol{\theta}_{\mathrm{i}}$. The corresponding reference angular velocity $\boldsymbol{\Omega}_{\mathrm{i}}$ generated by the outer-loop controller is then fed into the inner-loop system. The inner-loop SMC law subsequently ensures accurate tracking of the reference angular velocity $\boldsymbol{\Omega}_{\mathrm{i}}$. Specifically, the inner loop is responsible for angular velocity regulation, while the outer loop focuses on achieving precise attitude tracking.

The outer-loop sliding mode controller is developed to follow the desired attitude angular velocity $\boldsymbol{\Omega}_{\mathrm{i}}$. Its associated sliding surface is defined as

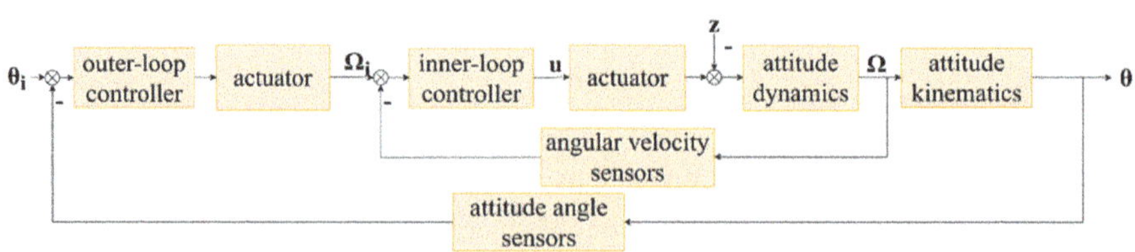

Fig. 8.1 The control block diagram of the control system for nominal SMC

$$\mathbf{s}_o = \boldsymbol{\theta}_e + \mathbf{K}_o \int_0^t \boldsymbol{\theta}_e dt, \ \mathbf{s}_o \in \Re^3 \tag{8.10}$$

where $\boldsymbol{\theta}_e = \boldsymbol{\theta}_i - \boldsymbol{\theta}$ denotes the tracking error of attitude angle and $\mathbf{K}_o = \text{diag}\{k_{o1}, k_{o2}, k_{o3}\}$ is the gain matrix. To ensure system stability, the control error is driven toward an ideal sliding surface by appropriately designing $\mathbf{K}_o$, thereby achieving the desired dynamic performance.

From Eq. (8.10), it can be obtained that

$$\dot{\mathbf{s}}_o = \dot{\theta}_e + \mathbf{K}_o\boldsymbol{\theta}_e = \dot{\theta}_i - \dot{\theta} + \mathbf{K}_o\boldsymbol{\theta}_e = \dot{\theta}_i\mathbf{R}(\boldsymbol{\theta})(\boldsymbol{\Omega}_e - \boldsymbol{\Omega}_i) + \mathbf{K}_o\boldsymbol{\theta}_e \tag{8.11}$$

Thus, the control law of the outer-loop can be designed as

$$\boldsymbol{\Omega}_i = \mathbf{R}^{-1}(\boldsymbol{\theta})\left(\dot{\theta}_c + \mathbf{K}_o\boldsymbol{\theta}_e\right) + \mathbf{R}^{-1}(\boldsymbol{\theta})\rho_1\mathbf{s}_o \tag{8.12}$$

where $\rho_1 > 0$.

Theorem 8.1 *Considering Eq. (8.4), if the outer-loop sliding surface is designed as Eq. (8.10), then the system achieves asymptotic stability under the proposed outer-loop virtual control law given by Eq. (8.12).*

Proof The candidate Lyapunov function is chosen as

$$V_1 = \frac{1}{2}\mathbf{s}_o^T\mathbf{s}_o \tag{8.13}$$

Thus, one can obtain that

$$\begin{aligned}\dot{V}_1 &= \mathbf{s}_o^T\dot{\mathbf{s}}_o = \mathbf{s}_o^T\left(\dot{\theta}_i + \mathbf{R}(\theta)\ (\Omega_e - \Omega_i) + \mathbf{K}_o\theta_e\right)\\ &= \mathbf{s}_o^T\left\{\dot{\theta}_i - \mathbf{R}(\theta)\left[\mathbf{R}^{-1}(\theta)\left(\dot{\theta}_c + \mathbf{K}_o\theta_e + \rho_1\mathbf{s}_o\right)\right] + \mathbf{R}(\theta)\Omega_e + \mathbf{K}_o\theta_e\right\}\\ &= -\rho_1\|\mathbf{s}_o\|^2 + \mathbf{s}_o^T\mathbf{R}(\theta)\Omega_e\end{aligned} \tag{8.14}$$

From Eq. (8.14), to keep $\dot{V}_1 \leq 0$, the attitude angle error $\boldsymbol{\Omega}_e$ needs to be small enough. Therefore, a fast-converging inner-loop SMC is needed. □

The controller of the inner-loop is designed to ensure that $\boldsymbol{\Omega}$ accurately tracks $\boldsymbol{\Omega}_i$, with the objective of driving the tracking error between them to zero through an appropriately formulated control law $\mathbf{u}$. The inner-loop sliding surface can be derived as

$$\mathbf{s}_i = \boldsymbol{\Omega}_e + \mathbf{K}_i \int_0^t \boldsymbol{\Omega}_e dt, \ \mathbf{s}_i \in \Re^3 \tag{8.15}$$

where $\mathbf{K}_i = \text{diag}\{k_{i1}, k_{i2}, k_{i3}\}$ is the gain matrix. Combining Eqs. (8.6) and (8.15), $\dot{\mathbf{s}}_i$ can be derived as

$$\dot{\mathbf{s}}_{\mathrm{i}} = \dot{\Omega}_{\mathrm{e}} + \mathbf{K}_{\mathrm{i}}\boldsymbol{\Omega}_{\mathrm{e}} = \dot{\Omega}_{\mathrm{c}} + \mathbf{J}^{-1}\left(\boldsymbol{\Omega}^{\times}\mathbf{J}\Omega - \mathbf{u} - \mathbf{z}\right) + \mathbf{K}_{\mathrm{i}}\boldsymbol{\Omega}_{\mathrm{e}} \tag{8.16}$$

Thus, the control law of the inner-loop can be designed as

$$\mathbf{u} = \boldsymbol{\Omega}^{\times}\mathbf{J}\Omega + \mathbf{J}\left(\dot{\Omega}_{\mathrm{i}} + \mathbf{K}_{\mathrm{i}}\boldsymbol{\Omega}_{\mathrm{e}}\right) + \rho_2 \mathrm{sat}(\mathbf{s}_{\mathrm{i}}) + \mathbf{k}\mathbf{s}_{\mathrm{i}} \tag{8.17}$$

where $\rho_2 > \max|\mathbf{z}_{\mathrm{i}}|$, $\mathbf{k} = \mathrm{diag}\{k, k, k\}, k > 0$ and $\mathrm{sat}(s_j) = \begin{cases} \frac{s_j}{\Delta_j}, |s_j| \le \Delta_j \\ \mathrm{sgn}(s_j), |s_j| > \Delta_j \end{cases}, j = 1, 2, 3$ is the saturating function, where s_j is the jth component of the sliding surface. Thus, Eq. (8.16) can be rewritten as

$$\dot{\mathbf{s}}_{\mathrm{i}} = -\mathbf{J}^{-1}(\rho_2 \mathrm{sat}(\mathbf{s}_{\mathrm{i}}) + \mathbf{k}\mathbf{s}_{\mathrm{i}} + \mathbf{z}) \tag{8.18}$$

Theorem 8.2 *Considering Eq. (8.1), if the inner-loop sliding surface is designed as Eq. (8.15), then the system achieves asymptotic stability under the proposed inner-loop control law described in Eq. (8.17).*

Proof The candidate Lyapunov function is chosen as

$$V_2 = \frac{1}{2}\mathbf{s}_{\mathrm{i}}^{\mathrm{T}}\mathbf{J}\mathbf{s}_{\mathrm{i}} \tag{8.19}$$

Thus, one can obtain that

$$\dot{V}_2 = \mathbf{s}_{\mathrm{i}}^{\mathrm{T}}\mathbf{J}\dot{\mathbf{s}}_{\mathrm{i}} = -\rho_2\sum_{n=1}^{3}|\mathbf{s}_{\mathrm{i}}| - \mathbf{s}_{\mathrm{i}}^{\mathrm{T}}\mathbf{z} - \mathbf{k}\|\mathbf{s}_{\mathrm{i}}\|^2 \le -\mathbf{k}\|\mathbf{s}_{\mathrm{i}}\|^2 \tag{8.20}$$

By designing $\mathbf{k} \ge \frac{1}{2}\mathbf{J}_{\max}$, one can obtain that $\mathbf{k}\|\mathbf{s}_{\mathrm{i}}\|^2 \ge \frac{1}{2}\mathbf{s}_{\mathrm{i}}^{\mathrm{T}}\mathbf{J}\mathbf{s}_{\mathrm{i}}$, which is equivalent to $-\mathbf{k}\|\mathbf{s}_{\mathrm{i}}\|^2 \le -V_2$, which implies that $\dot{V}_2 \le -V_2$. Thus, one can obtain that $V_2 \le e^{-t}$, indicating that when $t \to \infty$, the Lyapunov function V_2 converges exponentially. This also implies that $\boldsymbol{\Omega}_{\mathrm{e}}$ also exhibits exponential convergence. □

In a double-loop control framework, the dynamic behavior of the inner loop directly influences the stability of the outer loop and consequently impacts the overall stability of the closed-loop system. To ensure system stability, a common approach is to design the inner loop to converge significantly faster than the outer loop. To achieve this, the parameter $\mathbf{K}_{\mathrm{i}} \ge \mathbf{K}_{\mathbf{o}}$ is set and $\mathbf{k}$, is assigned a sufficiently large value. This accelerates the elimination of $\boldsymbol{\Omega}_{\mathrm{e}}$, thereby ensuring that $\dot{V}_1 \le 0$ holds, and guarantees that the convergence rate of the outer loop remains considerably slower than that of the inner loop.

8.3.2 An Interval-Based Sliding Mode Attitude Controller

Based on the preceding analysis, various uncertainties are inevitably present in practical scenarios. Therefore, to enhance the realism and applicability of the control design, it is essential to account for interval-type uncertainties in the system parameters. To this end, this chapter incorporates the IDWA method, which enables precise estimation of the bounds of uncertain parameters affected by interval uncertainties.

According to the IDWA method, for any variable $\mathbf{y}$ and the mapping between it and the interval uncertainty parameter $\mathbf{b}^{\mathrm{I}}$, expressed as $\mathbf{y} = \mathbf{F}(\mathbf{b}^{\mathrm{I}})$, a polynomial approximation $\hat{\mathbf{F}}(\mathbf{b}^{\mathrm{I}})$ can be used as

$$\hat{\mathbf{F}}(\mathbf{b}^{\mathrm{I}}) = \alpha_0 \mathbf{T}_0(x) + \cdots + \alpha_i \mathbf{T}_i(x) + \cdots + \alpha_{n+1} \mathbf{T}_{n+1}(x) \tag{8.21}$$

where $\mathbf{T}_i(x)$ is the orthogonal polynomial, and $x = \frac{2(b_j - \underline{b}_j)}{\overline{b}_j - \underline{b}_j} - 1 \in [-1, 1]$ is the normalized parameter value. The coefficient α_i is proposed to narrow the gap between $\hat{\mathbf{F}}(\mathbf{b}^{\mathrm{I}})$ and $\mathbf{F}(\mathbf{b}^{\mathrm{I}})$, which is calculated as

$$\alpha_i = \frac{2i + 1}{2} \int_{-1}^{1} \mathbf{F}(x) \mathbf{T}_k(x)\, \mathrm{d}x, \quad k = 0, 1, \ldots, n \tag{8.22}$$

Utilizing the numerical integration method, Eq. (8.22) can be solved.

By substituting Eq. (8.22) into Eq. (8.21), an approximation of $\mathbf{F}(\mathbf{b}^{\mathrm{I}})$ is derived. Since the new expression is simplified compared to the original one, it can be analyzed more easily. To obtain the extremum value of $\mathbf{F}(\mathbf{b}^{\mathrm{I}})$, the derivation expression roots must be found. The problem is equivalent to solving the roots x_{root} of the following expression:

$$\alpha_0 \hat{\mathbf{T}}_0(x) + \alpha_1 \hat{\mathbf{T}}_1(x) + \cdots + \alpha_{n+1} \hat{\mathbf{T}}_{n+1}(x) = \mathbf{0} \tag{8.23}$$

Therefore, the minimum point on $\mathbf{F}(\mathbf{b}^{\mathrm{I}})$ at the reference point b_j^{c} can be calculated as

$$b_{j\min}^{(1)} = b_j^{\mathrm{c}} + \frac{\overline{b}_j - b_j}{2} x_{j\min}, \quad j = 1, 2, \ldots, m \tag{8.24}$$

where

$$\mathbf{F}(x_{j\min}) = \min_{x_j \in \{-1, x_{\mathrm{root}} - 1\}} \mathbf{F}(x). \tag{8.25}$$

By computing Eq. (8.24), a new reference point $\mathbf{b}^{\mathrm{I}}_{(1)\min}$ is generated, which is expressed as

$$\mathbf{b}^{\mathrm{I}}_{(1)\min} = \left[b^{1}_{(1)\min}, b^{2}_{(1)\min}, \ldots, b^{m}_{(1)\min}\right]^{\mathrm{T}} \tag{8.26}$$

Following the above procedure, the maximum value point can also be computed. By calculating Eqs. (8.21)–(8.25), the reference point can be updated through iterative computation. The stop criterion of iteration is $\left\|\mathbf{b}_{(k+1)} - \mathbf{b}_{(k)}\right\| < \delta$, where δ is a small quantity. Once the extremum points are derived, the output variables can be calculated utilizing these points.

Following the above procedure, by incorporating interval uncertainties into the system equations, the corresponding system model can be reformulated in an interval-based form, which is expressed as

$$\begin{cases} \dot{\theta}_{j\min} = \mathbf{R}\left(\boldsymbol{\theta}_{j\min}\right)\boldsymbol{\Omega}\left(\mathbf{b}^{\mathrm{I}}_{(k)\min}\right) \\ \dot{\theta}_{j\max} = \mathbf{R}\left(\boldsymbol{\theta}_{j\max}\right)\boldsymbol{\Omega}\left(\mathbf{b}^{\mathrm{I}}_{(k)\max}\right) \end{cases} \tag{8.27}$$

Thus, by calculating the LB and UB of each dimension, the bounds of $\boldsymbol{\theta}$ are presented as

$$\begin{cases} \underline{\theta} = \left[\underline{\gamma}\ \underline{\phi}\ \underline{\varphi}\right]^{\mathrm{T}} \\ \overline{\theta} = \left[\overline{\gamma}\ \overline{\phi}\ \overline{\varphi}\right]^{\mathrm{T}} \end{cases} \tag{8.28}$$

Therefore, the output containing interval uncertainties can be determined.

8.4 Reliability-Constrained Optimal Attitude Controller

To design the optimal attitude controller, a reliability-constrained optimization problem is proposed. By incorporating the reliability results obtained from reliability analysis as constraints, it is possible to identify pairs of system parameters that minimize both the overall deviation between the interval response trajectory and the target trajectory over the entire time history, denoted by $\boldsymbol{\theta}_{\mathrm{eh}}$, as well as the maximum deviation between these trajectories, denoted by $\boldsymbol{\theta}_{\mathrm{emax}}$. Based on this formulation, the following multi-objective optimization problem can be established:

$$\begin{aligned} &\text{find } \rho_1, \rho_2 \\ &\min \quad \boldsymbol{\theta}_{\mathrm{eh}}, \boldsymbol{\theta}_{\mathrm{emax}} \\ &\text{s.t.} \quad Rel(T) \geq \text{poss} \end{aligned} \tag{8.29}$$

where ρ_1 and ρ_2 are designed parameters in the sliding mode controller. To address this problem efficiently, the CCMO algorithm (Tian et al. 2021), which has been introduced in Subsection 5.7.2, is employed.

8.5 Numerical Example

The previously proposed method will be applied in this section to address a realistic spacecraft attitude tracking control problem. The key parameters related to the satellite and control system are summarized as follows. The uncertain spacecraft model represented in Eq. (8.6) is set as $\mathbf{J}^{\mathrm{c}} = \begin{bmatrix} 1200 & 100 & -200 \\ 100 & 2200 & 300 \\ -200 & 300 & 2100 \end{bmatrix} \mathrm{kg} \cdot \mathrm{m}^2$ and $\Delta\mathbf{J} = 0.04|\mathbf{J}^{\mathrm{c}}|$. The center value and radius of the initial angular velocity are set to $\boldsymbol{\Omega}^{\mathrm{c}} = \begin{bmatrix} 0 \; 0 \; 0 \end{bmatrix}^{\mathrm{T}}$ and $\Delta\boldsymbol{\Omega} = \begin{bmatrix} 0 \; 0 \; 0 \end{bmatrix}^{\mathrm{T}}$. The initial attitude angle is defined as: $\boldsymbol{\theta}^{\mathrm{c}} = \begin{bmatrix} 0 \; 0 \; 0 \end{bmatrix}^{\mathrm{T}}$. The external disturbance is defined as: $\mathbf{z}^{\mathrm{c}} = \begin{bmatrix} \sin t \; \cos t \; \sin(2t) \end{bmatrix}^{\mathrm{T}}$.

The related parameter of the sliding mode controller is set as follows. For the outer-loop controller, $\mathbf{K}_{\mathrm{o}} = \mathrm{diag}\{0.3, 0.3, 0.3\}$. For the inner-loop controller, $\mathbf{K}_{\mathrm{i}} = \mathrm{diag}\{10, 10, 10\}$, $\mathbf{k} = \mathrm{diag}\{100, 100, 100\}$, while the saturation function parameter is set as: $\Delta_i = 0.01$, where $i = 1, 2, 3$.

To certify the efficacy of the proposed approach, the aforementioned IDWA approach will be demonstrated through two distinct load cases, each involving a different desired tracking trajectory. The first load case is designed to illustrate the basic performance and efficiency of IDWA. In this case, IDWA will be compared against the classical MCS method, highlighting its computational advantages and accuracy. Building upon the results from the first case, the second load case introduces a reliability constraint into the analysis. In addition to comparing IDWA with MCS, this scenario incorporates a reliability-based constraint phase. By enforcing these reliability requirements, a multi-objective optimization problem, which is identical in structure to the one formulated earlier in this chapter, will be established. Reliability data will then be extracted across specified time intervals to show both the performance and robustness of the IDWA approach under uncertainty.

In the first load case, and without loss of generality, only a single dimension of the attitude angle is considered for illustration, denoted as θ. The desired trajectory is defined as $\begin{cases} \underline{\theta} = 0.95 \cos t \\ \overline{\theta} = 1.05 \cos t \end{cases}$, and the system parameters are set as $\rho_1 = 5$, $\rho_2 = 5$. The simulation is conducted over a time period of 5 s. The results obtained using the IDWA method are presented in Fig. 8.2.

To further validate the performance of the introduced approach, a comparative study is conducted between the proposed approach and the MCS, where MCS is performed with 10,000 sample runs. For clarity and without loss of generality, only the UBs of the results from both methods are plotted. The comparison is illustrated in Fig. 8.3.

Figures 8.2 and 8.3 collectively illustrate the effectiveness of the introduced approach. As shown in Fig. 8.2, the designed controller successfully tracks the desired trajectory, achieving a sufficiently small tracking error at steady state. With the comparison of the MCS and the proposed method results in Fig. 8.3, it is obvious that the response curve produced by the proposed method consistently encloses the

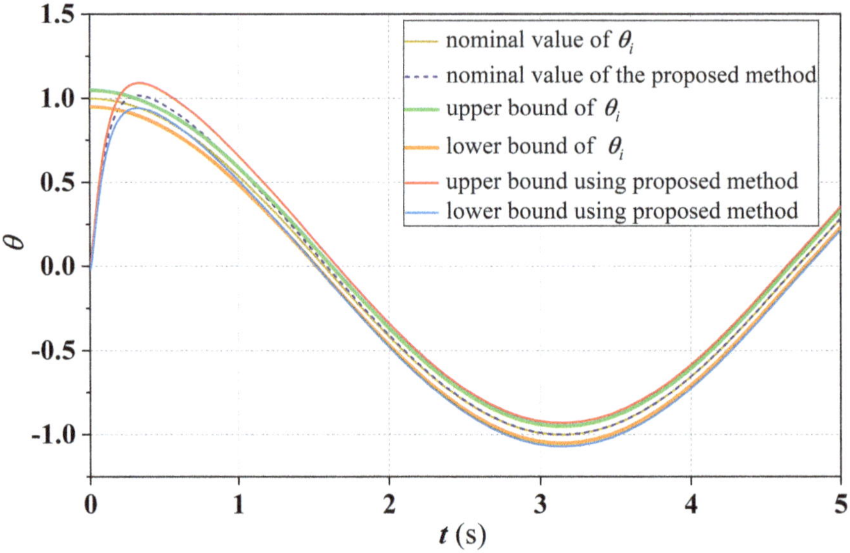

Fig. 8.2 The attitude angle using the proposed method

Fig. 8.3 UB of attitude angle using MCS and the proposed method

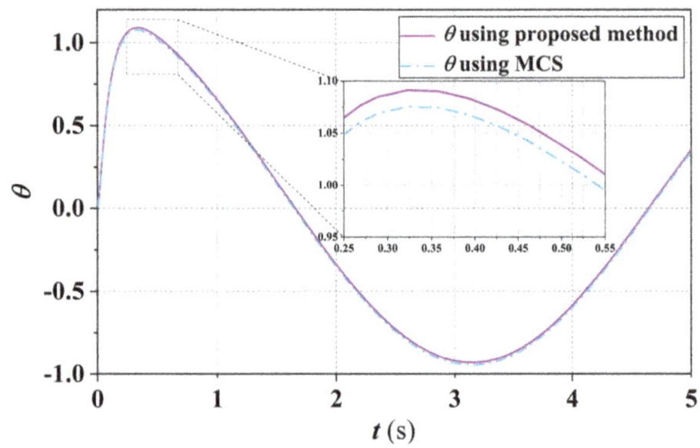

one obtained from MCS. This is primarily due to the fact that MCS, as a probabilistic simulation technique, provides a more accurate approximation of the true distribution. Although the proposed method exhibits a slightly larger error compared to MCS, the deviation remains acceptably small. Specifically, the maximum difference between the two methods is approximately 1.4% of the peak value obtained from MCS. This result confirms the accuracy and reliability of the IDWA method in handling interval uncertainties. Moreover, the proposed IDWA method exhibits a notable advantage in computational efficiency. For this load case, MCS requires 632 s of computation time, whereas the IDWA method requires only 12 s—approximately

1.9% of the MCS computation time. Therefore, considering its superior computational performance and acceptable accuracy, IDWA proves to be an effective and efficient tool for interval analysis in spacecraft attitude control systems.

To further evaluate the performance and robustness of the proposed method, Load Case 2 is introduced for further validation. In contrast to Load Case 1, it utilizes LB and UB that gradually converge, which can be expressed as $\begin{cases} \underline{\theta} = \cos t - 0.05e^{-t} \\ \overline{\theta} = \cos t + 0.05e^{-t} \end{cases}$. This time-dependent bound structure is designed to more rigorously test the tracking capability of the IDWA method. The results are again compared with those obtained from MCS, which involves 10,000 simulation runs over a total simulation time of 5 s. Distinct from the first load case, Load Case 2 incorporates reliability constraints to identify optimal system parameters. Traditional approaches that rely on constant threshold functions often fail to adequately capture the dynamic behavior across different stages of system response. To overcome this limitation, a time-dependent function that mirrors the trend of the target trajectory is employed as the threshold, replacing the conventional fixed threshold. This allows for a more accurate fit to the transient and steady-state responses of the system. In this chapter, the threshold is set as $\theta^{cr} = 2.331 - 2.71^{0.493t}$. Moreover, as the influence of the reliability constraint diminishes once the system converges, the optimization process focuses on the transient phase of the system response, the time interval from 0.5 s to 1.5 s. As previously mentioned, two performance indices are selected as the optimization objectives, and a reliability constraint of 96.5% is imposed. The CCMO algorithm is employed to solve this problem, configured with a population size of 100 and a maximum of 10,000 iterations. The resulting Pareto front is illustrated in Fig. 8.4.

In Fig. 8.4, the midpoint of the Pareto front, labeled as Point No. 50, is selected as the optimal solution since it achieves a balanced trade-off between the two optimization objectives $\theta_{\text{eh}} = 0.054$ and $\theta_{\text{emax}} = 0.185$. The corresponding controller parameters obtained at this point are $\rho_1 = 2.03$ and $\rho_2 = 3.33$. For comparative

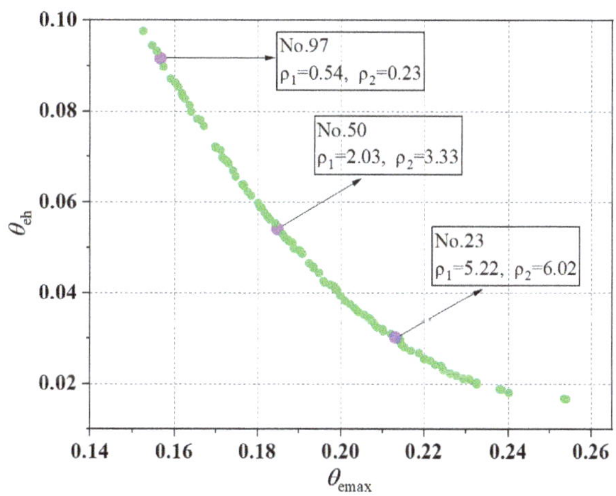

Fig. 8.4 Pareto front using the introduced approach

analysis, two additional solutions are chosen from the Pareto front—Point No. 23 and Point No. 97. Point No. 23 represents a solution with a relatively smaller average error, with objective values given by $\theta_{\text{eh}} = 0.031$ and $\theta_{\text{emax}} = 0.212$, and associated controller parameters: $\rho_1 = 5.22$, $\rho_2 = 6.02$. On the other hand, Point No. 97 corresponds to a solution that minimizes the maximum error, with objective values $\theta_{\text{eh}} = 0.092$ and $\theta_{\text{emax}} = 0.157$, and controller parameters $\rho_1 = 0.54$ and $\rho_2 = 0.23$. These three representative solutions reflect different performance characteristics in terms of tracking accuracy. The system responses corresponding to these points are illustrated in Figs. 8.5 and 8.6, and. 8.7.

Figures 8.5, 8.6 and 8.7 illustrate that the system response varies significantly with different parameter settings. However, regardless of the chosen parameters, the response obtained using IDWA does not achieve full convergence. This behavior can be attributed to two main factors. First, the desired trajectory θ_{i} converges according to the exponential term e^{-t}, which inherently prevents the tracking trajectory from reaching a complete steady state. Second, the IDWA method assumes a uniform distribution of uncertain variables within their respective intervals, which may introduce certain modeling inaccuracies and deviate from real-world conditions. Despite

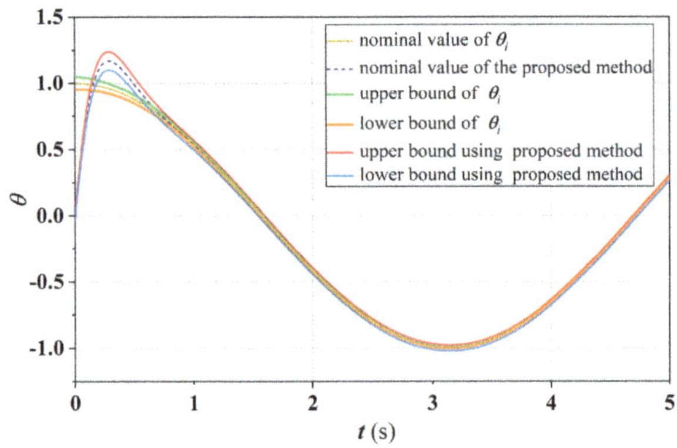

Fig. 8.5 No. 23 solution set in Pareto front

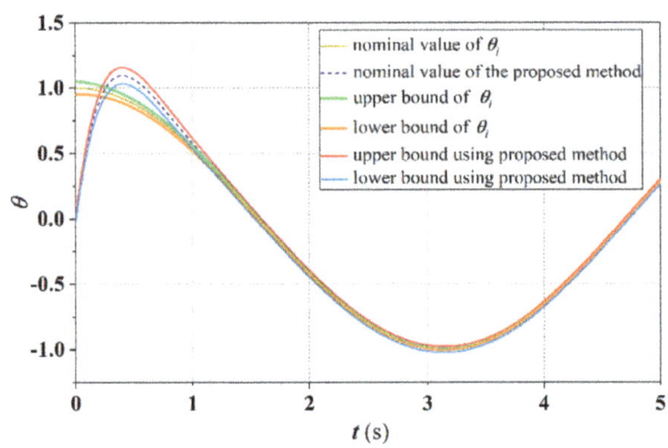

Fig. 8.6 No. 50 solution set in Pareto front

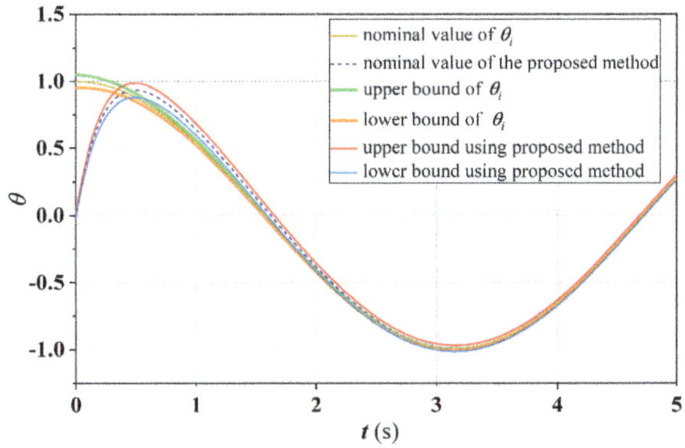

Fig. 8.7 No. 97 solution set in Pareto front

these limitations, the results show that the IDWA-based responses still exhibit a clear trend toward convergence, demonstrating their effectiveness in capturing the overall system behavior under interval uncertainties.

As shown in Fig. 8.5, 8.6 and 8.7, the system corresponding to Point No. 23 demonstrates a rapid response during the transient phase, which leads to a larger peak error θ_{emax}. However, its swift convergence results in a smaller average error θ_{eh} over the entire time period is smaller over the entire time horizon. In contrast, the system linked to Point No. 97 exhibits a slower response, yielding a lower maximum error θ_{emax}, but suffering from a reduced convergence speed, which increases the overall average error θ_{eh}. Therefore, considering its balanced performance in both transient response and steady-state accuracy, Point No. 50 is selected as the representative solution for further analysis.

First, the optimized parameters obtained from the multi-objective optimization are substituted into the MCS framework. The results generated by IDWA are then compared with those obtained from MCS. In this validation process, MCS is executed 10,000 times, which serves as a benchmark for evaluating the performance of the IDWA approach.

As shown in Fig. 8.8, the MCS results exhibit slightly better convergence compared to those obtained using IDWA. However, the difference between the two methods is relatively small, indicating that IDWA maintains a high level of accuracy despite its lower computation cost. Specifically, the computation time for IDWA is only 9 s, whereas MCS requires 536 s, making IDWA approximately 1.7% of the computational cost of MCS. Furthermore, since reliability constraints were incorporated into the IDWA-based optimization, Fig. 8.9 presents a comparison between the optimized IDWA response curve and the defined threshold function.

Furthermore, to more effectively evaluate the reliability performance of the IDWA method, a reliability analysis was also conducted using MCS for comparison. The results show that the reliability obtained from MCS is 0.994, whereas the reliability computed by IDWA is 0.972, as illustrated in Fig. 8.9. As observed in the figure, the reliability of the proposed method gradually converges to 0.972 during the system evolution. Although the reliability value from IDWA is slightly lower than that of

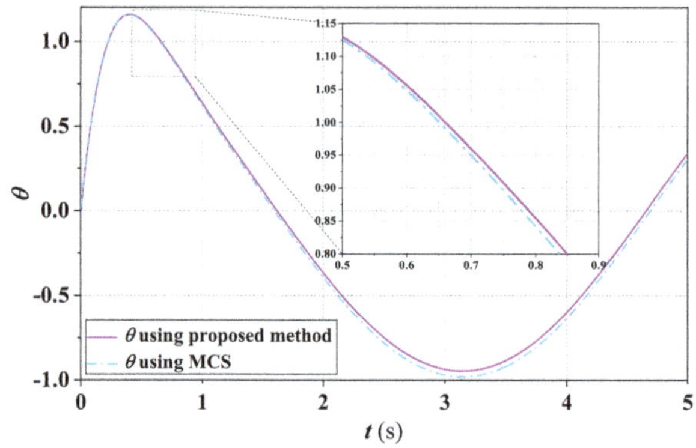

Fig. 8.8 Comparison between MCS and the proposed method

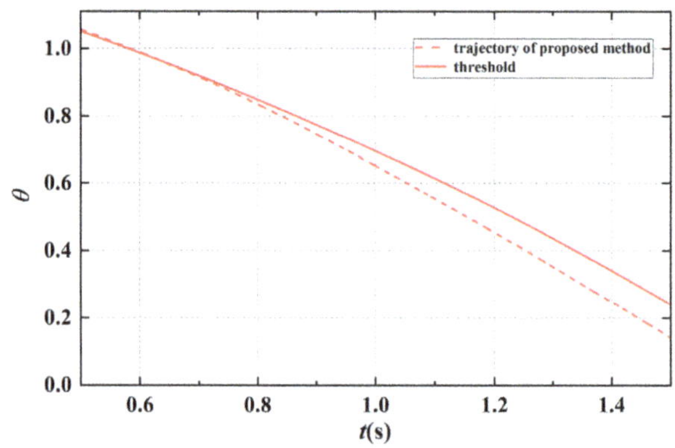

Fig. 8.9 Comparison between the proposed method and threshold

MCS, both results exceed the imposed reliability constraint of 0.965. This indicates that the IDWA-based solution satisfies the reliability requirement and achieves a level of performance that is reasonably close to that of the benchmark MCS method, further validating its effectiveness under interval uncertainties (Fig. 8.10).

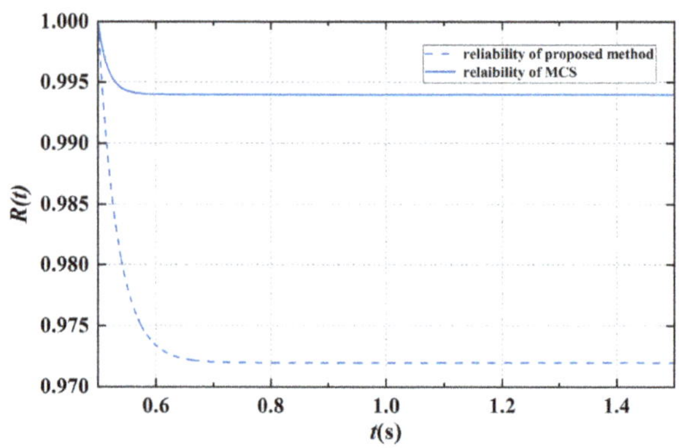

Fig. 8.10 Comparison of reliability between the proposed method and MCS

8.6 Summary

In this chapter, an interval uncertainty-driven and time-dependent reliability-constrained SMC framework was established to address spacecraft attitude tracking under data-sparse uncertainties. Unlike traditional probabilistic approaches, the proposed methodology integrates the IDWA method, grounded in orthogonal polynomials, to extend the nominal SMC into the interval domain. This extension enables the controller to effectively track the bounds of desired trajectories despite parameter variations. A contribution of this work is the introduction of an NPTDR measure, which employs a dynamic, non-constant threshold to rigorously constrain system safety across different operational phases. To balance performance and safety, the control parameters were tuned via a CCMO algorithm. Numerical validations on two distinct load cases confirmed that this interval-based strategy achieves approximation accuracy comparable to MCSs but with significantly superior computational efficiency.

References

Bello A, Olfe KS, Rodriguez J, Ezquerro JM, Lapuerta V (2023) Experimental verification and comparison of fuzzy and PID controllers for attitude control of nanosatellites. Adv SPACE Res 71:3613–3630. https://doi.org/10.1016/j.asr.2022.05.055

Fravolini ML, Costante G, Yucelen T, Napolitano MR (2020) Interval prediction models for data-driven design of aerial vehicle's robust adaptive controllers. J Guid CONTROL Dyn 43:1512–1528. https://doi.org/10.2514/1.G004937

Hu Y, Lu Z, Ling KV, Liao W, Zhang X (2022) Multiplexed MPC attitude control of a moving mass satellite using dual-rate piecewise affine model. Aerosp Sci Technol 128. https://doi.org/10.1016/j.ast.2022.107778

Liu X, Zhao T, Cao J, Li P (2023) Design of an interval type-2 fuzzy neural network sliding mode robust controller for higher stability of magnetic spacecraft attitude control. ISA Trans 137:144–159. https://doi.org/10.1016/j.isatra.2023.01.011

Nguyen NP, Oh H, Moon J (2022) Continuous nonsingular terminal sliding-mode control with integral-type sliding surface for disturbed systems: Application to attitude control for quadrotor uavs under external disturbances. IEEE Trans Aerosp Electron Syst 58:5635–5660. https://doi.org/10.1109/TAES.2022.3177580

Nixon ME, Shtessel YB (2022) Adaptive sliding mode control of a perturbed satellite in a formation antenna array. IEEE Trans Aerosp Electron Syst 58:4595–4614. https://doi.org/10.1109/TAES.2022.3164410

Sari NN, Jahanshahi H, Fakoor M (2019) Adaptive fuzzy PID control strategy for spacecraft attitude control. Int J FUZZY Syst 21:769–781. https://doi.org/10.1007/s40815-018-0576-2

Tian Y, Zhang T, Xiao J, Zhang X, Jin Y (2021) A coevolutionary framework for constrained multiobjective optimization problems. IEEE Trans Evol Comput 25:102–116. https://doi.org/10.1109/TEVC.2020.3004012

Wang L, Zhou Z, Liu J (2022) Interval-based optimal trajectory tracking control method for manipulators with clearance considering time-dependent reliability constraints. Aerosp Sci Technol 128. https://doi.org/10.1016/j.ast.2022.107745

Wang L, Liu J, Zhou Z, Li Y (2023) A two-stage dimension-reduced dynamic reliability evaluation (TD-DRE) method for vibration control structures based on interval collocation and narrow bounds theories. ISA Trans 136:622–639. https://doi.org/10.1016/j.isatra.2022.10.033

Yadegari H, Beyramzad J, Khanmirza E (2022) Magnetorquers-based satellite attitude control using interval type-II fuzzy terminal sliding mode control with time delay estimation. Adv SPACE Res 69:3204–3225. https://doi.org/10.1016/j.asr.2022.01.018

Yin S, Xiao B, Ding SX, Zhou D (2016) A review on recent development of spacecraft attitude fault tolerant control system. IEEE Trans Ind Electron 63:3311–3320. https://doi.org/10.1109/TIE.2016.2530789

Zhou B, Zhang K-K (2024) Smooth time-varying feedback control of nonholonomic systems with applications to the attitude control of underactuated spacecraft. Guid Navig Control 04. https://doi.org/10.1142/S2737480724500043

Zhou B (2015) Global stabilization of periodic linear systems by bounded controls with applications to spacecraft magnetic attitude control. Automatica 60:145–154. https://doi.org/10.1016/j.automatica.2015.07.003

Zhou B (2019) On stability and stabilization of the linearized spacecraft attitude control system with bounded inputs. Automatica 105:448–452. https://doi.org/10.1016/j.automatica.2019.03.011

Part III
Set Theory-Based Control for Rigid-Flexible Coupling Spacecrafts

Chapter 9
Set Theory-Based Optimal Attitude-Vibration Control for Rigid-Flexible Coupling Satellite

Abstract This chapter presents a control framework designed for rigid-flexible coupling (RFC) satellites subject to multi-source uncertainties. Adopting a set-theoretical perspective, specifically interval analysis, the complex RFC dynamics are modeled by quantifying uncertain physical parameters as bounded intervals. To efficiently compute the uncertainty propagation without high computational costs, an interval-based Linear Quadratic Regulator (LQR) is developed to simultaneously stabilize the satellite's attitude and suppress flexible vibrations. Furthermore, to guarantee operational safety under dynamic conditions, the aforementioned interval-based time-dependent reliability (ITDR) index is integrated as a rigorous constraint within a multi-objective optimization scheme. Numerical simulations validate that this set-based methodology effectively mitigates the effects of uncertainties and coupling disturbances, achieving a superior balance between control performance and system reliability compared to traditional approaches.

9.1 Introduction

As technology and science for space exploration progress, spacecraft are increasingly trending towards large-scale designs. These structures are characteristically defined as RFC systems, which create pressing demands for effective methodologies in attitude and vibration control (AVC). Models based purely on rigid body dynamics are inadequate for accurately representing such systems (Subramanian and Mahadevan 2023), rendering the completion of AVC unfeasible. Thus, establishing robust dynamic modeling frameworks for RFC spacecrafts is a prerequisite for implementing advanced control methods to achieve this objective.

A common representation used to characterize satellite systems is the rigid-flexible assembly model characterized by a central hub and flexible appendages (Wu et al. 2018). This fundamental paradigm in rigid-flexible coupled dynamics is frequently employed to create or reduce the order of complex spacecraft system models and is instrumental in performing AVC (He et al. 2022; Azimi and Joubaneh 2020; Liu et al. 2019; Hu et al. 2021; Murilo et al. 2021). Presently, the RFC dynamic

C. Yang and Y. Xia, *Set Theory-Based Spacecraft Dynamics and Control*, Intelligent Control and Learning Systems 26, https://doi.org/10.1007/978-981-95-9179-4_9

model sees mature application in large-scale spacecraft, serving as a reliable tool for managing AVC in operational scenarios (Zhou et al. 2022; Yang and Xia 2023).

The interval attitude-vibration dynamic model is established by first quantifying the uncertain RFC behavior and state observation as UBB parameters. We introduce the interval LQR. This proposed regulator optimizes the uncertain control input and performance index. A significant advantage is its ability to conveniently compute the sensitivity of the state response to the uncertain parameters that is essential for the following time-dependent reliability assessment.

Leveraging the mentioned uncertain state response and computed sensitivity in the form of partial derivatives, an ITDR constraint is obtained. This constraint is designed to more accurately reflect the correspondence between the uncertain state responses and defined safety thresholds. The multi-objective optimal AVC problem under reliability constraints is then formulated by incorporating both the nominal cost function and the uncertain state response into a bi-objective optimization framework. An advanced algorithm is employed to solve this problem. Finally, an RFC spacecraft serves as an application case to demonstrate the efficacy and precision of the presented uncertain control methodology.

9.2 Deterministic Rigid–Flexible Coupling Satellite Model

Figure 9.1 illustrates an RFC spacecraft model designed to represent a rigid central cube connected to a huge flexible solar antenna. The model comprises a central core body with a flexible peripheral extension, which is a beam featuring a significant end mass (Wu et al. 2018). The central cylindrical rotor has a radius of R_r and a moment of inertia designated as J_r. The flexible beam is attached to the mentioned cylindrical rotor at its fixed end; this beam possesses a length L, a linear density μ, a banding stiffness EI, and a damping coefficient K_c. (The term I represents the sectional moment of inertia of the beam, and the term E represents the Young modulus.) The rotor itself is characterized by a viscous friction coefficient b_m. The control torque τ that acts upon this rotor is considered the exclusive input for the controller of the RFC spacecraft system, which will be detailed subsequently. At the tip of the beam, there is a concentrated mass m_L with its own inertia moment J_L.

Aiming at constructing the RFC spacecraft model, establishing two coordinate systems is necessary. The first is the non-inertial frame of reference xy, which is attached to the hub of the satellite and translates or spins concurrently, the same as the central rotor. The second is the inertial coordinate system, denoted as x_0y_0.

The rigid displacement, also known as the satellite's attitude, is expressed as $\theta(t)$, an angle between these two frames of reference. The elastic deflection of the beam should be defined, as its flexible deformation is expected to be larger than $\theta(t)$, particularly when including the tip mass. Assuming $w(x, t)$ represents the separation between a specific deflection and the position of static equilibrium within the xy frame, expressed as a function affected by the position x and the current instant t.

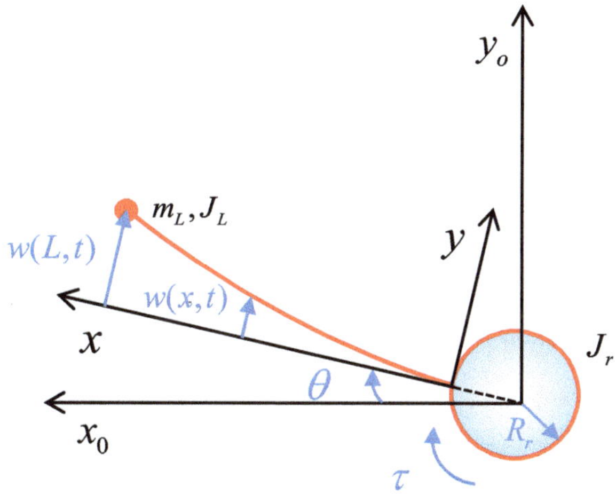

Fig. 9.1 RFC spacecraft model

Within the dynamics of the RFC spacecraft, the beam is represented as a conventional Euler–Bernoulli beam model. The subsequent force analysis is conducted using an infinitesimal approach, leading to the following relationships:

$$\begin{cases} \dfrac{\partial Q}{\partial x} = \mu \mathrm{d}x \dfrac{\partial^2 w(x,t)}{\partial t^2} \\ Q = \dfrac{\partial M}{\partial x} \end{cases} \tag{9.1}$$

Here, Q and M, respectively, represent the shear force and moment, applied to a differential Euler–Bernoulli beam segment.

Additionally, the beam's moment–curvature relationship provides

$$M = -EI \frac{\partial^2 w(x,t)}{\partial x^2} \tag{9.2}$$

Integrating Eqs. (9.1) and (9.2) produces the fundamental equation governing the transverse motion:

$$EI \frac{\partial^4 w(x,t)}{\partial x^4} + \mu \frac{\partial^2 w(x,t)}{\partial t^2} = 0 \tag{9.3}$$

To solve Eq. (9.3), it is requisite to apply the boundary conditions for both tips of the beam, which are given as

$$\begin{cases} w(0,t) = 0 \\ \dfrac{\partial w(0,t)}{\partial x} = 0 \\ \dfrac{\partial^2 w(L,t)}{\partial x^2} = 0 \\ EI\dfrac{\partial^3 w(x,t)}{\partial x^3} - \mu\dfrac{\partial^2 w(x,t)}{\partial t^2} = 0 \end{cases} \tag{9.4}$$

The boundary value problem defined previously is solved by applying the separation of variables method, expressed as

$$w(x,t) = X(x)T(t) \tag{9.5}$$

Inserting Eq. (9.5) into Eq. (9.3) while respecting the conditions in Eq. (9.4) yields the subsequent characteristic equation:

$$\cos(\beta L)\cosh(\beta L) = -1 \tag{9.6}$$

Solving Eq. (9.6) produces an unlimited series of spatial natural modes, denoted as β_i. The index i corresponds to the i-th order modal shape, $X(x)$, which is solved by the following equation:

$$X(x) = A_i[\cosh(\beta x) - \cos(\beta x) + \kappa_i(\sinh(\beta x) - \sin(\beta x))] \tag{9.7}$$

in which κ_i is defined as:

$$\kappa_i = \frac{\sin(\beta L) - \sinh(\beta L)}{\cos(\beta L) - \cosh(\beta L)} \tag{9.8}$$

The deflection of the flexible beam, $w(x,t)$, is determined through the superposition of n_m vibration modes. This is calculated as the product of the matrix of modal shape $\mathbf{\Phi}(x)$ and the instantaneous generalized displacements $\eta(t)$, expressed as:

$$\mathbf{w}(x,t) = [\mathbf{\Phi}(x)]^{\mathrm{T}}\mathbf{\Upsilon}(t) \text{ or } w(x,t) = \sum_{i=1}^{n_{\mathrm{m}}} \varphi_i(x)\eta_i(t) \tag{9.9}$$

In this formulation, $\mathbf{\Phi}(x)$ represents the matrix $\left[\phi_1(x), \phi_2(x), ..., \phi_{n_m}(x)\right]$ and $\boldsymbol{\eta}(t)$ is the vector $[\eta_1(t), \eta_2(t), ..., \eta_{n_m}(t)]^{\mathrm{T}}$. The term n_m signifies the total quantity of modal shapes included in the analysis.

Utilizing the Lagrange method, the dynamic equation for the RFC spacecraft model is then constructed. This derivation relies on the system's energy terms in the Lagrangian, respectively, denoted as E_k and E_p, which are detailed as follows:

Here is the paraphrased version of the content from the image. All technical variables, terminology, and equations are preserved as requested:

$$E_{\mathrm{k}} = E_{\mathrm{k}}^{\mathrm{rotor}} + E_{\mathrm{k}}^{\mathrm{rod}} + E_{\mathrm{k}}^{\mathrm{mass}} \tag{9.10}$$

$$E_{\mathrm{p}} = \frac{1}{2}EI\int_{0}^{L}\left(\frac{\partial^2 w}{\partial x^2}\right)^2 \mathrm{d}x \tag{9.11}$$

A detailed calculation of the kinetic energy E_k from Eq. (9.10) is provided as

$$\begin{aligned} E_{\mathrm{k}} = {} & \frac{1}{2}\dot{\theta}\left(J_{\mathrm{r}} + \mu\left(\int_{0}^{L} w^2\mathrm{d}x + \frac{1}{3}(R+L)^3 - \frac{1}{3}R^3\right) + m_{\mathrm{L}}\left(w_{\mathrm{L}}^2 + (R+L)^2\right) + J_{\mathrm{L}}\right) \\ & + \frac{1}{2}\dot{\theta}\left(2\mu\int_{0}^{L}\dot{w}(R+w(x))\mathrm{d}x + 2m_{\mathrm{L}}\dot{w}_{\mathrm{L}} + J_{\mathrm{L}}\frac{\partial \dot{w}}{\partial x} + \frac{1}{2}\left(\mu\int_{0}^{L}\dot{w}^2\mathrm{d}x + m_{\mathrm{L}}\frac{\partial \dot{w}}{\partial x}\right)\right) \end{aligned} \tag{9.12}$$

Furthermore, the RFC model incorporates the Rayleigh dissipation function $\mathcal{R}$, which is calculated as

$$\mathcal{R} = \frac{1}{2}b_{\mathrm{m}}\dot{\theta}^2 + \frac{1}{2}K_{\mathrm{e}}EI\int_{0}^{L}\left(\frac{\partial^2 \dot{w}}{\partial x^2}\right)^2 \mathrm{d}x \tag{9.13}$$

Within the Lagrange method, defined as $\mathcal{L} = E_k - E_p$, obeys the fundamental equation:

$$\frac{\mathrm{d}}{\mathrm{d}t}\left(\frac{\partial \mathcal{L}}{\partial \chi_i}\right) - \frac{\partial \mathcal{L}}{\partial \chi_i} + \frac{\partial \mathcal{R}}{\partial \chi_i} = \Gamma_i \tag{9.14}$$

In this formulation, $\boldsymbol{\chi} = \left[\theta\ \eta_1(t)\ \cdots\ \eta_{n_{\mathrm{m}}}(t)\right]^{\mathrm{T}}$ represents the generalized coordinates, and $\boldsymbol{\Gamma} = \left[\tau\ 0\ \cdots\ 0\right]^{\mathrm{T}}$ is the general force vector, both within the Lagrange method. The analysis is based on the system dynamics near equilibrium; consequently, the non-linearized dynamical model of the RFC spacecraft is permissible to be linearized. This leads to the dynamical formula for the RFC spacecraft model, which is derived to be

$$\begin{cases} \ddot{\theta}\left[I_{\mathrm{t}} + \mathbf{q}^{\mathrm{T}}\mathbf{C}_{\mathrm{rr}}\mathbf{q}\right] + \ddot{\mathbf{q}}^{\mathrm{T}}\mathbf{M}_{\mathrm{rf}} + \dot{\theta}\left[2\mathbf{q}^{\mathrm{T}}\mathbf{C}_{\mathrm{rr}}\dot{\mathbf{q}} + b_m\right] = \tau \\ \ddot{\theta}\mathbf{M}_{\mathrm{rf}} + \mathbf{M}_{\mathrm{ff}}\ddot{\mathbf{q}} - \dot{\theta}^2\mathbf{C}_{\mathrm{rr}}q + \mathbf{K}_{\mathrm{ff}}\mathbf{q} + \mathbf{B}_{\mathrm{ff}}\dot{\mathbf{q}} = \mathbf{0} \end{cases} \tag{9.15}$$

where

$$\begin{cases} I_{\mathrm{t}} = J_{\mathrm{r}} + \dfrac{1}{3}\mu\left((R+L)^3 - R^3\right) + m_{\mathrm{L}}(R+L)^2 + J_{\mathrm{L}} \\ \mathbf{B}_{\mathrm{ff}} = K_{\mathrm{e}}EI\displaystyle\int_0^L \left(\frac{\mathrm{d}^2\boldsymbol{\Phi}}{\mathrm{d}x^2}\right)^{\mathrm{T}} \frac{\mathrm{d}^2\boldsymbol{\Phi}}{\mathrm{d}x^2}\mathrm{d}x \\ \mathbf{C}_{\mathrm{rr}} = \mu\displaystyle\int_0^L \boldsymbol{\Phi}^{\mathrm{T}}\boldsymbol{\Phi}\mathrm{d}x + m_{\mathrm{L}}(\boldsymbol{\Phi}(L))^{\mathrm{T}}\boldsymbol{\Phi}(L) \\ \mathbf{K}_{\mathrm{ff}} = EI\displaystyle\int_0^L \left(\frac{\mathrm{d}^2\boldsymbol{\Phi}}{\mathrm{d}x^2}\right)^{\mathrm{T}} \frac{\mathrm{d}^2\boldsymbol{\Phi}}{\mathrm{d}x^2}\mathrm{d}x \\ \mathbf{M}_{\mathrm{ff}} = \mu\displaystyle\int_0^L \boldsymbol{\Phi}^{\mathrm{T}}\boldsymbol{\Phi}\mathrm{d}x + m_{\mathrm{L}}(\boldsymbol{\Phi}(L))^{\mathrm{T}}\boldsymbol{\Phi}(L) + \frac{1}{2}J_{\mathrm{L}}\frac{\mathrm{d}(\boldsymbol{\Phi}(L))^{\mathrm{T}}}{\mathrm{d}x}\frac{\mathrm{d}\boldsymbol{\Phi}(L)}{\mathrm{d}x} \\ \mathbf{M}_{\mathrm{rf}} = \mu\displaystyle\int_0^L \boldsymbol{\Phi}(R+x)\mathrm{d}x + m_{\mathrm{L}}(R+L)\boldsymbol{\Phi}(L) + \frac{1}{2}J_{\mathrm{L}}\frac{\mathrm{d}(\boldsymbol{\Phi}(L))^{\mathrm{T}}}{\mathrm{d}x}\frac{\mathrm{d}\boldsymbol{\Phi}(L)}{\mathrm{d}x} \end{cases} \tag{9.16}$$

Based on the relationships in Eq. (9.15), the nominal state-space equation for the RFC spacecraft model is able to be finalized and expressed as follows:

$$\dot{\mathbf{z}}(t) = \mathbf{A}\mathbf{z}(t) + \mathbf{B}\mathbf{u}(t) \tag{9.17}$$

Here

$$\begin{cases} \mathbf{A} = \begin{bmatrix} \mathbf{0} & \mathbf{I} \\ -\mathbf{M}^{-1}\mathbf{K} & -\mathbf{M}^{-1}\mathbf{D} \end{bmatrix} \\ \mathbf{B} = \begin{bmatrix} \mathbf{0} \\ \mathbf{M}^{-1}\mathbf{N} \end{bmatrix} \\ \mathbf{u} = \tau(t) \\ \mathbf{z}(t) = \begin{bmatrix} \theta(t) & \eta_1(t) & \cdots & \eta_{n_{\mathrm{m}}}(t) & \dot{\theta}(t) & \dot{\eta}_1(t) & \cdots & \dot{\eta}_{n_{\mathrm{m}}}(t) \end{bmatrix}^{\mathrm{T}} \end{cases} \tag{9.18}$$

and

$$\begin{cases} \mathbf{M} = \begin{bmatrix} I_\mathrm{t} & \mathbf{M}_\mathrm{rf}^\mathrm{T} \\ \mathbf{M}_\mathrm{rf} & \mathbf{M}_\mathrm{ff} \end{bmatrix} \\ \mathbf{D} = \begin{bmatrix} b_\mathrm{m} & \mathbf{0} \\ \mathbf{0} & \mathbf{B}_\mathrm{ff} \end{bmatrix} \\ \mathbf{K} = \begin{bmatrix} 0 & \mathbf{0} \\ \mathbf{0} & \mathbf{K}_\mathrm{ff} \end{bmatrix} \\ \mathbf{N} = \begin{bmatrix} 1 \\ \mathbf{0} \end{bmatrix} \end{cases} \tag{9.19}$$

With the component matrices defined in Eq. (9.19), the constitution of the nominal state-space equation for the RFC spacecraft dynamics is now complete. This model, however, does not yet incorporate uncertainties. When the system is subject to parameters with uncertainty, it can't be precisely formulated in this manner, nor can it be addressed with the baseline solution technique previously discussed. Therefore, the rest of this chapter will endeavor to solve the uncertain RFC spacecraft model and associated control issue.

9.3 Interval LQR Control Method Based on IDWA

For the controller of the RFC spacecraft, considering the uncertain system matrices $\mathbf{A}(\mathbf{b}^\mathrm{I})$ and $\mathbf{B}(\mathbf{b}^\mathrm{I})$, the nominal expression is adapted into interval notation as

$$\mathbf{Q} - \mathbf{S}^\mathrm{I}\mathbf{B}(\mathbf{b}^\mathrm{I})\mathbf{R}^{-1}(\mathbf{B}(\mathbf{b}^\mathrm{I}))^\mathrm{T}\mathbf{S}^\mathrm{I} + (\mathbf{A}(\mathbf{b}^\mathrm{I}))^\mathrm{T}\mathbf{S}^\mathrm{I} + \mathbf{S}^\mathrm{I}\mathbf{A}(\mathbf{b}^\mathrm{I}) = \mathbf{0} \tag{9.20}$$

In accordance with the IDWA method mentioned in the previous chapters, each row of $\mathbf{B}_s$ represents the input uncertainty for one sample point from a total of N_s sample points. By employing these parameters in the ARE in sequence, $\mathbf{S}(t)$ is calculated as follows:

Here is the paraphrased version of the content from the image, maintaining all technical variables, equations, and terminology:

$$\begin{aligned} &\mathbf{Q} - \mathbf{S}\left(\hat{\mathbf{b}}_k^{(i)}\right)\mathbf{B}\left(\hat{\mathbf{b}}_k^{(i)}\right)\mathbf{R}^{-1}\left(\mathbf{B}\left(\hat{\mathbf{b}}_k^{(i)}\right)\right)^\mathrm{T}\mathbf{S}\left(\hat{\mathbf{b}}_k^{(i)}\right) \\ &+ \left(\mathbf{A}\left(\hat{\mathbf{b}}_k^{(i)}\right)\right)^\mathrm{T}\mathbf{S}\left(\hat{\mathbf{b}}_k^{(i)}\right) + \mathbf{S}\left(\hat{\mathbf{b}}_k^{(i)}\right)\mathbf{A}\left(\hat{\mathbf{b}}_k^{(i)}\right) = \mathbf{0},\ i = 1, 2, \ldots, n_\mathrm{P} \end{aligned} \tag{9.21}$$

Following this procedure, yields the interval bound $\mathbf{S}(t)$ as

$$\mathbf{S} \in \mathbf{S}^\mathrm{I} = [\underline{\mathbf{S}}, \overline{\mathbf{S}}] = \mathbf{S}^\mathrm{I}_\mathrm{s} \bigcup \mathbf{S}^\mathrm{I}_\mathrm{a} = [\underline{\mathbf{S}_\mathrm{s}}, \overline{\mathbf{S}_\mathrm{s}}] \bigcup [\underline{\mathbf{S}_\mathrm{a}}, \overline{\mathbf{S}_\mathrm{a}}] \tag{9.22}$$

Here, the subscript index 'a' and 's', respectively, signify approximate points and the sample within the IDWA method.

Moreover, the nominal optimal feedback gain $\mathbf{P}$ is reformulated into the form of interval uncertainty $\mathbf{P}^{\mathrm{I}}$, which can be described as:

$$\mathbf{P}^{\mathrm{I}} = -\mathbf{R}^{-1}\left(\mathbf{B}\left(\mathbf{b}^{\mathrm{I}}\right)\right)^{\mathrm{T}}\mathbf{S}^{\mathrm{I}} \tag{9.23}$$

Utilizing the introduced IDWA method, the uncertain control gain interval is acquired via the following computation:

$$\mathbf{P}\left(\hat{\mathbf{b}}_k^{(i)}\right) = -\mathbf{R}^{-1}\left(\mathbf{B}\left(\hat{\mathbf{b}}_k^{(i)}\right)\right)^{\mathrm{T}}\mathbf{S}\left(\hat{\mathbf{b}}_k^{(i)}\right) \tag{9.24}$$

Analogously, the interval bound of $\mathbf{P}(t)$ for the feedback gain with uncertainty is determined by reiterating the process:

$$\mathbf{P} \in \mathbf{P}^{\mathrm{I}} = \left[\underline{\mathbf{P}}, \overline{\mathbf{P}}\right] = \mathbf{P}_{\mathrm{s}}^{\mathrm{I}} \bigcup \mathbf{P}_{\mathrm{a}}^{\mathrm{I}} = \left[\underline{\mathbf{P}_{\mathrm{s}}}, \overline{\mathbf{P}_{\mathrm{s}}}\right] \bigcup \left[\underline{\mathbf{P}_{\mathrm{a}}}, \overline{\mathbf{P}_{\mathrm{a}}}\right] \tag{9.25}$$

Employing the control law $\mathbf{u} = -\mathbf{P}\mathbf{x}$ in Eq. (9.23) results in the equation as:

$$\mathbf{R}\mathbf{u} = -\mathbf{B}^{\mathrm{T}}\mathbf{S}\mathbf{x} \tag{9.26}$$

The deterministic performance metric J, originally defined in $J = \int_0^{\infty}\left(\mathbf{x}^{\mathrm{T}}\mathbf{Q}\mathbf{x} + \mathbf{u}^{\mathrm{T}}\mathbf{R}\mathbf{u}\right)\mathrm{d}t$, can thus be derived by incorporating $\mathbf{u} = -\mathbf{P}\mathbf{x}$ along with Eqs. (5.33), (9.23), and (9.26) as:

$$J = \int_0^{\infty} \mathbf{x}^{\mathrm{T}}\left(2\mathbf{Q} + \mathbf{A}^{\mathrm{T}}\mathbf{S} + \mathbf{S}\mathbf{A}\right)\mathbf{x}\mathrm{d}t \tag{9.27}$$

Leveraging the IDWA method and using the following expression, the uncertain control cost interval can be computed as

$$J\left(\hat{\mathbf{b}}_k^{(i)}\right) = \int_0^{\infty} \left(\mathbf{z}\left(\hat{\mathbf{b}}_k^{(i)}\right)\right)^{\mathrm{T}}\left(2\mathbf{Q} + \left(\mathbf{A}\left(\hat{\mathbf{b}}_k^{(i)}\right)\right)^{\mathrm{T}}\mathbf{S}\left(\hat{\mathbf{b}}_k^{(i)}\right) + \mathbf{S}\left(\hat{\mathbf{b}}_k^{(i)}\right)\mathbf{A}\left(\hat{\mathbf{b}}_k^{(i)}\right)\right)\mathbf{z}\left(\hat{\mathbf{b}}_k^{(i)}\right)\mathrm{d}t \tag{9.28}$$

Following the same procedure, reiterating the previous process determines the interval bound for the uncertain cost of control $J(t)$:

$$J \in J^{\mathrm{I}} = \left[\underline{J}, \overline{J}\right] = J_{\mathrm{s}}^{\mathrm{I}} \bigcup J_{\mathrm{a}}^{\mathrm{I}} = \left[\underline{J_{\mathrm{s}}}, \overline{J_{\mathrm{s}}}\right] \bigcup \left[\underline{J_{\mathrm{a}}}, \overline{J_{\mathrm{a}}}\right] \tag{9.29}$$

In conclusion, this methodology successfully yields accurate interval estimators for the Riccati equation solution, $[\underline{\mathbf{S}}, \overline{\mathbf{S}}]$; the gain, $\mathbf{P}^{\mathrm{I}} = [\underline{\mathbf{P}}, \overline{\mathbf{P}}]$; and the cost of control, $J^{\mathrm{I}} = [\underline{J}, \overline{J}]$.

9.4 Interval Uncertainty-Based Multi-objective Optimization for Attitude-Vibration Control

9.4.1 Interval Uncertainty-Based Multi-objective Optimization Problem

Achieving superior performance in an RFC spacecraft heavily relies on the appropriate selection of its LQR controller parameters. Consequently, this work proposes an uncertain multi-objective optimization problem structured to concurrently minimize both the control costs (J) and the norm of the interval deflection at the beam tip ($||\Delta w(L, t)||$), all while adhering to specified reliability constraints. The formal expression of this optimization problem is:

$$\begin{aligned} &\text{find} && \mathbf{Q}, \mathbf{R} \\ &\min && J \\ &\min && \|\Delta w(L, t)\| \\ &\text{s.t.} && Rel(T) \geq \text{poss} \\ & && \underline{\mathbf{b}} \leq \mathbf{b} \leq \overline{\mathbf{b}} \end{aligned} \tag{9.30}$$

9.4.2 Solution Tool: C-TAEA

The constrained two-archive evolutionary algorithm (C-TAEA) (Li et al. 2022) is a novel algorithm for multi-objective optimization with constraints selected to resolve the issue outlined above for the RFC spacecraft. This algorithm is particularly adept at handling such optimization problems as it is fundamentally designed to simultaneously address convergence, diversity, and feasibility (Li et al. 2019), which outperforms GA in some aspects (Yang et al. 2019). The algorithm functions by operating two collaborative archives in parallel: one is a convergence-oriented archive (CA) that serves as the internal driving force, promoting the population's progression toward the Pareto front, and the other is a diversity-oriented archive (DA) that runs concurrently to ensure the population diversity is maintained.

The continuous interaction and updating of both the DA and CA are the central part of the full optimization cycle. The DA's role is to survey unexplored areas in CA, including not feasible areas, thereby supplementing CA's performance and injecting information of diversity into the process. To maximize this complementary synergy, the optimization employs a restricted mate selection mechanism. This mechanism adaptively chooses appropriate mating parents on the basis of the population's current state of the population (Li et al. 2022).

In this chapter, each population individual is defined by its location points, which are derived from the optimization objectives. The entire population undergoes iterative updates to converge on the optimal solution. This update process is driven by the driving force from the CA parameters influencing the population, combined with the diversified behavior (influenced by the feasible region) originating from the DA parameters. Consequently, implementing C-TAEA for this optimization process effectively demonstrates the algorithm's efficiency for this specific multi-objective optimization problem in RFC spacecraft control.

9.4.3 Flowchart of the Proposed Method

The complete analysis procedure is bifurcated into two primary segments: the multi-objective optimization with constraints process and the control system performance evaluation, which leverages the parameter sets produced during the optimization's iterations. Each optimization round is dedicated to updating the CA and DA populations. This step generates the relevant system parameters, including the weight matrices $\mathbf{Q}$ and $\mathbf{R}$, which are then supplied to the uncertain control system model to calculate its system responses. The metrics derived from these responses are subsequently verified against the specified optimization constraint requirements. The control system analysis component is subdivided into three elements: the satellite model, the IDWA method, and the analysis for reliability. The coupling mechanism between them can be listed as follows: First, the satellite model is formulated as an order-extended state-space, which allows for the extraction of partial derivative parameters. These parameters are then utilized within the assessment for reliability. Concurrently, the IDWA method is applied to the nominal control model to formulate the attitude control system with interval uncertainties and to settle its responses in the presence of uncertainties. Finally, the resulting interval boundaries of the response are designated as the target of the reliability calculation, which employs the ITDR method in conjunction with the previously calculated parameters of the partial derivative. The flowchart of the proposed method is illustrated in Fig. 9.2.

9.5 Numerical Example

To assess the satellite's vibration response, its flexible beam is characterized as a typical Euler–Bernoulli beam, while the rigid displacement angle between the coordinate systems is employed to define the attitude situation. Table 9.1 enumerates the parameters for this nominal RFC model of the satellite.

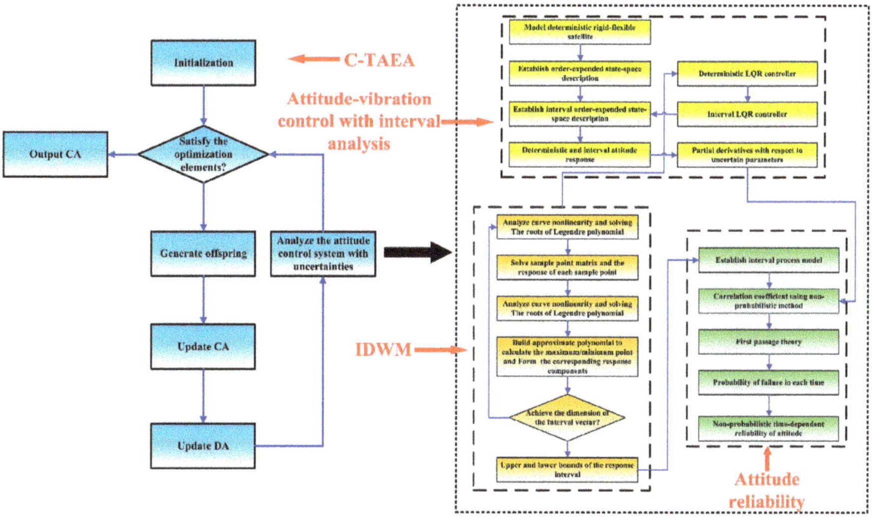

Fig. 9.2 Flowchart of the proposed method

Table 9.1 Nominal parameters for constructing the RFC model

Notation	Description	Value/unit
R_r	Rotor radius	0.05 m
J_c	Rotor inertia moment	0.3 kg m^2
L	Beam length	1.5 m
μ	Beam linear density	0.54 kg/m
K_e	Beam damping coefficient	0.03
EI	Beam bending stiffness	18.4 N m^2
b_m	Rotor viscous damping	0.15
m_L	Tip mass	0.35 kg
J_L	Tip mass moment of inertia	0.04 kg m^2

The configuration and metrics of an RFC spacecraft are usually impacted to some degree by the complex environment in space, leading to deviations from its on-ground characterization. The satellite's structure inherently contains uncertain parameters, which in turn affect the efficacy of its attitude-vibration control.

Consequently, the variables of specific components are treated as UBB parameters. For this chapter, the inertia moment of the satellite's central rotor, J_c, and the rotor's viscous damping, b_m, are designated as interval uncertain parameters. This choice is made because they are anticipated to have a significant influence on the satellite's attitude and rod deflection. The specified parameters of uncertainty are $J_c^I = [0.27, 0.33]$ and $b_m^I = [0.14, 0.16]$.

The RFC spacecraft model can thus be constructed. Utilizing the core state variables of the dynamics, the basic state is established as $\mathbf{z}(t)$ =

$[\theta(t), \eta_1(t), \eta_2(t), \dot{\theta}(t), \dot{\eta}_1(t), \dot{\eta}_2(t)]^{\mathrm{T}}$ to facilitate the analysis of the system's dynamic responses. To concurrently calculate the system state's partial derivatives, an order-expanded interval state space is constructed, which computes the results by employing time-dependent reliability. Consequently, the basic state is converted to $\mathbf{Z}(t) = [\theta(t), \eta_1(t), \eta_2(t), \dot{\theta}(t), \dot{\eta}_1(t), \dot{\eta}_2(t), \frac{\partial \mathbf{z}(t)}{\partial b_k}]^{\mathrm{T}}$. Here, b_k corresponds to the uncertain parameters J_c and b_m, noting that $\frac{\partial \mathbf{z}(t)}{\partial b_k}$ represent the partial derivatives vector.

The IDWA method presented in this chapter is then utilized to establish the RFC system, enabling the computation of the uncertainty band of the responses. Extending the IDWA method to the attitude LQR controller facilitates the proposal of the interval LQR control approach.

Employing the particular weight matrices $\mathbf{Q}$ and $\mathbf{R}$ allows for the derivation of the associated control law. This subsequently permits obtaining both the nominal control cost J and the norm of the interval deflection at the beam tip $||\Delta w(L, t)||$. These two results serve as optimal objectives within the multi-objective optimization, representing the inherent trade-offs in the control process. The aim is to attain the desired control performance, namely, a reduced response variation and lower control effort.

With the optimization objectives defined, the constrained optimization is performed. The uncertain response of the satellite attitude angle $\theta(t)$ is readily obtained via analysis of interval uncertainty. The accuracy of the resulting LB and UB of the $\theta(t)$ dynamic response is then verified by comparison with 10,000 MCS runs. Furthermore, the quantified reliability is computed by applying the ITDR method, which utilizes the partial derivatives solved in the order-expanded state. This reliability assessment outcome is imposed as an optimization constraint. Thus, to ensure the system is more reliable, the reliability threshold for the attitude angle $\theta(t)$ exceeding 37° is established at 0.9.

The optimization is configured with an initial population size of 50 and a maximum generations of 5000. Executing this process generates the Pareto set, which is illustrated in Fig. 9.3. From this front, two representative solution sets are chosen to evaluate the attitude-vibration control performance. Table 9.2 provides a detailed list of their corresponding solution ID, decision variables, and optimization objectives. Following the acquisition of these typical optimal solutions, the analysis proceeds to the deflection of the flexible beam and the attitude angle of the spacecraft. Figure 9.4 exhibits the two clusters of interval responses for the angle $\theta(t)$. To confirm the constraint satisfaction, the computed reliability metrics for the system outputs $\theta(t)$ (shown in Fig. 9.4) for these two solution sets are acquired using the ITDR method.

With the 37° threshold, the calculated reliabilities of the spacecraft are enumerated in Table 9.3; all values surpass the initial 0.9 setting. All results are benchmarked against 10,000 MCS runs using corresponding parameters, and these reliability results are also included in Table 9.3. Additionally, Table 9.3 presents the relative errors against MCS to validate the computational accuracy. Besides this, a

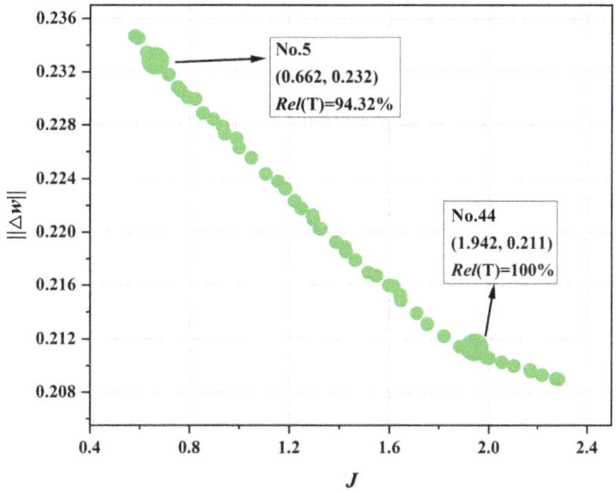

Fig. 9.3 Pareto front of the control output

Table 9.2 The selected solution ID of optimized Q and R values and their corresponding system parameters

Solution ID	**Q**(1, 1)	**Q**(2, 2)	**Q**(3, 3)	**Q**(4, 4)	**Q**(5, 5)	$\mathbf{Q}_2(1, 1)$	**R**	J	$\lVert \Delta w(L, t) \rVert$
No. 5	0.1240	0.8546	0.7206	0.4141	0.1814	0.1004	2.5012	0.662	0.233
No. 44	0.2984	0.2495	0.2023	0.9535	0.9789	0.1002	2.5020	1.942	0.211

non-probabilistic time-independent method (NPTIM) is applied for reliability assessment. The comparison confirms that the proposed interval method yields more realistic reliability estimates than those from NPTIM. The discrepancy between the proposed method and the actual scenario is extremely small, which validates the precision of the proposed approach to a certain extent.

The discrepancy concerning the reliability outcomes originates primarily from the preceding interval uncertainty analysis process and the reliability assessment process itself. The sampling points in the IDWA do not encompass the entire set of sample points; therefore, even with dimension-by-dimension optimization, the precision loss inherent in interval analysis is fundamentally a reflection of the algorithm's underlying principle.

Furthermore, when applying ITDR methods, the inherent limitations of numerical computation mandate a time-stepping solution, which consequently yields reliability estimates of compromised precision. In principle, the numerically feasible region also fails to perfectly delineate the effective computational domain. This is because the dependence analysis between variables typically generates an ellipsoidal feasible region, causing extreme points in the lower-left and upper-right regions of Fig. 5. 3 to be omitted. Whereas, due to the feasible region's specific geometry, the effect of these omitted points is generally minor. Beyond its high accuracy, the proposed method's efficiency significantly reduces time consumption compared to MCS. On

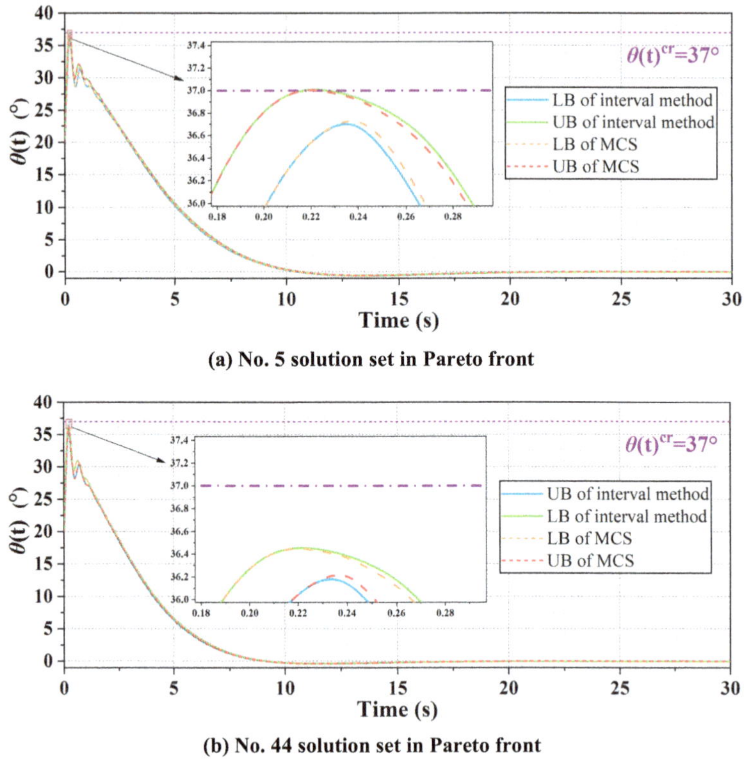

Fig. 9.4 θ response envelope of two typical solutions utilizing the interval analysis and MCS

Table 9.3 Reliability assessment of two solution sets using different model

Solution index	No. 5 (%)	No. 44 (%)
MCS	97.80	100
Interval-based approach	94.32	100
NPTIM	92.72	100
Error Relative to Benchmark (MCS)	3.48	-

a 2.20 GHz 16-core processor with 16 GB RAM, a single MCS control process averaged 82.5 s. In contrast, the introduced interval approach requires only 7.9 s. The introduced interval approach demonstrates significant computational efficiency.

Now that the optimal parameters have been determined, the controlled response of the satellite is obtained. This computation yields the nominal attitude angle response along with the bounds of the interval attitude response $\theta(t)$, considering the uncertainty variables. To ulteriorly validate the precision of the introduced interval-based approach, the response envelopes calculated from MCS are benchmarked against the associated solution sets, as illustrated in Fig. 9.4. The remarkable coincidence observed between the response interval envelope from the interval-based approach and those from MCS serves to assess the precision of the approach.

The optimization problem is established by two objectives: the vector norm of the interval deflection at the beam tip $||\Delta w(L, t)||$ and the nominal control cost J. Their respective responses over time history and within the defined uncertainty ranges are shown subsequently. When analyzing the satellite's operation under uncertain optimal attitude control, these two optimization objectives hold a level of importance comparable to that of the attitude angle. The terms $w(L, t)$ and $\Delta w(L, t)$ represent the nominal and interval deflection at the beam tip, respectively.

Figure 9.5 illustrates the LB and UB for deflection considering uncertainty along the whole flexible beam for each of the four solution sets. Following this, Fig. 9.6 details the deflection response curve of the tip of the flexible beam, also presenting the LB and UB. To more clearly demonstrate the deformation's effect, the display is limited to the initial 3 s of the response. This is because the deflection at an arbitrary location on the beam trends toward zero after this interval. This outcome confirms that the proposed method is effective for controlling flexible deformation. It also validates the accuracy of the forecast for the vibration response with uncertainty throughout the control process.

Figure 9.7 presents the nominal and uncertainty of control costs for the RFC spacecraft. A particularly high degree of coincidence is evident between the bounds of the control costs derived from the interval-based approach and the results from MCS, which substantiates the precision of the interval-based approach for this system.

Furthermore, the general tendency of the control costs shows its amplitude promptly converging to zero over time, a fact that verifies the effectiveness of the uncertain control system and the relative stability of the system after regulation. This transformation in control costs is also relative to the satellite's attitude response illustrated in Fig. 9.4: Enhanced control effectiveness results in effective suppression of the attitude angle's amplitude after a time lag, and vice versa.

By maintaining the same ordinate scale across the two panels of Fig. 9.7, it is clear that both the nominal and uncertain result of control costs increase as illustrated from panel (a)–(b). This escalation in cost correlates with a progressively better system response (regarding the AVC responses depicted in Fig. 9.4 through Fig. 9.6).

These collective results show that the introduced interval-based control approach successfully fulfills the control of attitude vibration while satisfying the reliability constraint. This fulfills the system's required performance index and meets the control demands for the RFC spacecraft.

9.6 Summary

This chapter establishes a comprehensive AVC framework tailored for RFC satellites subject to multi-source interval uncertainties. Central to this methodology is the application of the IDWA technique grounded in polynomial chaos expansions, which is not only employed to accurately quantify the uncertain dynamics of the coupled

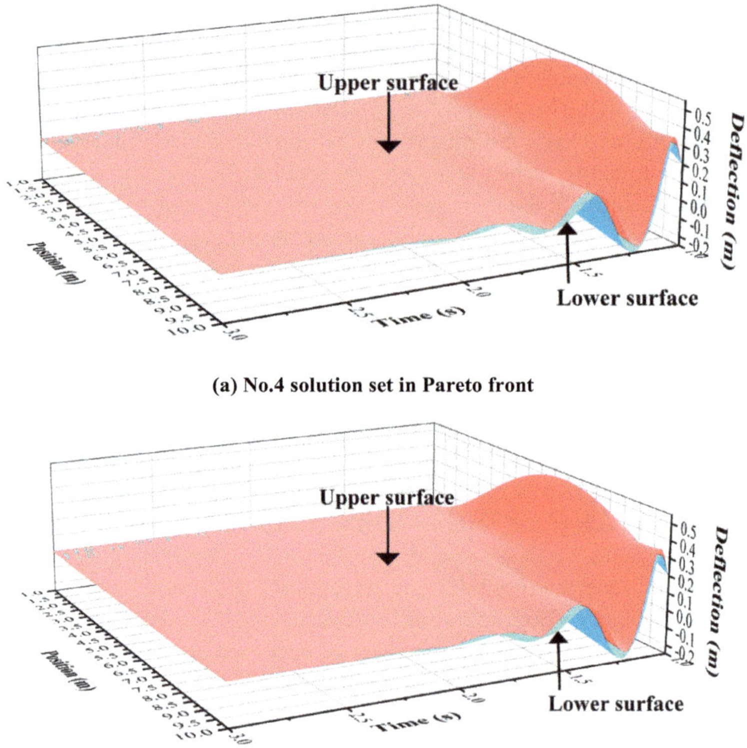

(a) No.4 solution set in Pareto front

(b) No.44 solution set in Pareto front

Fig. 9.5 Uncertain flexible beam deflection with two typical solutions

rigid-flexible system but also serves as the computational basis for designing an interval-based LQR. Furthermore, to rigorously guarantee system safety over time, an ITDR index is integrated as a critical constraint. To address the trade-offs between performance and safety, the C-TAEA is utilized to solve the complex multi-objective optimization problem. Consequently, the proposed strategy successfully achieves optimal control for the entire system, meeting stringent response requirements even under significant parameter perturbation.

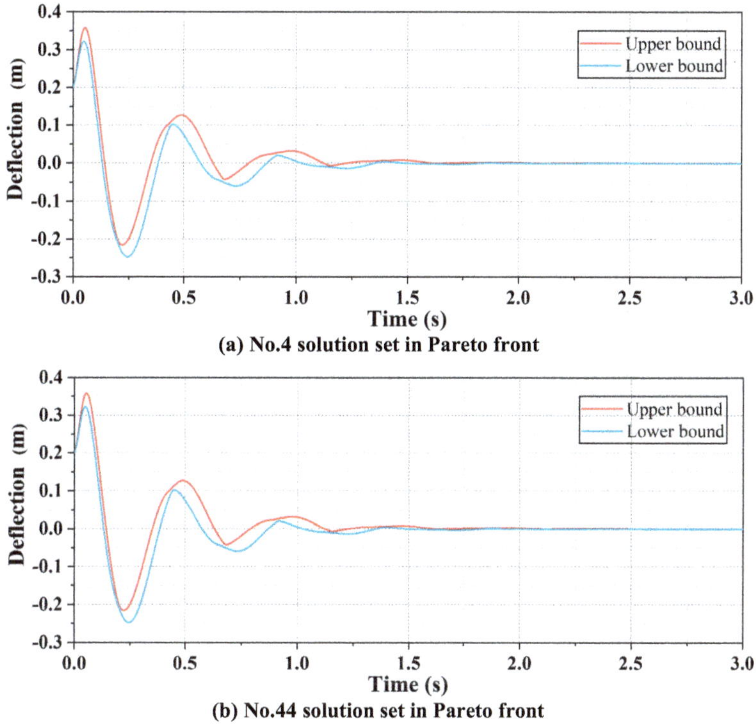

Fig. 9.6 Uncertain flexible deflection at beam tip with two typical solutions

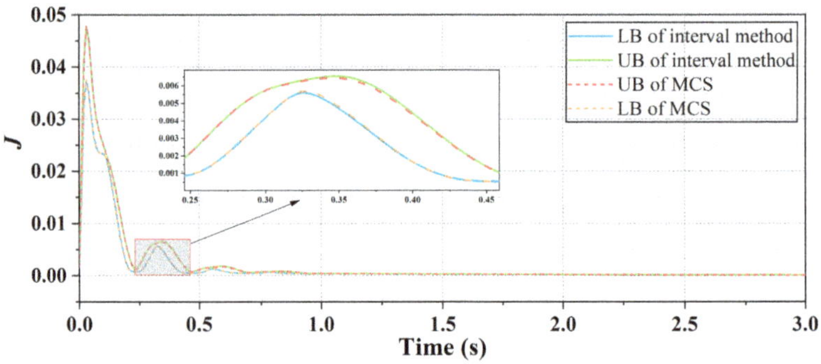

(a) No. 4 solution set in Pareto front

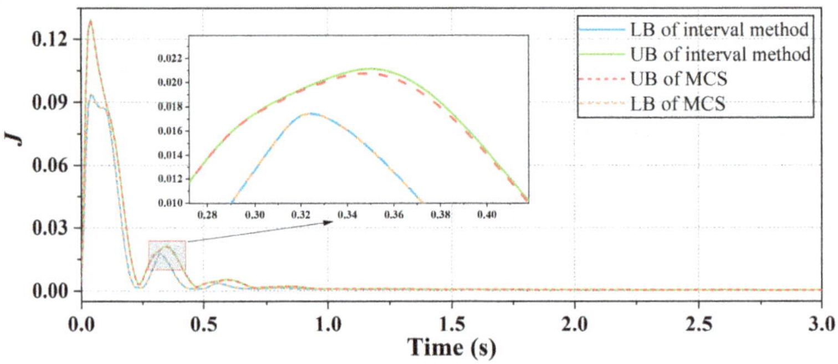

(b) No. 44 solution set in Pareto front

Fig. 9.7 Control costs of two typical solutions utilizing the interval analysis and MCS

References

Azimi M, Joubaneh EF (2020) Dynamic modeling and vibration control of a coupled rigid-flexible high-order structural system: a comparative study. Aerosp Sci Technol 102. https://doi.org/10.1016/j.ast.2020.105875

He G, Cao D, Cao Y, Huang W (2022) Dynamic modeling and orbit maneuvering response analysis for a three-axis attitude stabilized large scale flexible spacecraft installed with hinged solar arrays. Mech Syst SIGNAL Process 162. https://doi.org/10.1016/j.ymssp.2021.108083

Hu Y, Geng Y, Biggs JD (2021) Simultaneous spacecraft attitude control and vibration suppression via control allocation. J Guid CONTROL Dyn 44:1853–1861. https://doi.org/10.2514/1.G005834

Li K, Chen R, Fu G, Yao X (2019) Two-archive evolutionary algorithm for constrained multiobjective optimization. IEEE Trans Evol Comput 23:303–315. https://doi.org/10.1109/TEVC.2018.2855411

Li W, Gong W, Ming F, Wang L (2022) Constrained multi-objective evolutionary algorithm with an improved two-archive strategy. Knowl-BASED Syst 246. https://doi.org/10.1016/j.knosys.2022.108732

Liu L, Wang X, Sun S, Cao D, Liu X (2019) Dynamic characteristics of flexible spacecraft with double solar panels subjected to solar radiation. Int J Mech Sci 151:22–32. https://doi.org/10.1016/j.ijmecsci.2018.10.067

Murilo A, de Deus Peixoto PJ, Gadelha de Souza LC, Lopes RV (2021) Real-time implementation of a parameterized model predictive control for attitude control systems of rigid-flexible satellite. Mech Syst SIGNAL Process 149. https://doi.org/10.1016/j.ymssp.2020.107129

Subramanian A, Mahadevan S (2023) Probabilistic physics-informed machine learning for dynamic systems. Reliab Eng Syst Saf 230. https://doi.org/10.1016/j.ress.2022.108899

Wu S, Chu W, Ma X, Radice G, Wu Z (2018) Multi-objective integrated robust H ∞ control for attitude tracking of a flexible spacecraft. ACTA Astronaut 151:80–87. https://doi.org/10.1016/j.actaastro.2018.05.062

Yang C, Xia Y (2023) Interval uncertainty-oriented optimal control method for spacecraft attitude control. IEEE Trans Aerosp Electron Syst 59:5460–5471. https://doi.org/10.1109/TAES.2023.3257777

Yang C, Liang K, Zhang X, Geng X (2019) Sensor placement algorithm for structural health monitoring with redundancy elimination model based on sub-clustering strategy. Mech Syst SIGNAL Process 124:369–387. https://doi.org/10.1016/j.ymssp.2019.01.057

Zhou W, Zhang K, Wu S, Tan S, Wu Z (2022) Distributed cooperative control for vibration suppression of a flexible satellite. Aerosp Sci Technol 128. https://doi.org/10.1016/j.ast.2022.107750

Chapter 10
Bounded Model Predictive Control for Rigid-Flexible Coupling Satellite

Abstract This chapter introduces a bounded model predictive control (BMPC) framework tailored for the attitude and vibration control (AVC) of rigid-flexible coupling (RFC) satellites subject to multi-source uncertainties. Addressing the complexity of high-dimensional satellite systems, the chapter begins by establishing a nominal RFC dynamic model and proposing an interval uncertainty-based discretization method to process unknown-but-bounded (UBB) system parameters. A novel BMPC strategy is then developed, rooted in the interval dimension-wise analysis (IDWA) technique, to efficiently predict the bounds of control inputs, attitude angles, and flexible deflections. To further ensure operational safety, interval-based time-dependent reliability (ITDR) assessment is integrated into the control framework, evaluating system performance under various operational situations. Numerical validations confirm that this interval-based approach achieves approximation accuracy comparable to Monte Carlo simulations (MCSs) but with significantly reduced computational time, establishing a solid foundation for real-time control of uncertain satellite systems.

10.1 Introduction

Satellite-related technologies are in a state of continuous evolution. A large satellite system, which typically comprises a central stiff hub and large flexible appendages, is commonly represented as a classical RFC dynamic model (Ghorbani et al. 2022). A significant portion of contemporary research concentrates on controlling singular satellite features, for instance, orbit (Zhou et al. 2023), attitude (Cao and Zhao 2017; Kumar et al. 2025), or vibration (Chen et al. 2023; Zhou et al. 2022). The effectiveness of managing attitude and vibration is intimately linked to the satellite's self-safe state. Consequently, AVC is a critical factor in the safety and cost analysis of RFC satellites (Zhang et al. 2017; Wang et al. 2026). This indicates that an attitude-vibration coupling model is capable of accurately characterizing complex satellite systems. Moreover, a model of this nature can be effectively utilized in the on-orbit environment to perform necessary control tasks.

C. Yang and Y. Xia, *Set Theory-Based Spacecraft Dynamics and Control*, Intelligent Control and Learning Systems 26, https://doi.org/10.1007/978-981-95-9179-4_10

To address the pressing demands in satellite control, this chapter focuses on a specific set of problems: model analysis, AVC, and the associated safety performance for satellites operating under multi-source uncertainties in various complex operation situations. A nominal RFC dynamic model is first developed to perform the AVC task. By treating the uncertain parameters within the satellite system as UBB numbers, an interval uncertainty-based discrete method for the state-space representation is proposed. This discretization is a prerequisite for applying the uncertain MPC to the system. An IDWA method is then utilized to construct an uncertain control system based on the RFC dynamic model. Leveraging the IDWA method, the bounded model predictive control (BMPC) method is proposed to manage the uncertain AVC of RFC satellites. This BMPC approach is specifically designed to accommodate different constraints corresponding to different operation situations. An ITDR metric, introduced via the first passage method, is also incorporated. Finally, an RFC satellite model is employed to describe the validation of the introduced BMPC method for AVC across two operational situations.

10.2 Discrete State-Space Equation with Interval Uncertainty

A foundational requirement of the MPC method is its implementation on a discrete system. Consequently, this section focuses on the interval uncertainty-based discretization of the continuous state-space equation with uncertainties. For the continuous LTI state-space model, the solutions are provided as (Wigren 2023):

$$\mathbf{x}(t) = \exp(\mathbf{A}(t - t_0))\mathbf{x}(0) + \int_{t_0}^{t} \exp(\mathbf{A}(t - \tau))\mathbf{B}\mathbf{u}(\tau)d\tau \tag{10.1}$$

In this equation, t_0 signifies the previous time of the time variable t; τ is the integral variable; and $\exp(\mathbf{A}(t))$ represents the system's state transfer matrix.

The conversion to a discrete-time representation begins by letting the discrete state as $\mathbf{x}(k) = \mathbf{x}(t)|_{t=kT_s}$ and the discrete input as $\mathbf{u}(k) = \mathbf{u}(t)|_{t=kT_s}$, where T_s denotes the time step. This results in the discrete system:

$$\mathbf{x}(k+1) = \exp(\mathbf{A}T_s)\mathbf{x}(k) + \int_{kT_s}^{(k+1)T_s} \exp(\mathbf{A}[(k+1)T_s - \tau])\mathbf{B}\mathbf{u}(\tau)d\tau \tag{10.2}$$

Assuming $\mathbf{u}(\tau)$ is constant and therefore independent of the integral variable τ, it can be denoted as $\mathbf{u}(k)$. Through the variable substitution $\hat{\tau} = (k+1)T_s - \tau$,, Eq. (10.22) is transformed as follows:

$$\mathbf{x}(k+1) = \exp(\mathbf{A}T_s)\mathbf{x}(k) + \int_{0}^{T_s} \exp(\mathbf{A}\overline{\tau})\mathbf{B}d\overline{\tau}\mathbf{u}(k) \tag{10.3}$$

By defining the state matrix as $\mathbf{G} = \exp(\mathbf{A}T_s)$ and the input matrix as $\mathbf{H} = \int_0^{T_s} \exp(\mathbf{A}\hat{\tau})\mathbf{B}d\hat{\tau}$, the discrete state-space equation can be formulated in a compact expression as follows:

$$\mathbf{x}(k+1) = \mathbf{G}\mathbf{x}(k) + \mathbf{H}\mathbf{u}(k) \tag{10.4}$$

where $\mathbf{x}(k) = \left[x_1(k)\ x_2(k) \cdots x_m(k) \right]^T \in \mathbb{R}^{m\times 1}$ and $\mathbf{u}(k) = \left[u_1(k)\ u_2(k) \cdots u_n(k) \right]^T \in \mathbb{R}^{n\times 1}$.

Based on interval notation, the nominal discrete state-space equation is recast as an interval uncertain model:

$$\mathbf{x}^I(k+1) = \mathbf{G}\big(\mathbf{b}^I\big)\mathbf{x}^I(k) + \mathbf{H}\big(\mathbf{b}^I\big)\mathbf{u}^I(k) \tag{10.5}$$

where $\mathbf{b}^{\mathrm{I}}$ denotes the interval uncertainty vector.

Solving the interval discrete state-space equation involves identifying the LB/UB of $\mathbf{x}^{\mathrm{I}}(k)$ at each time step while accounting for uncertainties (Yang and Yu 2025).

An approximation for the interval matrices can be derived using a first-order Taylor series expansion.

$$\mathbf{A}\big(\mathbf{b}^I\big) = \mathbf{A}(\mathbf{b}^{\mathrm{c}}) + \Delta\mathbf{A}\big(\mathbf{b}^I\big),\ \mathbf{B}\big(\mathbf{b}^I\big) = \mathbf{B}\big(\mathbf{b}^c\big) + \Delta\mathbf{B}\big(\mathbf{b}^I\big) \tag{10.6}$$

Here, $\mathbf{b}^{\mathrm{c}}$ represents the nominal vector of the uncertain parameters $\mathbf{b}$. The uncertain part is formulated as follows:

$$\Delta\mathbf{A}\big(\mathbf{b}^{\mathrm{I}}\big) = \sum_{q=1}^{m_u} \Big(\partial\mathbf{A}(\mathbf{b}^{\mathrm{c}}) \Big/ \partial b_q\Big)\Delta b_q \eta,\ \Delta\mathbf{B}\big(\mathbf{b}^{\mathrm{I}}\big) = \sum_{q=1}^{m_u} \Big(\partial\mathbf{B}(\mathbf{b}^{\mathrm{c}}) \Big/ \partial b_q\Big)\Delta b_q \eta \tag{10.7}$$

In this context, m_u signifies the total number of uncertainties; b_q denotes the qth interval component of $\mathbf{b}$; Δb_q is the radius of b_q; and $\eta = [-1, 1]$ represents the standard interval.

A hierarchical expression defines the exponential function of a matrix (Dwivedi and Sahai 2020):

$$\exp(\mathbf{A}) = \sum_{j=0}^{\infty} \mathbf{A}^n \big/ j! \tag{10.8}$$

Based on this definition, the uncertain state matrix $\mathbf{G}\big(\mathbf{b}^{\mathrm{I}}\big)$ for the uncertain discrete state-space is obtained as follows:

$$\mathbf{G}\big(\mathbf{b}^I\big) = \exp\Big(\mathbf{A}\big(\mathbf{b}^I\big)T_s\Big) = \sum_{j=0}^{\infty} \Big[\Big(\mathbf{A}(\mathbf{b}^c) + \Delta\mathbf{A}\big(\mathbf{b}^I\big)\Big)T_s\Big]^j \Big/ j! \tag{10.9}$$

A simplification is achieved by omitting terms that contain the high-order components of $\Delta\mathbf{A}\big(\mathbf{b}^{\mathrm{I}}\big)$. This approximation summarizes the uncertain $\mathbf{G}\big(\mathbf{b}^{\mathrm{I}}\big)$ as

$$\begin{aligned}\mathbf{G}\left(\mathbf{b}^{I}\right) &= \sum_{j=0}^{\infty}\frac{(\mathbf{A}(\mathbf{b}^{c})T_{s})^{j}}{j!} + \boldsymbol{\Gamma}(T_{s}) \\ &= \sum_{j=0}^{\infty}\frac{(\mathbf{A}(\mathbf{b}^{c})T_{s})^{j}}{j!} \\ &+ \sum_{j=1}^{\infty}\frac{\Delta\mathbf{A}\left(\mathbf{b}^{I}\right)T_{s}(\mathbf{A}(\mathbf{b}^{c})T_{s})^{j-1} + \mathbf{A}(\mathbf{b}^{c})T_{s}\Delta\mathbf{A}\left(\mathbf{b}^{\mathrm{I}}\right)T_{s}(\mathbf{A}(\mathbf{b}^{c})T_{s})^{j-2} + \cdots}{j!}\end{aligned} \tag{10.10}$$

Consequently, the decomposed form $\mathbf{G}\left(\mathbf{b}^{\mathrm{I}}\right) = \mathbf{G}(\mathbf{b}^{\mathrm{c}}) + \Delta\mathbf{G}\left(\mathbf{b}^{\mathrm{I}}\right)$ can be expressed as

$$\mathbf{G}\left(\mathbf{b}^{I}\right) = \exp\left(\mathbf{A}\left(\mathbf{b}^{c}\right)T_{s}\right) + \boldsymbol{\Gamma}(T_{s}) \tag{10.11}$$

where the uncertain component $\Delta\mathbf{G}\left(\mathbf{b}^{\mathrm{I}}\right)$ is derivable as $\Gamma(T_{s})$.

With $\mathbf{G}\left(\mathbf{b}^{\mathrm{I}}\right)$ acquired, an expression for $\exp\left(\mathbf{A}\left(\mathbf{b}^{\mathrm{I}}\right)\overline{\tau}\right)$ is available. This allows the uncertain input matrix $\mathbf{H}\left(\mathbf{b}^{\mathrm{I}}\right)$ to be initially represented in the following form:

$$\mathbf{H}\left(\mathbf{b}^{I}\right) = \int_{0}^{T_{s}}\left(\exp\left(\mathbf{A}\left(\mathbf{b}^{c}\right)\overline{\tau}\right) + \boldsymbol{\Gamma}(\overline{\tau})\right)\left(\mathbf{B}\left(\mathbf{b}^{c}\right) + \Delta\mathbf{B}\left(\mathbf{b}^{I}\right)\right)d\overline{\tau} \tag{10.12}$$

By neglecting the high-order uncertain terms, the expression $\mathbf{H}\left(\mathbf{b}^{\mathrm{I}}\right) = \mathbf{H}(\mathbf{b}^{\mathrm{c}}) + \Delta\mathbf{H}\left(\mathbf{b}^{\mathrm{I}}\right)$ can be written as

$$\begin{aligned}\mathbf{H}\left(\mathbf{b}^{I}\right) &= \int_{0}^{T_{s}}\exp\left(\mathbf{A}\left(\mathbf{b}^{c}\right)\overline{\tau}\right)\mathbf{B}\left(\mathbf{b}^{c}\right)d\overline{\tau} \\ &+ \int_{0}^{T_{s}}\left(\exp\left(\mathbf{A}\left(\mathbf{b}^{c}\right)\overline{\tau}\right)\Delta\mathbf{B}\left(\mathbf{b}^{I}\right) + \boldsymbol{\Gamma}(\overline{\tau})\mathbf{B}\left(\mathbf{b}^{c}\right)\right)d\overline{\tau}\end{aligned} \tag{10.13}$$

where $\mathbf{H}\left(\mathbf{b}^{I}\right) = \mathbf{H}(\mathbf{b}^{c}) + \Delta\mathbf{H}\left(\mathbf{b}^{I}\right)$, and the uncertain part $\Delta\mathbf{H}\left(\mathbf{b}^{I}\right)$ is derived from $\int_{0}^{T_{s}}\left(\exp(\mathbf{A}(\mathbf{b}^{c})\overline{\tau})\Delta\mathbf{B}\left(\mathbf{b}^{I}\right) + \boldsymbol{\Gamma}(\overline{\tau})\mathbf{B}(\mathbf{b}^{c})\right)d\overline{\tau}$.

This derivation leads to the state-space equation with interval uncertainties for the uncertain system response $\mathbf{x}^{\mathrm{I}}(k) = \mathbf{x}^{\mathrm{c}}(k) + \Delta\mathbf{x}^{\mathrm{I}}(k)$ and uncertain inputs $\mathbf{u}^{\mathrm{I}}(k) = \mathbf{u}^{\mathrm{c}}(k) + \Delta\mathbf{u}^{\mathrm{I}}(k)$, which is obtained as

$$\begin{aligned}\mathbf{x}^{c}(k+1) + \Delta\mathbf{x}^{I}(k+1) &= \left(\mathbf{G}\left(\mathbf{b}^{c}\right) + \Delta\mathbf{G}\left(\mathbf{b}^{I}\right)\right)\left(\mathbf{x}^{c}(k) + \Delta\mathbf{x}^{I}(k)\right) \\ &+ \left(\mathbf{H}\left(\mathbf{b}^{c}\right) + \Delta\mathbf{H}\left(\mathbf{b}^{I}\right)\right)\left(\mathbf{u}^{c}(k) + \Delta\mathbf{u}^{I}(k)\right)\end{aligned} \tag{10.14}$$

Thus, by retaining only the first-order term, the nominal discrete state-space equation can be rewritten as

$$\mathbf{x}^c(k+1) = \mathbf{G}(\mathbf{b}^c)\mathbf{x}^c(k) + \mathbf{H}(\mathbf{b}^c)\mathbf{u}^c(k) \tag{10.15}$$

This allows for the derivation of the corresponding interval uncertain discrete state-space equation:

$$\begin{aligned}\Delta\mathbf{x}^I(k+1) =& \mathbf{G}(\mathbf{b}^c)\Delta\mathbf{x}^I(k) + \Delta\mathbf{G}(\mathbf{b}^I)\mathbf{x}^c(k)\\ &+ \mathbf{H}(\mathbf{b}^c)\Delta\mathbf{u}^I(k) + \Delta\mathbf{H}(\mathbf{b}^I)\mathbf{u}^c(k)\end{aligned} \tag{10.16}$$

In this equation, the vectors are defined as $\mathbf{x}^c(k), \Delta\mathbf{x}^I(k) \in \mathbb{R}^{m\times1}$ and $\mathbf{u}^c(k), \Delta\mathbf{u}^I(k) \in \mathbb{R}^{n\times1}$. The system matrices are $\mathbf{G}(\mathbf{b}^c), \Delta\mathbf{G}(\mathbf{b}^I) \in \mathbb{R}^{m\times m}$ and $\mathbf{H}(\mathbf{b}^c), \Delta\mathbf{H}(\mathbf{b}^I) \in \mathbb{R}^{m\times n}$.

Remark 10.1: An order-expanded state-space equation is formulated aiming at simultaneously obtaining the nominal and uncertain state responses of the interval discrete state-space equation, as follows:

$$\begin{aligned}\begin{bmatrix}\mathbf{x}^c(k+1)\\ \Delta\mathbf{x}^I(k+1)\end{bmatrix} =& \begin{bmatrix}\mathbf{G}(\mathbf{b}^c) & \mathbf{0}\\ \Delta\mathbf{G}(\mathbf{b}^I) & \mathbf{G}(\mathbf{b}^c)\end{bmatrix}\begin{bmatrix}\mathbf{x}^c(k)\\ \Delta\mathbf{x}^I(k)\end{bmatrix}\\ &+ \begin{bmatrix}\mathbf{H}(\mathbf{b}^c) & \mathbf{0}\\ \Delta\mathbf{H}(\mathbf{b}^I) & \mathbf{H}(\mathbf{b}^c)\end{bmatrix}\begin{bmatrix}\mathbf{u}^c(k)\\ \Delta\mathbf{u}^I(k)\end{bmatrix}\end{aligned} \tag{10.17}$$

which can be furthermore rewritten as the following equations:

$$\mathbb{X}^I(k+1) = \mathbb{G}(\mathbf{b}^I)\mathbb{X}^I(k) + \mathbb{H}(\mathbf{b}^I)\mathbb{U}^I(k) \tag{10.18}$$

where $\mathbb{X}^I(k) = \begin{bmatrix}\mathbf{x}^c(k)\\ \Delta\mathbf{x}^I(k)\end{bmatrix} \in \mathbb{R}^{2m\times1}$, $\mathbb{U}^I(k) = \begin{bmatrix}\mathbf{u}^c(k)\\ \Delta\mathbf{u}^I(k)\end{bmatrix} \in \mathbb{R}^{2n\times1}$, $\mathbb{G}(\mathbf{b}^I) = \begin{bmatrix}\mathbf{G}(\mathbf{b}^c) & \mathbf{0}\\ \Delta\mathbf{G}(\mathbf{b}^I) & \mathbf{G}(\mathbf{b}^c)\end{bmatrix} \in \mathbb{R}^{2m\times2m}$ and $\mathbb{H}(\mathbf{b}^I) = \begin{bmatrix}\mathbf{H}(\mathbf{b}^c) & \mathbf{0}\\ \Delta\mathbf{H}(\mathbf{b}^I) & \mathbf{H}(\mathbf{b}^c)\end{bmatrix} \in \mathbb{R}^{2m\times2n}$

10.3 Bounded Model Predictive Control

10.3.1 Nominal Model Predictive Control Method

MPC is an approach employed for systems containing constraints (Dai et al. 2023). The future input sequence $\tilde{\mathbf{U}}_k$ specifies the predictive control law across the prediction horizon N, as defined in Dai et al. (2023):

$$\tilde{\mathbf{U}}_k = \left[(\mathbf{u}(k|k))^T, (\mathbf{u}(k+1|k))^T, \ldots, (\mathbf{u}(k+N-1|k))^T\right]^T \in \mathbb{R}^{Nn\times1} \tag{10.19}$$

where $\mathbf{u}(k+r|k)$ represents the control input for time $k+r$ as forecasted at time k.

By applying this input sequence to the discrete system, the corresponding future state sequence $\tilde{\mathbf{X}}_k$ can be determined as follows:

$$\tilde{\mathbf{X}}_k = \left[(\mathbf{x}(k|k))^T, (\mathbf{x}(k+1|k))^T, \ldots, (\mathbf{x}(k+N|k))^T\right]^T \in \mathbb{R}^{(N+1)m\times 1} \tag{10.20}$$

where $\mathbf{x}(k+r|k)$ signifies the predicted state at time $k+r$ relying on information at time k.

The control costs J for the MPC is formulated as (Xie et al. 2023):

$$\begin{aligned} J = &\sum_{i=0}^{N-1}\left((\mathbf{x}(k+i|k))^T\mathbf{Q}\mathbf{x}(k+i|k) + (\mathbf{u}(k+i|k))^T\mathbf{R}\mathbf{u}(k+i|k)\right) \\ &+(\mathbf{x}(k+N|k))^T\mathbf{F}\mathbf{x}(k+N|k) \end{aligned} \tag{10.21}$$

In this function, $\mathbf{Q} \in \mathbb{R}^{m\times m}$ serves as the weight matrix on the trajectory tracking deviation, and $\mathbf{R} \in \mathbb{R}^{r\times r}$ is the penalizing matrix for control excursion and stationary control at steady-state conditions. $\mathbf{F}$ is designated as the terminal penalty term.

An iterative solution for the state variables at each future time instant is computed using the discrete state-space equations:

$$\begin{cases} \mathbf{x}(k|k) = \mathbf{x}(k) \\ \mathbf{x}(k+1|k) = \mathbf{G}\mathbf{x}(k|k) + \mathbf{H}\mathbf{u}(k|k) = \mathbf{G}\mathbf{x}(k) + \mathbf{H}\mathbf{u}(k|k) \\ \vdots \\ \mathbf{x}(k+N|k) = \mathbf{G}^N\mathbf{x}(k) + \mathbf{G}^{N-1}\mathbf{H}\mathbf{u}(k|k) + \cdots + \mathbf{G}\mathbf{H}\mathbf{u}(k+N-2|k) + \mathbf{H}\mathbf{u}(k+N-1|k) \end{cases} \tag{10.22}$$

This entire iterative process can be compactly outlined using the following matrix form:

$$\begin{aligned} \tilde{\mathbf{X}}_k &= \begin{bmatrix} \mathbf{I} \\ \mathbf{G} \\ \vdots \\ \mathbf{G}^N \end{bmatrix}\mathbf{x}(k) + \begin{bmatrix} \mathbf{0} & \mathbf{0} & \cdots & \cdots & \mathbf{0} \\ \mathbf{H} & \mathbf{0} & \cdots & \cdots & \mathbf{0} \\ \mathbf{G}\mathbf{H} & \mathbf{H} & \cdots & \cdots & \mathbf{0} \\ \vdots & \vdots & \ddots & \ddots & \vdots \\ \mathbf{G}^{N-1}\mathbf{H} & \mathbf{G}^{N-2}\mathbf{H} & \cdots & \mathbf{G}\mathbf{H} & \mathbf{H} \end{bmatrix}\tilde{\mathbf{U}}_k \\ &= \mathbf{M}\mathbf{x}(k) + \mathbf{C}\tilde{\mathbf{U}}_k,\ \mathbf{M} \in \mathbb{R}^{(N+1)m\times m}, \mathbf{C} \in \mathbb{R}^{(N+1)m\times Nn} \end{aligned} \tag{10.23}$$

Using this matrix formulation, the cost function J from (10.21) is subsequently rewritten as

$$J = \tilde{\mathbf{X}}_k^{\mathrm{T}}\hat{\mathbf{Q}}\tilde{\mathbf{X}}_k + \tilde{\mathbf{U}}_k^{\mathrm{T}}\hat{\mathbf{R}}\tilde{\mathbf{U}}_k = (\mathbf{x}(k))^{\mathrm{T}}\Omega\mathbf{x}(k) + 2(\mathbf{x}(k))^{\mathrm{T}}\Psi\tilde{\mathbf{U}}_k + \tilde{\mathbf{U}}_k^{\mathrm{T}}\Xi\tilde{\mathbf{U}}_k \tag{10.24}$$

In this expression, the matrices $\hat{\mathbf{Q}} = \text{diag}(\mathbf{Q}, ..., \mathbf{Q}, \mathbf{F}) \in \mathbb{R}^{(N+1)m\times(N+1)m}$ and $\hat{\mathbf{R}} = \text{diag}(\mathbf{R}, ..., \mathbf{R}) \in \mathbb{R}^{Nr\times Nr}$. The terms $\boldsymbol{\Omega}$, $\boldsymbol{\Psi}$, and Ξ represent the corresponding relationships: $\boldsymbol{\Omega} = \mathbf{M}^{\mathrm{T}}\hat{\mathbf{Q}}\mathbf{M}$, $\boldsymbol{\Psi} = \mathbf{M}^{\mathrm{T}}\hat{\mathbf{Q}}\mathbf{C}$, and $\Xi = \mathbf{C}^{\mathrm{T}}\hat{\mathbf{Q}}\mathbf{C} + \hat{\mathbf{R}}$, respectively.

Given that the MPC cost function is convex and its constraints are linear, the optimization problem for MPC can be formulated as a quadratic program (QP):

$$\begin{aligned}\tilde{\mathbf{U}}_k^{2*} = \arg \min_{\tilde{\mathbf{U}}_k\in\mathbb{R}^{Nn}} \sum_{i=0}^{N-1} \left[(\mathbf{x}(k+i|k))^T\mathbf{Q}\mathbf{x}(k+i|k) + (\mathbf{u}(k+i|k))^T\mathbf{R}\mathbf{u}(k+i|k)\right] \\ +(\mathbf{x}(k+N|k))^{\mathrm{T}}\mathbf{F}\mathbf{x}(k+N|k)\end{aligned} \tag{10.25}$$

$$\mathrm{s}\cdot\mathrm{t}\cdot\mathbf{x}(k+1) = \mathbf{G}\mathbf{x}(k) + \mathbf{H}\mathbf{u}(k) \tag{10.26}$$

$$\underline{\mathbf{u}}(k) \le \mathbf{u}(k) \le \overline{\mathbf{u}}(k) \tag{10.27}$$

Solving this QP optimization problem yields the optimal input sequence $\tilde{\mathbf{U}}_k^*$. The first sample from this sequence, $\mathbf{u}^*(k|k)$, is thus utilized as the control input $\mathbf{u}(k)$ to the RFC satellite system at the kth sampling time. The MPC process is completed as the prediction horizon N is iteratively updated forward.

10.3.2 Uncertain Model Predictive Control Method Based on IDWA

Based on the uncertainty transfer relationships, the matrices $\mathbf{M}$ and $\mathbf{C}$ from Eq. (10.23) are subsequently deduced into their interval forms, denoted as $\mathbf{M}(\mathbf{b}^{\mathrm{I}})$ and $\mathbf{C}(\mathbf{b}^{\mathrm{I}})$, respectively. This reformulation allows for the derivation of the interval form of Eq. (10.23) as follows:

$$\tilde{\mathbf{X}}_k^I = \mathbf{M}(\mathbf{b}^I)\mathbf{x}^I(k) + \mathbf{C}(\mathbf{b}^I)\tilde{\mathbf{U}}_k^I \tag{10.28}$$

where the interval state sequence interval input sequence can be, respectively, defined as

$$\tilde{\mathbf{X}}_k^I = \left[\left(\mathbf{x}^I(k|k)\right)^T, \left(\mathbf{x}^I(k+1|k)\right)^T, ..., \left(\mathbf{x}^I(k+N|k)\right)^T\right]^T \tag{10.29}$$

$$\tilde{\mathbf{U}}_k^I = \left[\left(\mathbf{u}^I(k|k)\right)^T, \left(\mathbf{u}^I(k+1|k)\right)^T, ..., \left(\mathbf{u}^I(k+N-1|k)\right)^T\right]^T \tag{10.30}$$

The interval cost function can be formulated as follows:

$$J^I = \left(\tilde{\mathbf{X}}_k^I\right)^T\hat{\mathbf{Q}}\tilde{\mathbf{X}}_k^I + \left(\tilde{\mathbf{U}}_k^I\right)^T\hat{\mathbf{R}}\tilde{\mathbf{U}}_k^I$$

$$= \left(\mathbf{M}(\mathbf{b}^I)\mathbf{x}^I(k) + \mathbf{C}(\mathbf{b}^I)\tilde{\mathbf{U}}_k^I\right)^T \hat{\mathbf{Q}}\left(\mathbf{M}(\mathbf{b}^I)\mathbf{x}^I(k) + \mathbf{C}(\mathbf{b}^I)\tilde{\mathbf{U}}_k^I\right) + \left(\tilde{\mathbf{U}}_k^I\right)^T \hat{\mathbf{R}}\tilde{\mathbf{U}}_k^I$$
$$= \left(\mathbf{x}^I(k)\right)^T \boldsymbol{\Omega}(\mathbf{b}^I)\mathbf{x}^I(k) + 2\left(\mathbf{x}^I(k)\right)^T \boldsymbol{\Psi}(\mathbf{b}^I)\tilde{\mathbf{U}}_k^I + \left(\tilde{\mathbf{U}}_k^I\right)^T \boldsymbol{\Xi}(\mathbf{b}^I)\tilde{\mathbf{U}}_k^I \quad (10.31)$$

In accordance with the IDWA method, the parameters corresponding to each row of $\boldsymbol{\Theta}_p$ are inserted into the interval cost function (Eq. (10.31)). This step provides the interval optimization objective for the QP optimization problem:

$$J^I = \left(\tilde{\mathbf{X}}_k^I\left(\hat{\mathbf{b}}_k^{(i)}\right)\right)^T \hat{\mathbf{Q}}\tilde{\mathbf{X}}_k^I\left(\hat{\mathbf{b}}_k^{(i)}\right) + \left(\tilde{\mathbf{U}}_k^I\left(\hat{\mathbf{b}}_k^{(i)}\right)\right)^T \hat{\mathbf{R}}\tilde{\mathbf{U}}_k^I\left(\hat{\mathbf{b}}_k^{(i)}\right)$$
$$= \left(\mathbf{x}\left(\hat{\mathbf{b}}_k^{(i)}\right)\right)^T \boldsymbol{\Omega}\left(\left(\hat{\mathbf{b}}_k^{(i)}\right)\right)\mathbf{x}\left(\hat{\mathbf{b}}_k^{(i)}\right) + 2\left(\mathbf{x}\left(\hat{\mathbf{b}}_k^{(i)}\right)\right)^T \boldsymbol{\Psi}\left(\hat{\mathbf{b}}_k^{(i)}\right)\tilde{\mathbf{U}}_k^I\left(\hat{\mathbf{b}}_k^{(i)}\right)$$
$$+ \left(\tilde{\mathbf{U}}_k^I\left(\hat{\mathbf{b}}_k^{(i)}\right)\right)^T \boldsymbol{\Xi}\left(\hat{\mathbf{b}}_k^{(i)}\right)\tilde{\mathbf{U}}_k^I\left(\hat{\mathbf{b}}_k^{(i)}\right) \quad (10.32)$$

By reiterating the former procedures, the interval variable $J(t)$ for the control costs with uncertainty is determined:

$$J \in J^I = \left[\underline{J}, \overline{J}\right] = J_{sp}^I \bigcup J_{ap}^I = \left[\underline{J_{sp}}, \overline{J_{sp}}\right] \bigcup \left[\underline{J_{ap}}, \overline{J_{ap}}\right] \quad (10.33)$$

in which the Lower indices 's' and 'a' designate the results from samplings and approximate samples, respectively. Furthermore, the QP optimization problem is constructed in the following form:

$$\left(\tilde{\mathbf{U}}_k^I\right)^* = \arg \min_{\tilde{\mathbf{U}}_k^I \in \mathbb{R}^{Nn}} \left(\mathbf{x}^I(k)\right)^T \boldsymbol{\Omega}(\mathbf{b}^I)\mathbf{x}^I(k)$$
$$+ 2\left(\mathbf{x}^I(k)\right)^T \boldsymbol{\Psi}(\mathbf{b}^I)\tilde{\mathbf{U}}_k^I + \left(\tilde{\mathbf{U}}_k^I\right)^T \boldsymbol{\Xi}(\mathbf{b}^I)\tilde{\mathbf{U}}_k^I \quad (10.34)$$

$$\mathrm{s \cdot t \cdot}\ \mathbf{x}^I(k+1) = \mathbf{G}(\mathbf{b}^I)\mathbf{x}^I(k) + \mathbf{H}(\mathbf{b}^I)\mathbf{u}^I(k) \quad (10.35)$$

$$\tilde{\mathbf{U}}_k^I = \left[\left(\mathbf{u}^I(k|k)\right)^T, \left(\mathbf{u}^I(k+1|k)\right)^T, \ldots, \left(\mathbf{u}^I(k+N-1|k)\right)^T\right]^T \quad (10.36)$$

$$\underline{\mathbf{u}}(k) \le \mathbf{u}^I(k) \le \overline{\mathbf{u}}(k) \quad (10.37)$$

$$\underline{\mathbf{b}} \le \mathbf{b} \le \overline{\mathbf{b}} \quad (10.38)$$

This formulation allows for the solution of the corresponding optimized input sequence $(\tilde{\mathbf{U}}_k^{\mathrm{I}})^*$:

$$\left(\tilde{\mathbf{U}}_k^I\right)^* = \left[\left(\mathbf{u}\left(\hat{\mathbf{b}}_k^{(i)}, k|k\right)\right)^T, \left(\mathbf{u}\left(\hat{\mathbf{b}}_k^{(i)}, k+1|k\right)\right)^T, \ldots, \left(\mathbf{u}\left(\hat{\mathbf{b}}_k^{(i)}, k+N-1|k\right)\right)^T\right]^T \quad (10.39)$$

From this sequence, the interval envelope of the optimal control input, which is implemented as the first sample $\mathbf{u}^*(k|k)$, is presented as

$$\mathbf{u}^* \in \left(\mathbf{u}^{\mathrm{I}}\right)^* = \left[\underline{\mathbf{u}}^*, \overline{\mathbf{u}^*}\right] = \left(\mathbf{u}_{\mathrm{sp}}^{\mathrm{I}}\right)^* \bigcup \left(\mathbf{u}_{\mathrm{ap}}^{\mathrm{I}}\right)^* = \left[\underline{\mathbf{u}_{\mathrm{sp}}^*}, \overline{\mathbf{u}_{\mathrm{sp}}^*}\right] \bigcup \left[\underline{\mathbf{u}_{\mathrm{ap}}^*}, \overline{\mathbf{u}_{\mathrm{ap}}^*}\right] \quad (10.40)$$

With this final step, the BMPC method is fully constituted.

10.4 Numerical Example

The implementation of the BMPC control approach is significantly influenced by the structural parameters of the RFC satellite, as the method relies on the complete constructed model of the system. To analyze the system responses, the appendage of the spacecraft is abstracted as a typical Euler–Bernoulli beam. In addition, the angles between coordinate systems are used to represent the attitude situation of the satellite's rigid part. The physical model specifications for the nominal RFC model are listed in Table 10.1. With the structural parameters defined above and the relationships from Eqs. (9.1)–(9.19), the system state $\mathbf{x}(t) = [\theta(t), \eta_1(t), \eta_2(t), \dot{\theta}(t), \dot{\eta}_1(t), \dot{\eta}_2(t)]^{\mathrm{T}}$ for the RFC satellite is established. This state vector is subsequently used to solve the system responses using the continuous state-space equation $\dot{\mathbf{x}}(t) = \mathbf{A}\mathbf{x}(t) + \mathbf{B}\mathbf{u}(t)$.

To implement the BMPC process, the aforementioned continuous state-space equation must be discretized. This conversion results in the discrete state-space equation $\mathbf{x}(k+1) = \mathbf{G}\mathbf{x}(k) + \mathbf{H}\mathbf{u}(k)$, which employs a sampling time step of $T_s = 0.05$ s. The corresponding non-continuous state vector is $\mathbf{x}(k)$. Uncertainties exist within the RFC satellite system as a result of complex operational situations. All such parameters in the system are treated as UBB numbers. To evaluate the proposed method, b_m and J_c are chosen as the interval uncertain parameters, with assumed values of $b_m^{\mathrm{I}} = [0.13, 0.17]$ and $J_c^{\mathrm{I}} = [0.26, 0.34]$, respectively.

Table 10.1 Nominal physical model for constructing the RFC satellite

Symbol	Parameter	Value
R_{r}	Rotor Radius	0.05 m
J_{c}	Inertia moment on the central rotor	$0.3\ \mathrm{kg \cdot m^2}$
L	Flexible beam length	1.5 m
μ	Flexible beam linear density	0.54 kg/m
K_{e}	Flexible beam damping coefficient	0.03
EI	Flexible beam banding stiffness	$10.4\ \mathrm{N \cdot m^2}$
b_m	Rotor viscous friction coefficient	$0.15\ \mathrm{m^2/s}$
m_{L}	Concentrated tip mass	0.25 kg
J_{L}	Tip inertia moment	$0.04\ \mathrm{kg \cdot m^2}$

To implement the MPC, the penalty matrices are configured as $\mathbf{Q} = \text{diag}(10, 1, 1, 1, 1, 1)$ and $\mathbf{R} = 5$, while the terminal penalty term $\mathbf{F}$ is set to 0. In accordance with Eq. (10.24), the nominal cost function for the RFC satellite's MPC system is determined at each kth sampling time. A prediction horizon of $N = 20$ is used, which is iteratively updated forward for every sampling time. This procedure yields the controlled system's state responses, such as the attitude angle and the flexible deformations of the beam.

In practical service, two principal operation situations can be abstracted from different mission modes, as illustrated in Fig. 10.1. One is elliptical orbits. In this scenario, orbit control is less energy-intensive, permitting more energy to be allocated for precise AVC without constraints. The other is circular orbits. Conversely, these orbits demand a large amount of energy for orbit control, which necessitates that the control inputs for AVC be constrained (Zhao et al. 2023).

To analyze the effect of the uncertainty variables on the RFC satellite, the introduced BMPC method is simulated under the two previously defined operation situations. The effectiveness of this approach is consistently assessed by validating all simulation results against a 10,000-run MCS. Employing the IDWA method, two distinct QP optimization problems are formulated: one for unconstrained control input and another for constrained control input. The resulting sets of interval control inputs, $\mathbf{u}^{\mathrm{I}}(k)$, for both cases are presented in Fig. 10.2. In each situation, the LB and UB curves for $\mathbf{u}^{\mathrm{I}}(k)$ are verified by MCS, which confirms the effectiveness of the proposed BMPC. Notably, a comparison between plots (a) and (b) in Fig. 10.2 demonstrates that the constrained interval control input is strictly held within the maximum and minimum input saturation levels, thereby satisfying the requirements of that

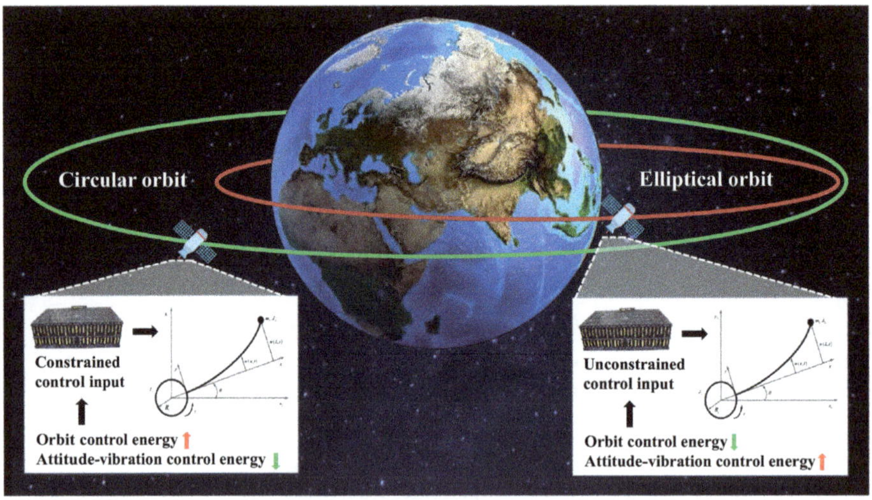

Fig. 10.1 Diagram of the two operational modes and their corresponding orbit types

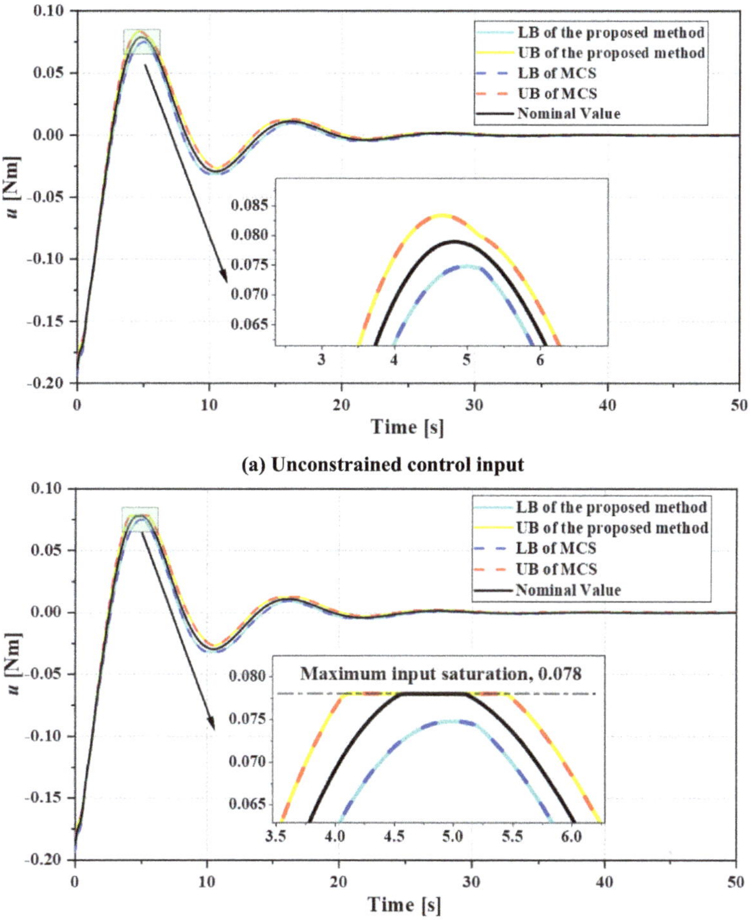

(a) Unconstrained control input

(b) Constrained control input

Fig. 10.2 Comparison of uncertain control inputs for the BMPC and MCS in two situations

specific operation situation. The analysis also yields the corresponding bounded responses for the attitude $\theta(t)$, which are displayed in Fig. 10.3. Similar to the input verification, the LB and UB curves for the bounded attitude responses $\theta(t)$ are validated against MCS for each respective situation.

The DPTDR for the different operation situations is computed adopting a lower threshold of -7.6°, as illustrated in Fig. 10.3. In the scenario where the control inputs are constrained, the LB of the corresponding interval attitude angle $\theta(t)$ breaches the lower threshold to a greater degree, which consequently yields a lower reliability result.

Figure 10.4 illustrates the LB and UB of the uncertain flexible deformation across various positions for both operation situations, while Fig. 10.5 specifically details the bounded responses at the free end. Validating these LB and UB results for the

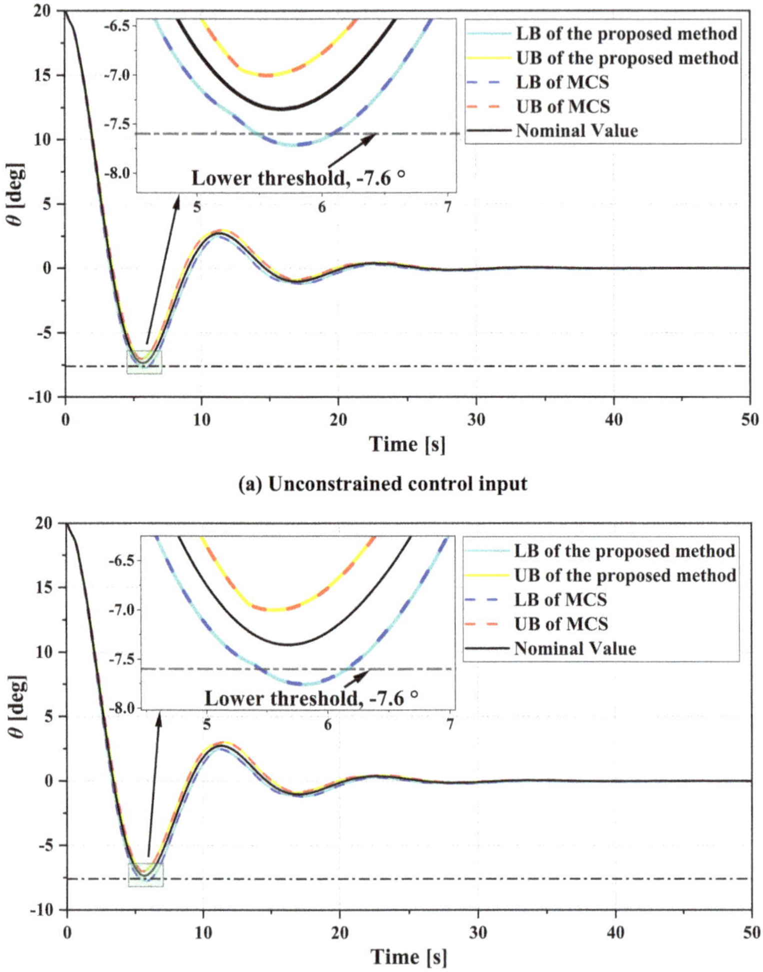

(a) Unconstrained control input

(b) Constrained control input

Fig. 10.3 Comparison of uncertain attitude angle for the BMPC and MCS in two situations

bounded vibration responses against MCS confirms the validity of the presented BMPC approach. The introduction of constrained control inputs notably impacts the interval vibration response, particularly at the point of input saturation. In addition, Fig. 10.6 exhibits the LB and UB for the bounded cost function in both situations, with its effectiveness also being corroborated by MCS. A key advantage of BMPC over the nominal MPC method is its superior ability to demonstrate how uncertain parameters influence the control system inputs.

In addition to achieving relatively high accuracy, the proposed BMPC method is highly effective at reducing time consumption when compared to MCS. The computations in this chapter were calculated on a computer workstation with a 3.0 GHz

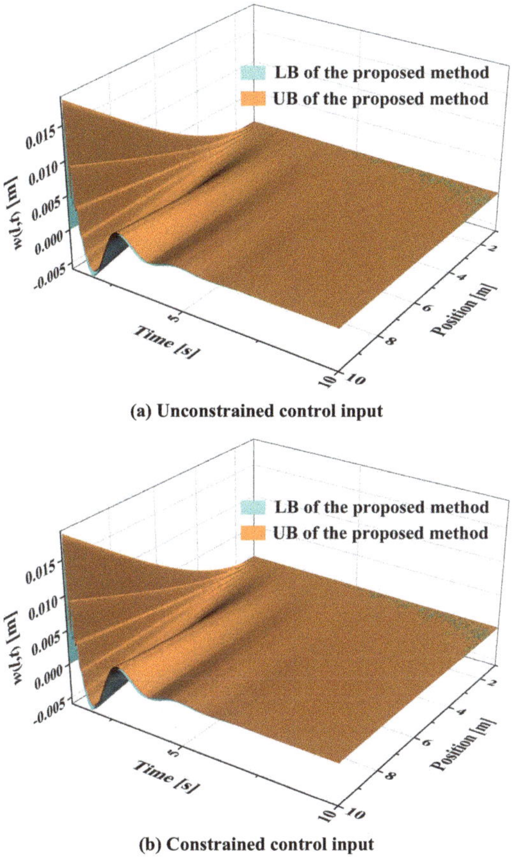

Fig. 10.4 Flexible beam deflection envelope for the BMPC in two situations

CPU (12 cores) and 32 GB RAM. Across the two operation situations, the average solving time for MCS was approximately 32,455 s, whereas the average time for the proposed BMPC method was only about 69 s. This substantial advantage in computation time clearly illustrates the significant efficiency of the proposed BMPC method.

10.5 Summary

This chapter introduced a BMPC method for designing the control system of an RFC satellite subject to uncertainties and reliability constraints. The foundation of this approach was an RFC mathematical model of the satellite. A method for discretizing

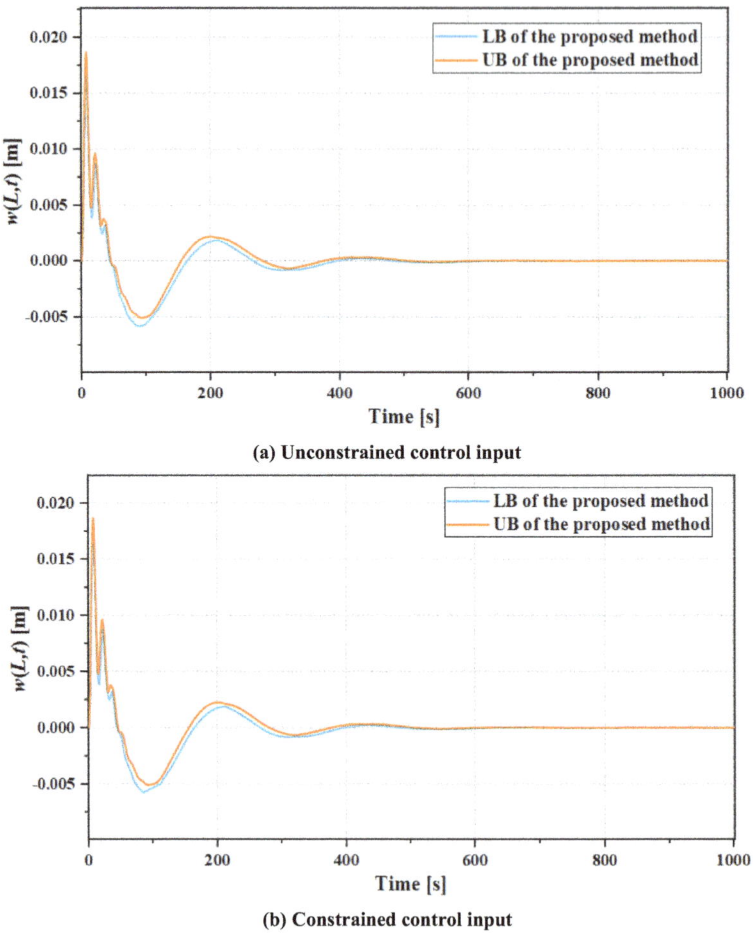

Fig. 10.5 Flexible deflection envelope at beam tip for the BMPC in two situations

the continuous state-space equation under uncertainty was also proposed. Furthermore, a novel BMPC method founded on the IDWA method was developed to achieve the uncertain AVC of the RFC satellite. This technique successfully yielded the interval uncertainty-based control input and system response. ITDR was incorporated into the BMPC system to assess two operational situations.

The proposed BMPC method for the RFC spacecraft (with its reliability constraint) was validated against MCS in these two operation situations, demonstrating the method's capability to handle different constraints. In addition to maintaining a certain degree of accuracy in the uncertain control system's response, the method's time consumption was greatly reduced compared to MCS, further proving its validity. This chapter establishes a solid foundation for applying the proposed BMPC method to the broader field of satellite control with uncertainties.

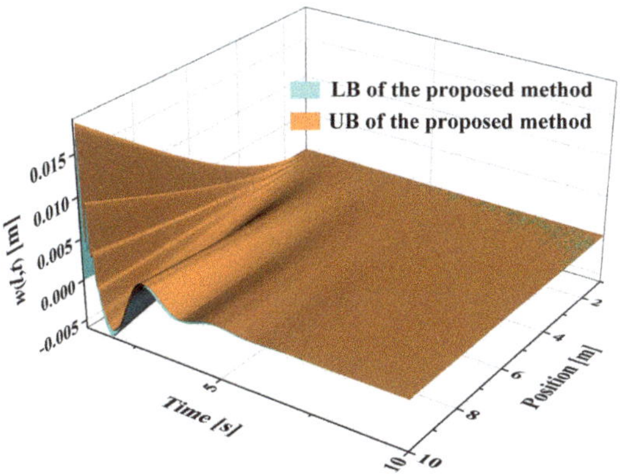

(a) Unconstrained control input

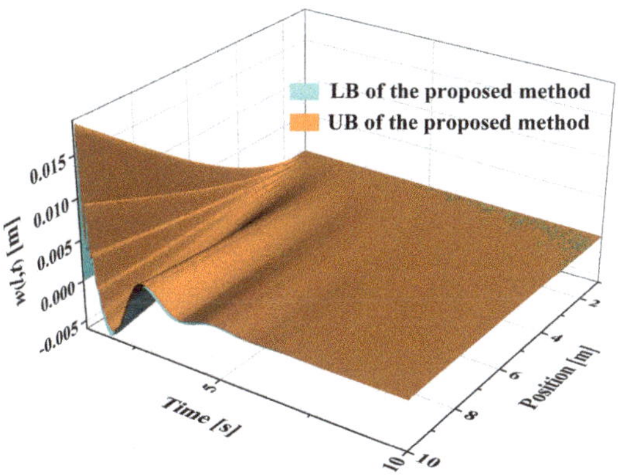

(b) Constrained control input

Fig. 10.4 Flexible beam deflection envelope for the BMPC in two situations

CPU (12 cores) and 32 GB RAM. Across the two operation situations, the average solving time for MCS was approximately 32,455 s, whereas the average time for the proposed BMPC method was only about 69 s. This substantial advantage in computation time clearly illustrates the significant efficiency of the proposed BMPC method.

10.5 Summary

This chapter introduced a BMPC method for designing the control system of an RFC satellite subject to uncertainties and reliability constraints. The foundation of this approach was an RFC mathematical model of the satellite. A method for discretizing

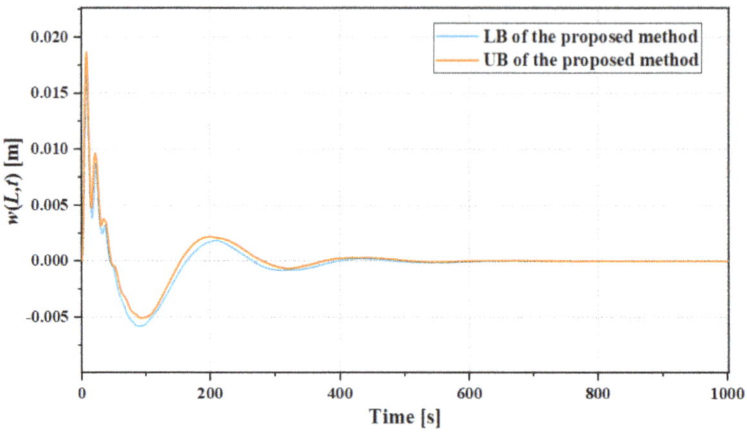

(a) Unconstrained control input

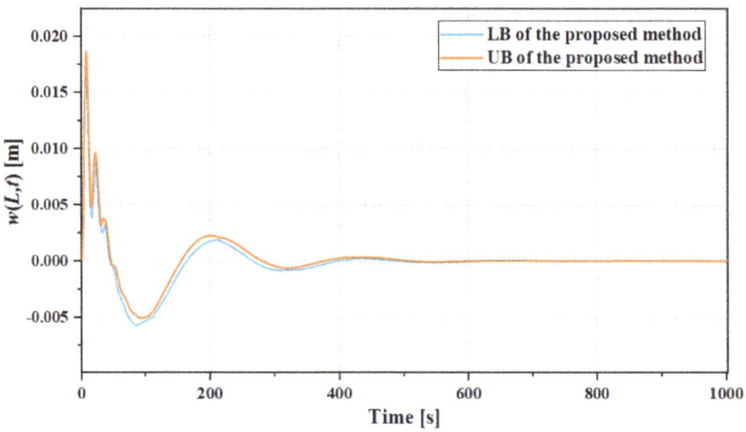

(b) Constrained control input

Fig. 10.5 Flexible deflection envelope at beam tip for the BMPC in two situations

the continuous state-space equation under uncertainty was also proposed. Furthermore, a novel BMPC method founded on the IDWA method was developed to achieve the uncertain AVC of the RFC satellite. This technique successfully yielded the interval uncertainty-based control input and system response. ITDR was incorporated into the BMPC system to assess two operational situations.

The proposed BMPC method for the RFC spacecraft (with its reliability constraint) was validated against MCS in these two operation situations, demonstrating the method's capability to handle different constraints. In addition to maintaining a certain degree of accuracy in the uncertain control system's response, the method's time consumption was greatly reduced compared to MCS, further proving its validity. This chapter establishes a solid foundation for applying the proposed BMPC method to the broader field of satellite control with uncertainties.

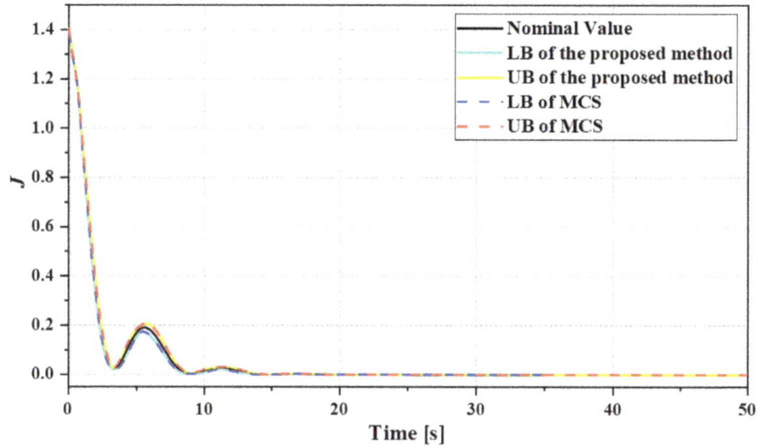

(a) Unconstrained control input

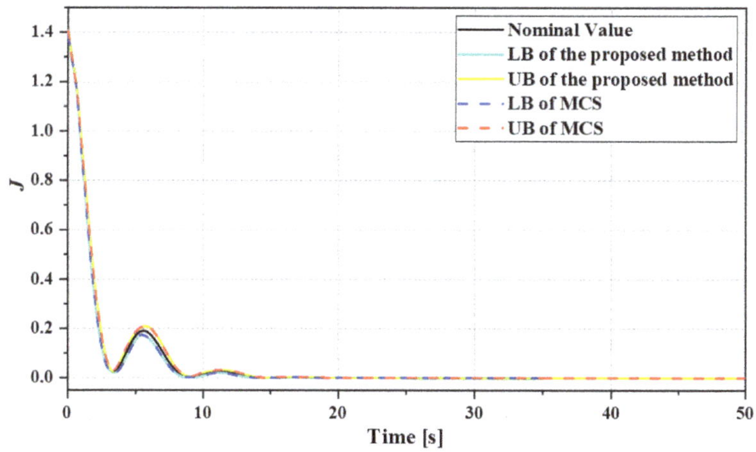

(b) Constrained control input

Fig. 10.6 Control costs for the BMPC and MCS in two situations

References

Cao S, Zhao Y (2017) Anti-disturbance fault-tolerant attitude control for satellites subject to multiple disturbances and actuator saturation. Nonlinear Dyn 89:2657–2667. https://doi.org/10.1007/s11071-017-3614-y

Chen L, Zhang J, Quan Q (2023) Quadcopter attitude control with vibration reduction by additive-state-decomposition dynamic inversion design with a notch filter. Nonlinear Dyn 111:8313–8327. https://doi.org/10.1007/s11071-023-08272-6

Dai L, Ma Y, Gao R, Wu J, Xia Y (2023) Cloud-based computational model predictive control using a parallel multiblock ADMM approach. IEEE Internet Things J 10:10326–10343. https://doi.org/10.1109/JIOT.2023.3238508

Dwivedi R, Sahai V (2020) On certain properties and expansions of zeta matrix function, digamma matrix function and polygamma matrix function. Quaest Math 43:97–105. https://doi.org/10.2989/16073606.2018.1539046

Ghorbani H, Vatankhah R, Farid M (2022) General planar motion modeling and control of a smart rigid-flexible satellite considering large deflections. Nonlinear Dyn 108:911–939. https://doi.org/10.1007/s11071-022-07242-8

Kumar TK, Etagi A, Giri DK (2025) Attitude control of Earth-pointing satellites employing novel hybrid actuator configurations. Aerosp Sci Technol 163:110333. https://doi.org/10.1016/j.ast.2025.110333

Wang H, Baoyin H, Lu S, Xue Z, Gao Y (2026) Observer-based adaptive fixed-time robust control for detumbling a large flexible defunct satellite. Acta Astronaut 238:127–139. https://doi.org/10.1016/j.actaastro.2025.09.090

Wigren T (2023) Recursive identification of a nonlinear state space model. Int J Adapt Control Signal Process 37:447–473. https://doi.org/10.1002/acs.3531

Xie H, Dai L, Sun Z, Xia Y (2023) Maximal admissible disturbance constraint set for tube-based model predictive control. IEEE Trans Autom Control 68:6773–6780. https://doi.org/10.1109/TAC.2023.3241273

Yang C, Yu Q (2025) Multi-objective optimization-inspired set theory-based regularization approach for force reconstruction with bounded uncertainties. Comput Methods Appl Mech Eng 438:117814. https://doi.org/10.1016/j.cma.2025.117814

Zhang Y, Zang Y, Li M, Wang Y, Li W (2017) Active-passive integrated vibration control for control moment gyros and its application to satellites. J Sound Vib 394:1–14. https://doi.org/10.1016/j.jsv.2017.01.002

Zhao B, Wang J, Hu G, Colombi A, Liao W-H, Liang J (2023) Time-sharing orbit jump and energy harvesting in nonlinear piezoelectric energy harvesters using a synchronous switch circuit. Mech Syst Signal Process 200:110601. https://doi.org/10.1016/j.ymssp.2023.110601

Zhou W, Zhang K, Wu S, Tan S, Wu Z (2022) Distributed cooperative control for vibration suppression of a flexible satellite. Aerosp Sci Technol 128:107750. https://doi.org/10.1016/j.ast.2022.107750

Zhou S, Xia C, Zhang G (2023) Orbit design and control for periodically revisiting multiple formation satellites. Aerosp Sci Technol 135:108199. https://doi.org/10.1016/j.ast.2023.108199

www.ingramcontent.com/pod-product-compliance
Ingram Content Group UK Ltd.
Pitfield, Milton Keynes, MK11 3LW, UK
UKHW020632260926
475829UK00003BA/47

9789819591787